Applied Mathematical Sciences

EDITORIAL STATEMENT

The mathematization of all sciences, the fading of traditional scientific boundaries, the impact of computer technology, the growing importance of mathematical-computer modelling and the necessity of scientific planning all create the need both in education and research for books that are introductory to and abreast of these developments.

The purpose of this series is to provide such books, suitable for the user of mathematics, the mathematician interested in applications, and the student scientist. In particular, this series will provide an outlet for material less formally presented and more anticipatory of needs than finished texts or monographs, yet of immediate interest because of the novelty of its treatment of an application or of mathematics being applied or lying close to applications.

The aim of the series is, through rapid publication in an attractive but inexpensive format, to make material of current interest widely accessible. This implies the absence of excessive generality and abstraction, and unrealistic idealization, but with quality of exposition as a goal.

Many of the books will originate out of and will stimulate the development of new undergraduate and graduate courses in the applications of mathematics. Some of the books will present introductions to new areas of research, new applications and act as signposts for new directions in the mathematical sciences. This series will often serve as an intermediate stage of the publication of material which, through exposure here, will be further developed and refined. These will appear in conventional format and in hard cover.

MANUSCRIPTS

The Editors welcome all inquiries regarding the submission of manuscripts for the series. Final preparation of all manuscripts will take place in the editorial offices of the series in the Division of Applied Mathematics, Brown University, Providence, Rhode Island.

SPRINGER-VERLAG NEW YORK INC., 175 Fifth Avenue, New York, N. Y. 10010

Printed in U.S.A.

Applied Mathematical Sciences | Volume 37

Stephen H. Saperstone

Semidynamical Systems in Infinite Dimensional Spaces

Springer-Verlag
New York Heidelberg Berlin

Stephen H. Saperstone
George Mason University
Department of Mathematics
4400 University Drive
Fairfax, VA 22030
U.S.A.

Library of Congress Cataloging in Publication Data

Saperstone, Stephen H.
Semidynamical systems in infinite dimensional spaces.

(Applied mathematical sciences; v. 37)
Bibliography: p.
Includes indexes.
1. Differentiable dynamical systems. 2. Topological imbeddings. 3. Function spaces. I. Title. II. Series: Applied mathematical sciences (Springer-Verlag New York Inc.); v. 37.
QA614.8.S26 515.3'5 81-16681
AACR2

Printed in the United States of America.

9 8 7 6 5 4 3 2 1

ISBN 0-387-90643-6 Springer-Verlag New York Heidelberg Berlin
ISBN 3-540-90643-6 Springer-Verlag Berlin Heidelberg New York

To my parents

PREFACE

Where do solutions go, and how do they behave en route? These are two of the major questions addressed by the qualitative theory of differential equations. The purpose of this book is to answer these questions for certain classes of equations by recourse to the framework of semidynamical systems (or topological dynamics as it is sometimes called). This approach makes it possible to treat a seemingly broad range of equations from nonautonomous ordinary differential equations and partial differential equations to stochastic differential equations. The methods are not limited to the examples presented here, though.

The basic idea is this: Embed some representation of the solutions of the equation (and perhaps the equation itself) in an appropriate function space. This space serves as the phase space for the semidynamical system. The phase map must be chosen so as to generate solutions to the equation from an initial value. In most instances it is necessary to provide a "weak" topology on the phase space. Typically the space is infinite dimensional.

These considerations motivate the requirement to study semidynamical systems in non locally compact spaces. Our objective here is to present only those results needed for the kinds of applications one is likely to encounter in differential equations. Additional properties and extensions of abstract semidynamical systems are left as exercises. The power of the semidynamical framework makes it possible to characterize the asymptotic behavior of the solutions of such a wide class of equations.

A caveat is in order. The stability results obtained in many of the examples can be gotten directly without recourse to the abstract semidynamical system setting. Moreover, in some instances, sharper results can be obtained by utilizing special techniques and methods suitably adjusted to that particular equation. On the other hand, the generality of the semidynamical system approach allows for a greater understanding of the unifying concepts running through all of the examples.

The first three chapters are devoted to the theory of semidynamical systems. Virtually all of the results hold for a discrete time parameter as well as a continuous time parameter. Because of their simplicity some examples of discrete semidynamical systems are included to illustrate the variety of asymptotic behavior. The remainder of the book is devoted to applications of the theory. The range of applications reflects recent mathematical activity. The choice of examples, though, reflects my interests and bias as well.

The presentation is meant to be self contained (except for a few lapses in Chapters 4, 5, and 7, where references are supplied). Appendices on functional analysis and probability are provided for this purpose. Definitions of terms not found in the text can usually be found in one of the appendices. Each chapter concludes with a set of exercises and a section called "Notes and Comments." This provides the reader with the source of the results of that chapter. It also offers some commentary and related results. Most of the source material is from the late 1960's and 1970's. The reader should be familiar with real analysis on the level of Royden [1] and ordinary differential equations on the level of Hirsch and Smale [1]. A little knowledge of partial differential equations in Chapter 5 and Markov processes in Chapter 7 would be useful. The chapter dependence is as follows:

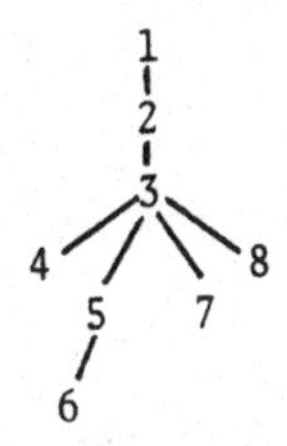

I want to acknowledge the contributions of many people. The initial impetus for this book came from Nam Bhatia. Much of the first chapter is based on his notes. Jim Yorke and Wei Shaw read portions of the manuscript. I am grateful for their helpful comments. Numerous colleagues have assisted me through their participation in seminars based on this material. Marshall Slemrod read the entire manuscript and provided invaluable suggestions which I feel improve the manuscript. A number of reviewers and referees also provided helpful criticisms and suggestions for improvement. Any remaining faults are mine.

The following people typed portions of the manuscript at various stages of its evolution: Pam Lambert, Mary Beth Minton, Nancy Dame, Carol Granis, and Susie Evers. Kate MacDougall typed the final camera-ready copy. I am grateful for their careful work and seemingly unbounded patience. Thanks are also due to Leon Booth, former Dean of CAS, for financial support in the preparation of the manuscript.

Fairfax, Virginia
August 1981

Stephen Saperstone

TABLE OF CONTENTS

Page

CHAPTER I. BASIC DEFINITIONS AND PROPERTIES 1

1. Introduction 1
2. Semidynamical Systems: Definitions and Conventions 2
3. A Glimpse of Things to Come; An Example from a Function Space 5
4. Solutions 7
5. Critical and Periodic Points 10
6. Classification of Positive Orbits 16
7. Discrete Semidynamical Systems 24
8. Local Semidynamical Systems; Reparametrization 25
9. Exercises 31
10. Notes and Comments 33

CHAPTER II. INVARIANCE, LIMIT SETS, AND STABILITY 35

1. Introduction 35
2. Invariance 36
3. Limit Sets: The Generalized Invariance Principle 39
4. Minimality 45
5. Prolongations and Stability of Compact Sets 52
6. Attraction: Asymptotic Stability of Compact Sets 56
7. Continuity of the Hull and Limit Set Maps in Metric Spaces 62
8. Lyapunov Functions: The Invariance Principle 77
9. From Stability to Chaos: A Simple Example 80
10. Exercises 92
11. Notes and Comments 95

CHAPTER III. MOTIONS IN METRIC SPACE 98

1. Introduction 98
2. Lyapunov Stable Motions 99
3. Recurrent Motions 105
4. Almost Periodic Motions 111
5. Asymptotically Stable Motions 121
6. Periodic Solutions of an Ordinary Differential Equation 125
7. Exercises 131
8. Notes and Comments 133

CHAPTER IV. NONAUTONOMOUS ORDINARY DIFFERENTIAL EQUATIONS 137

1. Introduction 137
2. Construction of the Skew Product Semidynamical System 140
3. Compactness of the Space $\mathcal{F}$ 151
4. The Invariance Principle for Ordinary Differential Equations 155
5. Limiting Equations and Stability 173

Page

CHAPTER IV (cont.)

6. Differential Equations without Uniqueness 189
7. Volterra Integral Equations 192
8. Exercises 202
9. Notes and Comments 205

CHAPTER V. SEMIDYNAMICAL SYSTEMS IN BANACH SPACE 209

1. Introduction 209
2. Nonlinear Semigroups and Their Generators 212
3. The Generalized Domain for Accretive Operators 225
4. Precompactness of Positive Orbits 231
5. Solution of the Cauchy Problem 244
6. Structure of Positive Limit Sets for Contraction Semigroups 253
7. Exercises 270
8. Appendix: Proofs of Theorems 2.4 and 2.16 273
9. Notes and Comments 279

CHAPTER VI. FUNCTIONAL DIFFERENTIAL EQUATIONS 283

1. Why Hereditary Dependence, Some Examples from Biology, Mechanics, and Electronics 283
2. Definitions and Notation: Functional Differential Equations with Finite or Infinite Delay. The Initial Function Space 285
3. Existence of Solutions of Retarded Functional Equations 292
4. Some Remarks on the Semidynamical System Defined by the Solution to an Autonomous Retarded Functional Differential Equation: The Invariance Principle and Stability 303
5. Some Examples of Stability of RFDE's 312
6. Remarks on the Asymptotic Behavior of Nonautonomous Retarded Functional Differential Equations 326
7. Critical Points and Periodic Solutions of Autonomous Retarded Functional Differential Equations 330
8. Neutral Functional Differential Equations 337
9. A Flip-Flop Circuit Characterized by a NFDE - The Stability of Solutions 351
10. Exercises 360
11. Notes and Comments 365

CHAPTER VII. STOCHASTIC DYNAMICAL SYSTEMS 369

1. Introduction 369
2. The Space of Probability Measures 370
3. Markov Transition Operators and the Semidynamical System 371
4. Properties of Positive Limit Sets 374
5. Critical Points for Markov Processes 378
6. Stochastic Differential Equations 380
7. The Invariance Principle for Markov Processes 384
8. Exercises 389
9. Notes and Comments 392

Table of Contents

	Page
CHAPTER VIII. WEAK SEMIDYNAMICAL SYSTEMS AND PROCESSES	393
1. Introduction	393
2. Weak Semidynamical Systems	395
3. Compact Processes	400
4. Uniform Processes	410
5. Solutions of Nonautonomous Ordinary Differential Equations Revisited - A Compact Process	411
6. Solutions of a Wave Equation - A Uniform Process	412
7. Exercises	422
8. Notes and Comments	423
APPENDIX A	424
0. Preliminaries	424
1. Commonly Used Symbols	424
2. Nets	425
3. Uniform Topologies	427
4. Compactness	428
5. Linear Spaces	429
6. Duality	431
7. Hilbert Spaces	432
8. Vector Valued Integration	433
9. Sobolev Spaces	435
10. Convexity	436
11. Fixed Point Theorems	437
12. Almost Periodicity	438
13. Differential Inequalities	438
APPENDIX B	440
1. Probability Spaces and Random Variables	440
2. Expectation	441
3. Convergence of Random Variables	443
4. Stochastic Processes; Martingales and Markov Processes	443
5. The Ito Stochastic Integral	446
REFERENCES	447
INDEX OF TERMS	465
INDEX OF SYMBOLS	473

CHAPTER I

BASIC DEFINITIONS AND PROPERTIES

1. Introduction

After the appropriate definitions in Section 2 we present in Section 3 a simple example of a semidynamical system in an infinite dimensional space. The example arises in conjunction with a Poisson process. The objective of Section 4 is to extend semidynamical systems to the negative time domain. This results in a classification of the maximal time domains of semidynamical systems. In Section 5 we characterize critical and periodic motions and prove an extension theorem for periodic motions. Examples are presented which distinguish semidynamical system behavior from its precursor, dynamical systems. Section 6 is devoted to both an algebraic and a topological classification of positive orbits. The chapter is closed following a reparametrization theorem. It permits us to treat global semidynamical systems throughout the book. This results in a considerable savings as regards the theory without losing the applicability of the results to many important examples.

Finally, because a main objective of the book is to develop examples and applications of semidynamical systems in infinite dimensional spaces, we avoid the requirement of local compactness of the phase space. We take exception to this in Chapter IV where our phase space turns out to be locally compact.

2. Semidynamical Systems: Definitions and Conventions

We begin by introducing the following notation.

$\mathbb{R}$ = the set of all real numbers.

$\mathbb{R}^+$ = the set of all nonnegative real numbers.

$\mathbb{R}^-$ = the set of all nonpositive real numbers.

$\mathbb{Z}$ = the set of all integers.

$\mathbb{Z}^+$ = the set of all nonnegative integers.

$\mathbb{N}$ = the set of all positive integers.

$\mathbb{Q}$ = the set of all rational numbers.

The set $\mathbb{R}$ will always be endowed with the Euclidean topology; all subsets of $\mathbb{R}$ will have the induced relative topology.

Definition 2.1. The pair (X,π) is called a *(continuous) semidynamical system* if X is a Hausdorff topological space and π is a mapping, $\pi : X \times \mathbb{R}^+ \to X$ which satisfies

(i) $\pi(x,0) = x$ for each $x \in X$ (initial value property),

(ii) $\pi(\pi(x,t),s) = \pi(x,t+s)$ for each $x \in X$ and $t,s \in \mathbb{R}^+$ (semigroup property), and

(iii) π is continuous on the product space $X \times \mathbb{R}^+$ (continuity property).

The modifier "continuous" for a semidynamical system is usually dropped. Except to distinguish them from discrete semidynamical systems, continuous semidynamical systems will always be referred to as semidynamical systems. The space $X \times \mathbb{R}^+$ is endowed with the product topology. The pair (X,π) is also commonly referred to as a *semiflow*. The space X is called the *phase space* and π the *phase map* of the semidynamical system (X,π). We often just refer to the map π as a semidynamical system or semiflow on the space X. The mapping $\pi^t: X \to X$ defined by $\pi^t(x) = \pi(x,t)$ is called a *t-transition*. It is clear that π^t is continuous.

Semidynamical systems represent a generalization of a dynamical system or a flow.

Definition 2.2. The pair (X,π) is called a *dynamical system* (or *flow*) if X is a Hausdorff topological space and π a mapping, $\pi: X \times \mathbb{R} \to X$ which satisfies

(i) $\pi(x,0) = x$ for each $x \in X$ (initial value property),

(ii) $\pi(\pi(x,t),s) = \pi(x,t+s)$ for each $x \in X$ and $t,s \in \mathbb{R}$ (group property), and

(iii) π is continuous on the product space $X \times \mathbb{R}$.

We note that a dynamical system incorporates "past" behavior, which is unspecified or even indeterminate for semidynamical systems.

Remark 2.3. If (X,π) is a dynamical system, then the t-transition π^t is a homeomorphism of X onto X. Indeed, for each $t \in \mathbb{R}$ the inverse of π^t is given by π^{-t}.

Let (X,π) be a semidynamical system. We shall often use the simpler notation xt in place of $\pi(x,t)$. Thus for example, properties (i) and (ii) of Definition 2.1 read $x0 = 0$ and $(xt)s = x(t+s)$, respectively. For sets $W \subset X$ and $T \subset \mathbb{R}^+$ we write WT for the set $\pi(W,T) = \{xt: x \in W, t \in T\}$. The symbol xT stands for $\{x\}T$ and Wt stands for $W\{t\}$. Consequently, the usage xt has a double meaning; namely, the image of (x,t) under π and the expression for the singleton $\{xt\}$.

Definition 2.4. Suppose (X,π) is a semidynamical system. If $M \subset X$ satisfies $M\mathbb{R}^+ = M$, then (M,π') is a semidynamical system and is called the *restriction* of (X,π) to M, where π' is the restricted mapping $\pi' = \pi|_{M\times \mathbb{R}^+}$.

Lemma 2.5. Suppose (X,π) is a semidynamical system. If for each $t \in \mathbb{R}^+$ the t-transition π^t is a homeomorphism from X onto X, we may define a mapping $\hat{\pi}: X \times \mathbb{R} \to X$ by

$$\hat{\pi}(x,t) = \begin{cases} \pi(x,t), & (x,t) \in X \times \mathbb{R}^+ \\ (\pi^{-t})^{-1}(x), & (x,t) \in X \times \mathbb{R}^-, \end{cases}$$

so that $(X,\hat{\pi})$ is a dynamical system.

Proof: Property (i) of Definition 2.2 is obvious. We will only verify property (ii) of Definition 2.2 in the case $s,t \in \mathbb{R}^-$. The other cases are similar. We have

$$\hat{\pi}(\hat{\pi}(x,t),s) = \hat{\pi}((\pi^{-t})^{-1}(x),s) = (\pi^{-s})^{-1}((\pi^{-t})^{-1}(x))$$
$$= (\pi^{-t}\circ\pi^{-s})^{-1}(x) = (\pi^{-t-s})^{-1}(x) = \hat{\pi}(x,t+s).$$

Finally we verify the continuity property (iii) of Definition 2.2. Let $\{x_\alpha\} \subset X$, $\{t_\alpha\} \subset \mathbb{R}$ be nets with $(x_\alpha,t_\alpha) \to (x,t) \in X \times \mathbb{R}$. Choose $s < t$. For sufficiently large α we must have

$t_\alpha - s \in \mathbb{R}^+$. Then

$$\begin{aligned}\hat{\pi}(x_\alpha, t_\alpha) &= \hat{\pi}(x_\alpha, s+(t_\alpha - s)) = \hat{\pi}(\hat{\pi}(x_\alpha, s), t_\alpha - s)\\ &\quad \pi(\hat{\pi}(x_\alpha, s), t_\alpha - s) \to \pi(\hat{\pi}(x,s), t-s)\\ &\quad \hat{\pi}(\hat{\pi}(x,s), t-s) = \hat{\pi}(x,t).\end{aligned}$$

□

Remark 2.6. If π^t is a homeomorphism, we may write $(\pi^t)^{-1} = \pi^{-t}$.

Definition 2.7. A semidynamical system (X,π) is said to *extend* to a dynamical system $(X,\hat{\pi})$ if $\hat{\pi}|_{X\times\mathbb{R}^+} = \pi$.

Corollary 2.8. Assume (X,π) is a semidynamical system so that π^t is a homeomorphism from X onto X for each $t \in \mathbb{R}^+$. Then (X,π) extends uniquely to a dynamical system $(X,\hat{\pi})$, where $\hat{\pi}$ is given in Lemma 2.5.

Proof: The proof is immediate in view of Lemma 2.5. □

3. A Glimpse of Things to Come; An Example from a Function Space

Let X denote the set of all bounded, real valued continuous functions on $\mathbb{R}$. Endow X with the topology generated by the norm $\|\phi\| = \sup_{\tau\in\mathbb{R}} |\phi(\tau)|$, $\phi \in X$. Fix $\lambda > 0$, $\alpha > 0$, and define $\pi\colon X \times \mathbb{R}^+ \to X$ by

$$\pi(\phi,t)(\tau) = e^{-\lambda t}\sum_{k=0}^{\infty}\frac{(\lambda t)^k}{k!}\phi(\tau - k\alpha), \quad \tau \in \mathbb{R}.$$

We see immediately that $\pi(\phi,0)(\tau) = \phi(\tau)$, $\tau \in \mathbb{R}$, so the initial value property (i) of Definition 2.1 is satisfied. Next,

$$\pi(\pi(\phi,t),s)(\tau) = e^{-\lambda s}\sum_{n=0}^{\infty}\frac{(\lambda s)^n}{n!}\left[e^{-\lambda t}\sum_{k=0}^{\infty}\frac{(\lambda t)^k}{k!}\phi(\tau-k\alpha-n\alpha)\right]$$

$$= e^{-\lambda(t+s)}\sum_{k=0}^{\infty}\frac{1}{k!}\left[k!\sum_{n=0}^{\infty}\frac{(\lambda s)^n}{n!}\frac{(\lambda t)^{k-n}}{(k-n)!}\phi(\tau-k\alpha)\right]$$

$$= e^{-\lambda(t+s)}\sum_{k=0}^{\infty}\frac{1}{k!}(\lambda t+\lambda s)^k\,\phi(\tau-k\alpha)$$

$$= \pi(\phi,t+s)(\tau).$$

Thus the semigroup property (ii) of Definition 2.1 is also satisfied. Finally we must verify the continuity property (iii) of Definition 2.1. In view of the semigroup property and the fact that each t-transition π^t is a bounded linear operator on X with norm $\|\pi^t\| \leq 1$, we have that for any (ϕ,t), $(\phi_0,t_0) \in X \times \mathbb{R}^+$,

$$\|\pi(\phi,t) - \pi(\phi_0,t_0)\| \leq \|\pi(\phi,t) - \pi(\phi_0,t)\| + \|\pi(\phi_0,t) - \pi(\phi_0,t_0)\|$$

$$\leq \|\pi^t(\phi - \phi_0)\| + \|\pi^{t_0}(\pi^{t-t_0}(\phi_0) - \phi_0)\|$$

$$\leq \|\phi - \phi_0\| + \|\pi(\phi_0,t-t_0) - \phi_0\| .$$

It will be sufficient to check the continuity of π at some point $(\phi_0,0) \in X \times \mathbb{R}^+$. For any $(\phi,t) \in X \times \mathbb{R}^+$ we have for each $\tau \in \mathbb{R}^+$

$$\|\pi(\phi,t) - \phi_0\| \leq \|\pi(\phi,t) - \pi(\phi_0,t)\| + \|\pi(\phi_0,t) - \phi_0\| .$$

We consider, in turn, each term on the right hand side of the last inequality.

$$|\pi(\phi,t)(\tau) - \pi(\phi_0,t)(\tau)| \leq \left|e^{-\lambda t}\sum_{k=0}^{\infty}\frac{(\lambda t)^k}{k!}[\phi(\tau-k\alpha) - \phi_0(\tau-k\alpha)]\right|$$

$$\leq \|\phi - \phi_0\| .$$

$$|\pi(\phi_0,t)(\tau) - \phi_0(\tau)| \le |e^{-\lambda t} \sum_{k=1}^{\infty} \frac{(\lambda t)^k}{k!}\phi_0(\tau-k\alpha)| + |e^{-\lambda t} - 1||\phi_0(\tau)|$$

$$\le e^{-\lambda t}(e^{-\lambda t} - 1)\|\phi_0\| + |e^{-\lambda t} - 1|\|\phi_0\|$$

$$\le 2|e^{-\lambda t} - 1|\|\phi_0\| .$$

It is clear now that $\|\pi(\phi,t) - \phi_0\| \to 0$ as $(\phi,t) \to (\phi_0,0)$. Thus π is continuous.

4. Solutions

As previously mentioned assume that we are given a semi-dynamical system (X,π).

Definition 4.1. A function $\phi: I \to X$ where I is a non-empty interval in $\mathbb{R}$ is called a *solution* of (X,π) if whenever $t \in I$, $s \in \mathbb{R}^+$ and $t+s \in I$, then $\pi(\phi(t),s) = \phi(t+s)$. The interval I is the domain of ϕ and according to our notation is represented by $\mathscr{D}(\phi)$. If $x \in X$, a solution ϕ with $0 \in \mathscr{D}(\phi)$ and $\phi(0) = x$ is called a *solution through* x. The function $\pi_x: \mathbb{R}^+ \to X$ given by $\pi_x(t) = xt$ is a solution through x and, indeed, is the unique solution through x with domain $\mathbb{R}^+$. We call this solution the *positive motion* through x.

Lemma 4.2. Every solution is continuous.

Proof: Suppose $\phi: I \to X$ is a solution and let $\{t_n\}$ be a sequence in I with $t_n \to t \in I$. Choose $s \in I$ with $s < t$. Then for sufficiently large $n \in \mathbb{N}$, $s \le t_n = s+(t_n-s)$. We obtain from the definition of a solution that

$$\phi(t_n) = \phi(s+(t_n-s)) = \pi(\phi(s),t_n-s) \to \pi(\phi(s),t-s) = \phi(t).$$

This shows that ϕ is continuous at t. □

The main purpose of this section is to extend positive motions to the negative time domain if possible.

Definition 4.3. A solution $\hat{\phi}$ is called an *extension of a solution* ϕ if $\mathscr{D}(\hat{\phi}) \supset \mathscr{D}(\phi)$ and $\hat{\phi} = \phi$ on $\mathscr{D}(\phi)$. A solution ϕ is called *maximal* if for every extension ψ of ϕ we have $\mathscr{D}(\psi) = \mathscr{D}(\phi)$ (and hence $\psi = \phi$ on $\mathscr{D}(\phi)$). We will use γ_x to denote a maximal solution through x.

Theorem 4.4. For each $x \in X$ there is a maximal solution through x.

Proof: π_x is a solution through x with $\mathscr{D}(\pi_x) = \mathbb{R}^+$. We show how to extend this to a maximal solution. Observe that the collection of all solutions through x is partially ordered by the relation $\prec$; namely, $\phi \prec \psi$ if and only if $\mathscr{D}(\phi) \subset \mathscr{D}(\psi)$ and $\phi = \psi$ on $\mathscr{D}(\phi)$. Now suppose $\{\phi_\alpha\}_{\alpha\in\Lambda}$ is a linearly ordered family of solutions through x. Set $I = \cup\{\mathscr{D}(\phi_\alpha): \alpha \in \Lambda\}$ and define $\phi: I \to X$ by $\phi(t) = \phi_\alpha(t)$ for $t \in \mathscr{D}(\phi_\alpha)$. It is easy to verify that $\phi(t)$ is well defined, and indeed is a solution through x with $\mathscr{D}(\phi) = I$. So the family $\{\phi_\alpha\}_{\alpha\in\Lambda}$ has an upper bound, ϕ. Zorn's lemma now implies that π_x admits a maximal extension through x. □

If ϕ is a maximal solution through x, then $\mathscr{D}(\phi) \supset \mathbb{R}^+$. No claims have been made though for the uniqueness of a maximal solution through x. Given right and left solutions through x their obvious meanings, it is clear that π_x is the unique right maximal solution through x. On the other hand, there need not be any left solutions through x. And even if they exist, they need not be unique. See Example 5.11 for such behavior. The following theorem provides a classification of maximal solutions. Its proof is obvious in view of the requirement that the domain of a maximal solution contains $\mathbb{R}^+$.

Theorem 4.5. Let γ_x be a maximal solution through $x \in X$. Then precisely one of the following holds.

(i) $\mathscr{D}(\gamma_x) = \mathbb{R}$.

(ii) $\mathscr{D}(\gamma_x) = [-a_x,\infty)$ for some $a_x \in \mathbb{R}^+$.

(iii) $\mathscr{D}(\gamma_x) = (-a_x,\infty)$ for some $a_x > 0$.

Definition 4.6. A maximal solution γ_x called *principal* if $\mathscr{D}(\gamma_x) = \mathbb{R}$.

We single out maximal solutions with domain $[-a,\infty)$, $a \in \mathbb{R}^+$. It is due to the structure of semidynamical systems that such solutions can exist. Examples of such domains are given in Chapter VI.

Definition 4.7. A point $x \in X$ is called a *start point* if $x \neq yt$ for every $y \in X$ and $t > 0$.

Theorem 4.8. If a maximal solution γ_x has domain $[-a,\infty)$, $a \in \mathbb{R}^+$, then $\gamma_x(-a)$ is a start point. Conversely, if x is a start point, then the positive motion through x is a maximal solution.

Proof: Suppose $\gamma_x(-a)$ is not a start point. Then there exists $y \in X$ and $\tau > 0$ so that $\gamma_x(-a) = y\tau$. Define $\phi: [-a-\tau,\infty) \to X$ by $\phi(t) = y(a+\tau+t)$ for $t \in [-a-\tau,\infty)$. It is straightforward to verify that ϕ is a solution. Since $\phi(-a) = \gamma_x(-a) = y\tau$, then $\phi = \gamma_x$ on $[-a,\infty)$. This is because $\pi_{y\tau}$ is the unique solution through $y\tau$ with domain $\mathbb{R}^+$. Consequently ϕ is an extension of γ_x. As this contradicts the maximality of γ_x, then $\gamma_x(-a)$ must be a start point. The converse statement follows immediately from the definition of a start point. □

Definition 4.9. If a maximal solution γ_x has domain $[-a,\infty)$, $a \geq 0$, then γ_x is said to *issue* from the start point $\gamma_x(-a)$.

5. Critical and Periodic Points

Let there be given a semidynamical system (X,π). The proof of the next proposition is easy and so is omitted.

Proposition 5.1. Fix $x \in X$ and $\tau \in \mathbb{R}^+$. Then the following are equivalent.

(i) $x\tau = x$.

(ii) $x(t+\tau) = xt$ for all $t \in \mathbb{R}^+$.

(iii) $x(n\tau) = x$ for all $n \in \mathbb{Z}^+$.

Definition 5.2. The point x (or equivalently, the positive motion π_x) is called *critical* if $x\tau = x$ for all $\tau \in \mathbb{R}^+$. Such a point x is also referred to as an *equilibrium* or *rest* point.

We now give a useful characterization of a critical point.

Theorem 5.3. Fix $x \in X$. Then the following are equivalent.

(i) x is a critical point.

(ii) $\mathscr{R}(\pi_x) = \{x\}$.

(iii) $x[a,b] = \{x\}$ for some $a < b$ in $\mathbb{R}^+$.

(iv) For each $c > 0$ there exists $\tau \in (0,c]$ such that $x\tau = x$.

Proof: The equivalence of (i) and (ii) is evident, as is the fact that (i) implies (iii). So first we show that (iv) follows from (iii). Now if (iii) is true then $(xa)(t-a) = x$

for $t \in [a,b]$. Let $c > 0$ and choose $\tau \in (0,c]$ such that $\tau \leq b-a$. As $xa = x$, then $x\tau = x$. This establishes the desired implication. Finally, to see that (iv) implies (i), we can choose a sequence $\{\tau_n\}$ in $\mathbb{R}^+ \setminus \{0\}$ with $\tau_n \to 0$ and $x\tau_n = x$ for every $n \in \mathbb{N}$. For each $t \in \mathbb{R}^+$ there exists a sequence $\{k_n\}$ in $\mathbb{Z}^+$ so that $k_n\tau_n < t \leq (k_n+1)\tau_n$ and $k_n\tau_n \to t$. Now use Proposition 5.1 to obtain $x(k_n\tau_n) = x$ for each $n \in \mathbb{N}$. Take the limit as $n \to \infty$ to get $xt = x$. □

Definition 5.4. The point x (or equivalently, the positive motion π_x) is called *periodic* if $x\tau = x$ for some $\tau > 0$. Such a number τ is called a *period* of x. A principal solution γ_x is called periodic with period τ if $\gamma_x(t+\tau) = \gamma_x(t)$ for all $t \in \mathbb{R}$.

Lemma 5.5. If $x \in X$ is periodic, then the set of all periods of x is a closed subset of $\mathbb{R}^+$.

Proof: Denote by P_x the set of all periods of x. Suppose $\{\tau_n\}$ is a sequence in P_x with $\tau_n \to \tau$, $\tau \in \mathbb{R}^+$. Then as $x = x\tau_n$ for each $n \in \mathbb{N}$ and $x\tau_n \to x\tau$, we have $x\tau = x$. Thus $\tau \in P_x$, so P_x is closed. □

In view of Lemma 5.5 if $x \in X$ is a periodic point, then the number T_x given by

$$T_x \overset{\text{def}}{=} \inf\{\tau > 0 : \tau \text{ is a period of } x\} \tag{5.1}$$

is a period of x.

Definition 5.6. If $x \in X$ is periodic, the number T_x given by (5.1) is called the *fundamental period* of x. If $T_x > 0$, x is called a *purely periodic* point. In this event the positive motion through x is called *T_x-periodic*. (It is customary to drop the modified "purely" when it is clear from the context that $T_x > 0$.)

Remark 5.7. It is clear that a critical point x is also a periodic point with the property that every $\tau \in \mathbb{R}^+$ is a period of x. On the other hand a purely periodic point cannot be critical in view of condition (iv) of Theorem 5.3.

Theorem 5.8. A necessary and sufficient condition for a point $x \in X$ to be purely periodic is that x has a positive period T such that $\{nT: n \in \mathbb{N}\}$ constitutes the full set of positive periods of x.

Proof: Sufficiency is obvious in view of the definition of T. Conversely, suppose x is purely periodic. Let T be the fundamental period of x. If P_x denotes the full set of positive periods of x, we must have $\{nT: n \in \mathbb{N}\} \subset P_x$. Now suppose $t \in P_x$ with $t \notin \{nT: n \in \mathbb{N}\}$. There exists $k \in \mathbb{N}$ so that $kT < t < (k+1)T$. As $x = x(kT) = xt = (x(kT))(t-kT) = x(t-kT)$, then $t-kT \in P_x$ with $0 < t-kT < T$. But this contradicts the fact that T is the fundamental period of x. Consequently $t \in \{nT: n \in \mathbb{N}\}$. □

In contrast to arbitrary positive motions which need not possess a unique extension, periodic positive motions admit a unique extension to a principle periodic solution.

Theorem 5.9. If π_x is T-periodic, then π_x admits a unique maximal extension to a principle T-periodic solution.

Proof: Define $\gamma_x: \mathbb{R} \to X$ by

$$\gamma_x(t) = \begin{cases} \pi_x(t), & t \in \mathbb{R}^+ \\ \pi_x(t+nT), & t < 0 \text{ and } n = [-t/T] + 1, \end{cases}$$

where $[s]$ denotes the greatest integer $\leq s$. Thus for $t < 0$, n is the unique positive integer for which $0 \leq t + nT \leq T$.

First we check that γ_x is indeed a solution through x with domain $\mathbb{R}$. Clearly, $\gamma_x(0) = x$ and $\gamma_x|_{\mathbb{R}^+} = \pi_x$. Now suppose $t < 0$ and $s \in \mathbb{R}^+$. We get $\pi(\gamma_x(t),s) = \pi(\pi_x(t+nT),s) = \pi_x(t+s+nT)$, where n is the unique positive integer for which $0 \leq t+nT \leq T$. If $0 \leq t+s+nT \leq T$, then $\pi_x(t+s+nT) = \gamma_x(t+s)$. On the other hand, if $t+s+nT > T$, choose an integer m in $[0,n]$ so that $0 \leq t+s+(n-m)T \leq T$. Then $\pi_x(t+s+nT) = \pi(x,t+s+nT) = \pi(\pi(x,mT),t+s+(n-m)T) = \pi(x,t+s+(n-m)T) = \pi_x(t+s+(n-m)T) = \gamma_x(t+s)$. This shows that γ_x is a solution which extends π_x. Since $\mathcal{D}(\gamma_x) = \mathbb{R}$ by construction of γ_x, it follows that γ_x is principal.

γ_x is T-periodic since for $t \in \mathbb{R}^+$ we have $\gamma_x(t) = \pi_x(t+T) = \gamma_x(t+T)$ while for $t < 0$ we have $\gamma_x(t) = \pi_x(t+nT) = \pi_x(t+T+(n-1)T) = \gamma_x(t+T)$.

In order to verify that the principal solution γ_x is unique, suppose that ψ_x is another T-periodic principal solution through x. Obviously, ψ_x and γ_x must agree with π_x on $\mathbb{R}^+$. So if $\psi_x \neq \gamma_x$, there must exist $t < 0$ so that $\psi_x(t) \neq \gamma_x(t)$. Choose $n \in \mathbb{Z}^+$ such that $t+nT \in \mathbb{R}^+$. Consequently, $\pi_x(t+nT) = \psi_x(t+nT) = \psi_x(t) \neq \gamma_x(t) = \gamma_x(t+nT) = \pi_x(t+nT)$, a contradiction. Thus γ_x is the unique T-periodic extension of π_x. □

It is customary to refer to the unique periodic extension in Theorem 5.9 as π_x. It is then called a *periodic motion*.

The next two examples illustrate the special behavior afforded semidynamical systems, but not dynamical systems.

Example 5.10. Let the phase space X be $\mathbb{R}^2$ with points denoted by $z = (x,y) \in \mathbb{R}^2$. Define the map $\pi: X \times \mathbb{R}^+ \to X$ by

$$\pi(z,t) = \begin{cases} (x+t,y), & x < 0, \quad x + t < 0 \\ (x-t,y), & x > 0, \quad x - t > 0 \\ (0,y), & x \leq 0, \quad x + t \geq 0 \\ (0,y), & x \geq 0, \quad x - t \leq 0 . \end{cases}$$

Then (X,π) is a semidynamical system with the representation given in Figure 5.1. Note the y-axis consists entirely of critical points. What makes this example so interesting is that positive motions through points z not on the y-axis reach a critical point in a finite amount of time. That is, if $z_0 = (x_0,y_0)$ with $x_0 \neq 0$, then $(0,y_0) = \pi(z_0,t_0)$ with $t_0 = |x_0|$. This cannot happen in the case of dynamical systems. Indeed, if (X,π) is a dynamical system, then no motion can reach a critical point in a finite amount of time. (z is a critical point for a dynamical system if $\pi(z,\tau) = z$ for all $\tau \in \mathbb{R}$.) Otherwise there exists a critical point $z \in X$, a point $u \in X$, $(u \neq z)$ and some $t > 0$ with $\pi(u,t) = z$. Thus $u = \pi(z,-t)$ which contradicts the assertion that z is critical.

Example 5.11. Again let the phase space X be $\mathbb{R}^2$, but now denote points $z \in \mathbb{R}^2$ in the polar form, $z = re^{i\theta}$, $r \in \mathbb{R}^+$, $\theta \in \mathbb{R}$, $i^2 = -1$. Define the map $\pi: X \times \mathbb{R}^+ \to X$ by

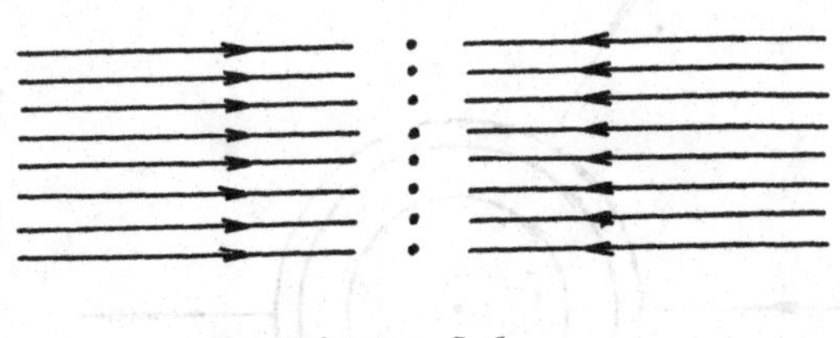

Figure 5.1

$$\pi(z,t) = \begin{cases} re^{i(\theta+t)}, & 0 \le r \le 1, \quad t \in \mathbb{R}^+ \\ (r-t)e^{i\theta}, & r > 1, \ 0 \le t \le r-1 \\ e^{i(\theta+t-r+1)}, & r > 1, \quad t > r-1. \end{cases}$$

Then (X,π) is a semidynamical system with the representation given in Figure 5.2. The point $z = 0$ is critical, and every point $z = re^{i\theta}$ with $r \in (0,1]$ is periodic with fundamental period 2π. Through each point $z = re^{i\theta}$ with $r > 1$ is a ray which merges into the periodic solution through $e^{i\theta}$ in time $r - 1$.

This example also illustrates the type of behavior one can have in semidynamical systems which cannot occur in dynamical systems. In particular, note the nonuniqueness of maximal solutions through each point of the unit circle, $r = 1$. Also observe that the positive motion through any point $z = re^{i\theta}$ with $r > 1$ gives rise to a so-called self-intersecting motion. This will be discussed in the next section.

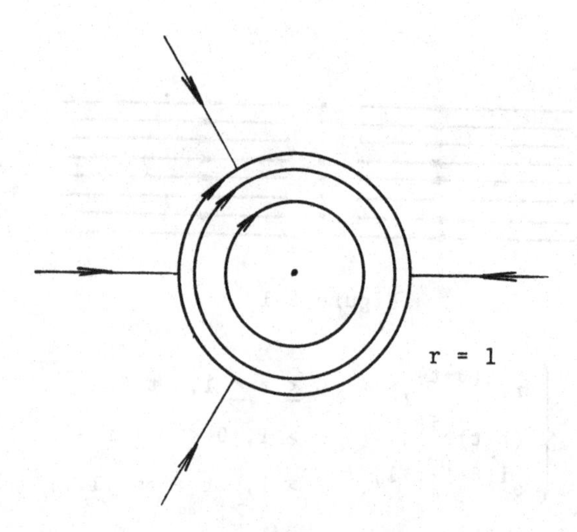

Figure 5.2

6. Classification of Positive Orbits

We begin with a definition which is motivated by Example 5.11. As in Sections 4 and 5, let there be given a fixed semiflow (X,π).

Definition 6.1. The positive motion π_x is called *self-intersecting* if $xt_1 = xt_2$ for some $t_1,t_2 \in \mathbb{R}^+$, $t_1 \neq t_2$.

The proposition which follows provides an algebraic characterization of self-intersecting positive motions. This important tool will enable us to obtain an algebraic classification of positive motions (Theorem 6.7) and a topological classification of the corresponding positive orbits (Theorem 6.11).

Proposition 6.2. For each $x \in X$ there exists $\lambda_x \in \mathbb{R}^+$ and a closed additive subgroup G_x of $\mathbb{R}$ such that $xt_1 = xt_2$ for some $t_1,t_2 \in \mathbb{R}^+$ with $t_1 \leq t_2$ if and only if $\lambda_x \leq t_1 \leq t_2$ and $t_2 - t_1 \in G_x$.

The proof of the proposition proceeds by a sequence three lemmas. The first lemma essentially states that the periods of a periodic point form a group.

Lemma 6.3. For each $x \in X$ and $t \in \mathbb{R}^+$ the set

$$G_x(t) \overset{\text{def}}{=} \{\tau \in \mathbb{R} : x(t+|\tau|) = xt\}$$

is a closed additive subgroup of $\mathbb{R}$.

Proof: First we show $G_x(t)$ is closed. Suppose $\{\tau_i\}$ is a sequence in $G_x(t)$ with $\tau_i \to \tau$. Thus $|\tau_i| \to |\tau|$. As $x(t+|\tau_i|) = xt$ and in view of the continuity of π, we obtain upon taking limits that $x(t+|\tau|) = xt$. Thus $\tau \in G_x(t)$.

Next we show $G_x(t)$ is an additive subgroup of $\mathbb{R}$. Clearly $0 \in G_x(t)$, and if $\tau \in G_x(t)$, then so does $-\tau \in G_x(t)$. Now suppose $\tau_1, \tau_2 \in G_x(t)$. We can assume without loss of generality that $|\tau_1| \geq |\tau_2|$. Then either $|\tau_1 + \tau_2| = |\tau_1| + |\tau_2|$ or $|\tau_1 + \tau_2| = |\tau_1| - |\tau_2|$. In the first case we have

$$x(t+|\tau_1+\tau_2|) = x(t+|\tau_1|+|\tau_2|) = (x(t+|\tau_2|))|\tau_1|$$
$$= (xt)|\tau_1| = x(t+|\tau_1|) = xt,$$

whereas in the second case we have

$$x(t+|\tau_1+\tau_2|) = x(t+|\tau_1|-|\tau_2|) = (xt)(|\tau_1|-|\tau_2|)$$
$$= (x(t+|\tau_2|))(|\tau_1|-|\tau_2|) = x(t+|\tau_1|) = xt.$$

Consequently, we have shown that $\tau_1 + \tau_2 \in G_x(t)$, and so $G_x(t)$ is an additive subgroup of $\mathbb{R}$.

The next lemma specifies the domain of periodicity of a possibly self-intersecting positive motion.

Lemma 6.4. For each $x \in X$ the set

$$I_x \overset{\text{def}}{=} \{t \in \mathbb{R}^+ : \ xt = x\tau \ \text{ for some } \ \tau > t\}$$

is either empty or is an interval of the form $[\lambda_x, \infty)$ in $\mathbb{R}^+$. In the latter case, there exists $\varepsilon = \varepsilon(x)$ so that $I_x = \{\lambda t \in \mathbb{R}^+ : xt = x(t+\varepsilon)\}$.

Proof: Suppose $I_x \neq \emptyset$. Choose $t_0 \in I_x$. Then $[t_0,\infty) \subset I_x$. In order to see this, first note that there exists $\varepsilon > 0$ so that $x(t_0+\varepsilon) = xt_0$. Then for any $t \geq t_0$ we obtain

$$xt = xt_0(t-t_0) = (x(t_0+\varepsilon))(t-t_0) = x(t+\varepsilon),$$

so that $t \in I_x$. Hence $[t_0,\infty) \subset I_x$, and I_x is an interval. Moreover, using the same ε chosen above, we have $x(t'+\varepsilon) = xt'$ for all $t' \in I_x$. That is, the choice of ε does not depend on $t' \in I_x$. Indeed, if $t' \in I_x$ with $t' < t_0$, there exists $\varepsilon' > 0$ so that $x(t'+\varepsilon') = xt'$. Hence $x(t'+n\varepsilon') = xt'$ for every $n \in \mathbb{Z}^+$. Choose $k \in \mathbb{Z}^+$ such that $t' + k\varepsilon' > t_0$. then

$$x(t'+\varepsilon) = (xt')\varepsilon = (x(t'+k\varepsilon'))\varepsilon = x(t'+k\varepsilon'+\varepsilon)$$

$$= x(t'+k\varepsilon') = xt'.$$

Finally, we observe that I_x is closed. In particular, let $\{t_i\}$ be a sequence in I_x with $t_i \to t$. Then $x(t_i+\varepsilon) = xt_i$ for each $i \in \mathbb{Z}^+$. Taking limits we obtain $x(t+\varepsilon) = xt$, which shows that $t \in I_x$. Now set $\lambda_x = \inf I_x$. We have $I_x = [\lambda_x,\infty)$ as promised. □

The final lemma demonstrates the invariance of the group of periods $G_x(t)$ with respect to the initial time t. The notation is that of the last two lemmas.

Lemma 6.5. $G_x(s) = G_x(t)$ for all $s,t \in I_x$.

Proof: From the proof of Lemma 6.4 we see that $x(s+|\tau|) = xs$ if and only if $x(t+|\tau|) = xt$ whenever $s,t \in I_x$. Hence $\tau \in G_x(s)$ if and only if $\tau \in G_x(t)$. □

We now turn to the proof of Proposition 6.2.

Proof: Let $x \in X$. If $xt_1 = xt_2$ holds only for $t_1 = t_2$, then choose $\lambda_x = 0$ and $G_x = \{0\}$. (Note that in this case $I_x = \emptyset$, but we define $\lambda_x = 0$ anyway.) On the other hand, if $xt_1 = xt_2$ for some $t_1 < t_2$ in $\mathbb{R}^+$, then choose λ_x as specified in Lemma 6.4, and $G_x = G_x(\lambda_x)$. It is straightforward to verify that λ_x and G_x have the desired properties. □

Before we can state the anticipated classification of positive motions, we require the following definition.

Definition 6.6. We say that the positive motion π_x *merges into critical point* y if there exists $\lambda_x \in \mathbb{R}^+$ so that $y \notin x[0,\lambda_x)$, but $x[\lambda_x,\infty) = y$. We say that the positive motion π_x *merges into the periodic motion* π_y if there exists $\lambda_x \in \mathbb{R}^+$ so that $y\mathbb{R}^+ \cap x[0,\lambda_x) = \emptyset$, but $y = x\lambda_x$ is a periodic point.

Theorem 6.7. For each $x \in X$ precisely one of the following alternatives is true (λ_x and G_x were established in Proposition 6.2):

(i) $G_x = \{0\}$; equivalently, π_x is nonself-intersecting.

(ii) $\lambda_x = 0$, $G_x = \mathbb{R}$; equivalently, π_x is critical.

(iii) $\lambda_x = 0$, $\{0\} \subsetneq G_x \subsetneq \mathbb{R}$; equivalently, π_x is purely periodic.

(iv) $\lambda_x > 0$, $G_x = \mathbb{R}$; equivalently, π_x merges into a critical point.

(v) $\lambda_x > 0$, $\{0\} \subsetneqq G_x \subsetneqq \mathbb{R}$; equivalently, π_x merges into a purely periodic motion.

The proof of this theorem is obvious and so, is omitted. We remark here that in cases (iii) and (v) the group G_x is a nontrivial infinite cyclic subgroup of $\mathbb{R}$.

Next we introduce the notion of the positive orbit through a point x.

Definition 6.8. The *positive orbit* through $x \in X$, denoted by $\gamma^+(x)$, is the range of the positive motion π_x through x. Positive orbits are called critical, periodic, or self-intersecting according to whether the corresponding motion is critical, periodic, or self-intersecting.

Remark 6.9. Observe that $\gamma^+(x)$ is the set $x\mathbb{R}^+$.

The classification of positive motions given in Theorem 6.7 translates into an equivalent topological classification of positive orbits. But first we need the following preliminary lemma.

Lemma 6.10. Let $x \in X$. $\gamma^+(x)$ is compact if and only if it is self-intersecting.

Proof: Suppose $\gamma^+(x)$ is self-intersecting. There exist $t_1 < t_2$ in $\mathbb{R}^+$ with $xt_1 = xt_2$. We claim $\gamma^+(x) = x[0,t_2]$, which shows $\gamma^+(x)$ is compact. In order to establish our claim, note that $x(t_1+\varepsilon+n(t_2-t_1)) = x(t_1+\varepsilon)$ holds for each $n \in \mathbb{Z}^+$ and $\varepsilon \geq 0$. For any $t > t_2$ we can choose ε and n so that $0 \leq \varepsilon < t_2-t_1$ and $t = t_1+\varepsilon+n(t_2-t_1)$. Thus

$xt = x(t_1+\varepsilon) \in x[0,t_2]$. This establishes $\gamma^+(x) = x[0,t_2]$.

Conversely, suppose $\gamma^+(x)$ is compact. As $\gamma^+(x)$ may be written $\gamma^+(x) = \bigcup_{n=0}^{\infty} x[0,n]$, it follows from the Baire category theorem that for some $n_0 \in \mathbb{N}$, $x[0,n_0]$ has nonempty interior relative to $\gamma^+(x)$. Let y be such an interior point, and write $y = xs_1$. The compactness of $\gamma^+(x)$ ensures that the sequence $\{xn\}$ has a cluster point $z = xs_2 \in \gamma^+(x)$. Consequently y is a cluster point of the sequence $\{x(n+s_1-s_2)\}$. (This sequence is defined for all integers $n \geq |s_1-s_2|$.) We can find arbitrarily large integers $n > n_0 + |s_1-s_2|$ so that $x(n+s_1-s_2) \in x[0,n_0]$. In other words $\gamma^+(x)$ is self-intersecting as there exists $t_1 \in (0,n_0)$, $t_1 \neq t_2 \overset{\text{def}}{=} n+s_1-s_2$ with $xt_2 = xt_1$.

Theorem 6.11. For each $x \in X$ precisely one of the following alternatives is true.

(i) $\gamma^+(x)$ is noncompact.

(ii) $\gamma^+(x)$ is homeomorphic to $\{x\}$.

(iii) $\gamma^+(x)$ is homeomorphic to the unit circle, S^1.

(iv) $\gamma^+(x)$ is homeomorphic to the unit interval, $[0,1]$.

(v) $\gamma^+(x)$ is homeomorphic to a figure-of-six.

This classification corresponds to that given in Theorem 6.7 in the same order.

Proof: Suppose $\gamma^+(x)$ is compact. Then it is self-intersecting, so precisely one of categories (ii), (iii), (iv) or (v) of Theorem 6.7 hold. Now consider the relation $\sim$ on $\mathbb{R}^+$ defined by $s \sim t$ if and only if $xs = xt$. Then $\sim$ is an equivalence relation. Define the corresponding quotient map, $h\colon \mathbb{R}^+ \to \mathbb{R}^+/\sim$, by $h(t) = \tilde{t}$, where $\tilde{t}$ is the equivalence class containing t. Since $\pi_x: \mathbb{R}^+ \to \gamma^+(x)$ is a continuous map onto $\gamma^+(x)$, there is a unique continuous one-to-one map $g: \mathbb{R}^+/\sim \to \gamma^+(x)$ onto $\gamma^+(x)$ so that $\pi_x = g \circ h$. Theorem

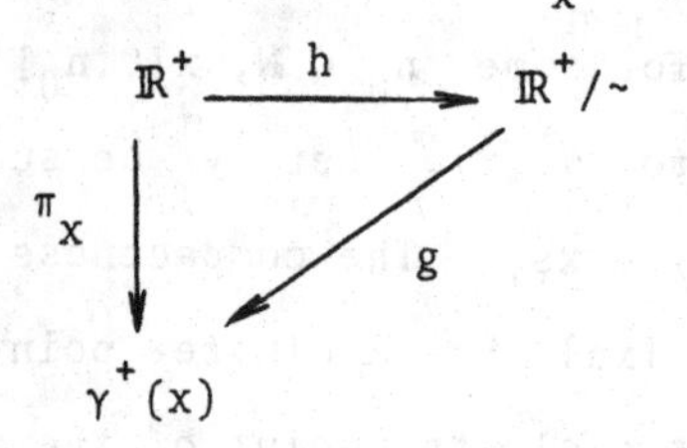

6.7 shows for self-intersecting π_x that $\mathbb{R}^+/\sim$ is homeomorphic to one of $\{x\}$, S^1, $[0,1]$, or a figure-of-six. Each of these spaces is compact, consequently g is a homeomorphism. As these four spaces are mutually nonhomeomorphic, the theorem is proved. □

We conclude this section with a discussion of orbits.

<u>Definition 6.12</u>. An *orbit* through $x \in X$ is the range of a maximal solution through x. We use the notation $\gamma(x)$ to denote an orbit through x. An orbit is called *principal, critical, periodic,* or *self-intersecting* according to whether the corresponding motion is principal, critical, periodic, or self-intersecting.

The following theorem provides a useful classification of orbits as they relate to start points (cf. Theorem 4.5). First, for each orbit $\gamma(x)$ through $x \in X$ define $a_x \in \mathbb{R}^+ \cup \{\infty\}$ by

$$a_x \overset{\text{def}}{=} \sup\{t \in \mathbb{R}^+ : \exists y \in \gamma(x) \text{ with } yt = x\}.$$

We omit the proof as the theorem is equivalent to Theorem 4.5.

<u>Theorem 6.13</u>. Let $\gamma(x)$ be an orbit through $x \in X$. Then precisely one of the following alternatives is true.

(i) There is a start point $y \in \gamma(x)$, in which case there exists $\tau \geq 0$ such that $y\tau = x$ and $\gamma^+(y) = \gamma(x)$.

(ii) There is no start point in $\gamma(x)$ and $a_x < \infty$.

(iii) There is no start point in $\gamma(x)$ and $a_x = \infty$.

<u>Corollary 6.14</u>. An orbit $\gamma(x)$ cannot contain more than one start point.

<u>Corollary 6.15</u>. If $\gamma(x)$ is relatively compact, then only alternatives (i) and (iii) of Theorem 6.13 are possible.

<u>Proof</u>: Suppose alternative (ii) occurs. There is a sequence $\{t_n\}$ in $\mathbb{R}^+$ with $t_n \uparrow a_x$ and a sequence $\{y_n\}$ in $\gamma(x)$ so that $y_n t_n = x$. As $\overline{\gamma(x)}$ is compact we may assume that $\{y_n\}$ has a cluster point $y \in \overline{\gamma(x)}$. Suppose $ya_x \neq x$. Let U be a neighborhood of ya_x which does not contain x. From the continuity of π there must be open sets V and W which contain y and a_x respectively and such that $VW \subset U$. There exists a positive integer n_0 so that $t_n \in W$ for all $n \geq n_0$. There also must exist $m \geq n_0$ so that $y_m \in V$. Hence $x = y_m t_m \in U$. As U was chosen to exclude x we have a contradiction. Thus $ya_x = x$. This means y is a start point, which again is a contradiction. □

<u>Definition 6.16</u>. The semidynamical system (X,π) is called *Lagrange stable* if $\overline{\gamma^+(x)}$ is compact for every $x \in X$.

7. Discrete Semidynamical Systems

Definition 7.1. The pair (X,π) is called a *discrete semidynamical system* if X is a Hausdorff topological space and π is a mapping, $\pi: X \times \mathbb{Z}^+ \to X$ which satisfies

(i) $\pi(x,0) = x$ for each $x \in X$,

(ii) $\pi(\pi(x,n),m) = \pi(x,m+n)$ for each $x \in X$ and $m,n \in \mathbb{Z}^+$, and

(iii) π is continuous.

If $\mathbb{Z}^+$ is replaced by $\mathbb{Z}$ in Definition 7.1, the pair (X,π) is then called a *discrete dynamical system*.

It is readily seen that a discrete semidynamical system determines a continuous mapping $F: X \to X$ given by $F(x) = \pi(x,1)$. Conversely, every continuous mapping $F: X \to X$ defines a phase map π for a discrete semidynamical system (X,π) by $\pi(x,n) = F^n(x)$, where F^n is the nth iterate of F. Thus the study of discrete semidynamical systems reduces to the study of the iterates of a continuous operator. This is a popular topic but we only treat those aspects having to do with stability and other related asymptotic properties. Also discrete semidynamical systems are worth singling out if only to provide some elementary yet useful examples which illustrate behavior not easily recognized in continuous systems.

The concept of a solution through $x \in X$ reduces to a sequence $\{x_i\}_{i=k}^{\ell}$ $(-\infty \leq k < \ell \leq +\infty)$ which agrees with the action of the mapping F; namely, $F^n(x_i) = x_{i+n}$ whenever $i+n \leq \ell$. If $\ell < +\infty$, we may always extend the solution beyond x_ℓ by setting $x_{\ell+1} = F(x_\ell)$, etc. In the other direction we need only find some $y \in X$ so that $F(y) = x_k$. Set $x_{k-1} = y$.

We can continue this process while such y exist to obtain a maximal solution through x. Start points have the same meaning in the discrete case so we obtain the following result, a restatement of Theorems 4.5 and 4.8.

Theorem 7.2. Let (X,π) be a discrete semidynamical system and γ_x a maximal solution through $x \in X$. Then either $\mathscr{D}(\gamma_x) = \mathbb{Z}$ or $\mathscr{D}(\gamma_x) = \{-m,-m+1,\ldots,-3,-2,-1\} \cup \mathbb{Z}^+$ for some positive integer m. In the latter case $\gamma_x(-m)$ is a start point.

Theorem 5.3 may be rephrased in an obvious way for the discrete case; so may Theorem 5.9. The algebraic classification of Theorem 6.7 also carries over to discrete semidynamical systems. The topological classification of Theorem 6.11 does not hold in the discrete case though because $\mathscr{D}(\gamma_x)$ is not connected. Instead we obtain the following.

Theorem 7.3. Let (X,π) be a discrete semidynamical system. The positive orbit $\gamma^+(x)$ is compact if and only if $\gamma^+(x)$ is a finite set.

8. Local Semidynamical Systems; Reparametrization

The theory of semidynamical systems concerns itself with the global behavior of solutions to certain classes of evolutionary equations. But it is known that these equations, e.g. ordinary differential equations, partial differential equations, Volterra integral equations, and functional differential equations, give rise to local semidynamical systems if indeed they possess solutions and positive uniqueness. Since many dynamic considerations are invariant to reparametrizations, it is of interest to know when a local semidynamical system can be reparametrized to yield a "global" one. The main theorem states this parametrization is possible provided the phase space is metric. Since most important examples of interest, including those listed above, possess metric phase spaces, we see that little generality is lost by treating only "global" semidynamical systems in the book.

<u>Definition 8.1</u>. A *local semidynamical system* is a pair (X,π) where X is a Hausdorff topological space and π is a mapping $\pi: \mathcal{D}_\pi \subset X \times \mathbb{R}^+ \to X$ which satisfies

(i) $\mathcal{D}_\pi$ is open in the product topology of $X \times \mathbb{R}^+$,

(ii) for every $x \in X$ there exists $\omega_x \in (0,\infty]$ such that $(x,t) \in \mathcal{D}_\pi$ if and only if $t \in [0,\omega_x)$,

(iii) $\pi(x,0) = x$ for every $x \in X$,

(iv) if $(x,t) \in \mathcal{D}_\omega$ and $(\pi(x,t),s) \in \mathcal{D}_\pi$, then $(x,t+s) \in \mathcal{D}_\pi$ and $\pi(\pi(x,t)s) = \pi(x,t+s)$,

(v) π is continuous on $\mathcal{D}_\pi$.

The adjective *global* in reference to a semidynamical system is actually redundant. It just serves to emphasize the distinction between a semidynamical system and a local semidynamical system. If for a local semidynamical system $\omega_x = \infty$ for every $x \in X$, then it is a global semidynamical system. The same notations and conventions will be used for local systems as was used for global ones.

<u>Lemma 8.2</u>. Suppose (X,π) is a local semidynamical system. The assignment $x \to \omega_x$ defines a lower semicontinuous map from X into $(0,\infty]$; that is, $\omega_x \leq \liminf_{y \to x} \omega_y$.

<u>Proof</u>: Suppose $t \in [0,\omega_x)$. Then xt is defined. In view of the continuity of π, yt is also defined for y near x.

Consequently for such y we must have $t < \omega_y$ so $t \leq \liminf_{y\to x} \omega_y$. Now take $t \uparrow \omega_x$ and we obtain the desired result. □

Lemma 8.3. Suppose (X,π) is a local semidynamical system. If for any $(x,t) \in X \times \mathbb{R}^+$, xt is defined, then $\omega_{xt} = \omega_x - t$.

Proof: We first show that $\omega_{xt} \leq \omega_x - t$. Take any $s \in [0,\omega_{xt})$. Then $(xt)s = x(t+s)$ is defined. Consequently $t+s < \omega_x$; that is, $s < \omega_x - t$. Now let $s \uparrow \omega_{xt}$ to obtain $\omega_{xt} \leq \omega_x - t$. In order to show $\omega_{xt} \geq \omega_x - t$ we will assume the contrary holds for some $(x,t) \in X \times \mathbb{R}^+$. Therefore $\omega_{xt} < \omega_x - t$, so by necessity $\omega_{xt} < \infty$. For any net $\{t_\alpha\}$ in $\mathbb{R}^+$ with $t_\alpha \uparrow \omega_{xt}$, $(xt)t_\alpha = x(t+t_\alpha)$ is defined. Thus $t+t_\alpha \to t + \omega_{xt} < \omega_x$ by assumption, so $x(t+t_\alpha)$ is also defined and $(xt)t_\alpha \to x(t+\omega_{xt})$. Set $y = x(t+\omega_{xt})$ and $x_\alpha = (xt)t_\alpha$. In view of Lemma 8.2 and the fact that $\omega_{x_\alpha} \leq \omega_{xt} - t_\alpha$, we have

$0 < \omega_y \leq \liminf_{x_\alpha\to y} \omega_{x_\alpha} \leq \liminf_\alpha (\omega_{xt} - t_\alpha) = \omega_{xt} - \omega_{xt} = 0$, a contradiction. □

We turn to the reparametrization of a local semidynamical system.

Definition 8.4. We say that two local semidynamical systems (X,π) and (X',π') are *isomorphic* provided there exists a homeomorphism h from X onto X' and a mapping ϕ from $\mathscr{D}_\pi$ into $\mathbb{R}^+$ which satisfies

(i) $\phi(x,0) = 0$ for every $x \in X$,

(ii) ϕ is continuous,

(iii) $\phi(x,\cdot)$ is a homeomorphism from $[0,\omega_x)$ onto $[0,\omega_{h(x)})$ for each $x \in X$, and

(iv) $h(\pi(x,t)) = \pi'(h(x),\phi(x,t))$ for every $(x,t) \in \mathscr{D}_\pi$.

ϕ is referred to as the *reparametrization* mapping.

Lemma 8.5. Suppose for each $x \in X$ there exists $\omega_x \in (0,\infty]$ such that

$$\mathcal{D} = \bigcup_{x \in X} \{\{x\} \times [0,\omega_x)\} \text{ is open in } X \times \mathbb{R}^+.$$

If there exists a continuous mapping $\phi: \mathcal{D} \to \mathbb{R}^+$ such that $\phi(x,\cdot): [0,\omega_x) \to \mathbb{R}^+$ is a homeomorphism onto $\mathbb{R}^+$ for each $x \in X$ with $\phi(x,0) = 0$, then the mapping $\eta: \mathcal{D} \to X \times \mathbb{R}^+$ defined by $\eta(x,t) = (x,\phi(x,t))$ is a homeomorphism onto $X \times \mathbb{R}^+$.

Proof: Clearly η is a one-to-one continuous mapping of $\mathcal{D}$ onto $X \times \mathbb{R}^+$. It will only be necessary to show that η is an open mapping. The basic open subsets of $\mathcal{D}$ are of the form $W \times [0,b)$ and $W \times (a,b)$, where W is a basic open subset of X and $W \times \{b\} \subset \mathcal{D}$. Using the facts that $\phi(x,\cdot)$ is a homeomorphism and $\phi(x,0) = 0$,

$$\eta(W\times[0,b)) = \bigcup_{\substack{x \in W \\ t \in [0,b)}} \eta(x,t) = \bigcup_{x \in W} \{\{x\} \times [0,\phi_x(b))\}.$$

As ϕ is continuous and $\mathcal{D}$ is open, then $\bigcup\{\{x\} \times [0,\phi(x,b)): x \in W\}$ is an open subset of $X \times \mathbb{R}^+$. Similarly for $0 < a < b$,

$$\eta(W \times (a,b)) = \bigcup_{x \in W} \{\{x\} \times (\phi(x,a),\phi(x,b))\},$$

which is also open in $X \times \mathbb{R}^+$. □

We are ready for the main result: necessary and sufficient conditions for a local semidynamical system to be reparametrized in the sense of Definition 8.4.

Theorem 8.6. Let (X,π) be a local semidynamical system. The following properties of (X,π) are equivalent.

(i) There exists a (global) semidynamical system (X,π') which is isomorphic to (X,π).

(ii) The closed sets $X \times \{0\}$ and $X \times \mathbb{R}^+ \setminus \mathcal{D}_\pi$ can be separated by a continuous function.

(iii) There exists a continuous function $f: X \to (0,1]$ with $f(x) \leq \omega_x$ for every $x \in X$.

Proof: We first show that (i) implies (ii). So let $\phi: \mathcal{D}_\pi \to \mathbb{R}^+$ be the reparametrization mapping according to Definition 8.4, and define $g: X \times \mathbb{R}^+ \to [0,1]$ by

$$g(x,t) = \begin{cases} \min\{1, 1/\phi(x,t)\}, & (x,t) \in \mathcal{D}_\pi \\ 0 \quad , & (x,t) \notin \mathcal{D}_\pi . \end{cases}$$

As ϕ is continuous on $\mathcal{D}_\pi$, then so is g continuous on $\mathcal{D}_\pi$. By hypothesis the homeomorphism h must be such that $\omega_{h(x)} = \infty$ for every $x \in X$. Consequently, if $\{(x_\alpha,t_\alpha)\}$ is a a net in $\mathcal{D}_\pi$ which converges to $(x,t) \in \bar{\mathcal{D}}_\pi \setminus \mathcal{D}_\pi$, then in view of property (ii) of Definition 8.1 we must have that $\phi(x_\alpha,t_\alpha) \to \infty$. Therefore g is continuous on all of $X \times \mathbb{R}^+$. Moreover $g^{-1}(0) = X \times \mathbb{R}^+ \setminus \mathcal{D}_\pi$, and $g(x,0) = 1$ for every $x \in X$. Thus g separates the closed sets $X \times \mathbb{R}^+ \setminus \mathcal{D}_\pi$ and $X \times \{0\}$.

Next we show that (ii) implies (iii). Given $g: X \times \mathbb{R}^+ \to [0,1]$ according to property (ii), define $f: X \to \mathbb{R}$ by

$$f(x) = \int_0^1 g(x,t)dt.$$

Clearly f is continuous with $0 \leq f(x) \leq 1$ for every $x \in X$. If for some $x \in X$ we have $\omega_x > 1$, then $f(x) < \omega_x$

is obvious. If for some $x \in X$ we have $\omega_x \leq 1$, then the separation property of g implies $g(x,t) = 0$ for $t > \omega_x$. Therefore

$$f(x) = \int_0^{\omega_x} g(x,t)dt \leq \omega_x .$$

Note that since $g(x,0) = 1$ for every $x \in X$, we must have $f(x) > 0$ on X. This establishes property (iii).

Finally we show that (iii) implies (i). Suppose there is given a continuous function $f\colon X \to (0,1]$ with $f(x) \leq \omega_x$ for every $x \in X$. The reparametrization is obtained by defining

$$\phi(x,t) = \int_0^t \frac{ds}{f(\pi(x,s))} .$$

Surely ϕ is continuous on $\mathscr{D}_\pi$ into $\mathbb{R}^+$ and $\frac{d}{dt}\phi(x,t)$ is positive for every $(x,t) \in \mathscr{D}_\pi$. Thus for each $x \in X$,

$$\theta_2 \overset{\text{def}}{=} \phi(\pi(x,t_1),t_2) = \int_0^{t_2} \frac{ds}{f(\pi(\pi(x,t_1),s))}$$

$$= \int_0^{t_2} \frac{ds}{f(\pi(x,t_1+s))} = \int_{t_1}^{t_1+t_2} \frac{ds}{f(\pi(x,s))} .$$

Therefore

$$\int_0^{t_1+t_2} \frac{ds}{f(\pi(x,s))} = \int_0^{t_1} \frac{ds}{f(\pi(x,s))} + \int_{t_1}^{t_1+t_2} \frac{ds}{f(\pi(x,s))} \overset{\text{def}}{=} \theta_1+\theta_2 ;$$

that is, $\phi(x,t_1+t_2) = \theta_1 + \theta_2$. Finally,

$$\pi'(x,\theta_1+\theta_2) = \pi\eta^{-1}(x,\theta_1+\theta_2) = \pi\eta^{-1}(x,\phi(x,t_1+t_2)) = \pi(x,t_1+t_2).$$

This completes the proof that (X,π') is a (global) semidynamical system. □

If $X \times \mathbb{R}^+$ is a normal topological space then property (ii) of Theorem 8.6 always holds. In particular if X is metric, we may reparametrize a local system to obtain a global one.

Corollary 8.7. Let (X,π) be a local semidynamical system with metric phase space X. There exists a (global) semidynamical system (X,π') which is isomorphic to (X,π).

9. Exercises

Unless otherwise stated assume a semidynamical system (X,π) is given.

9.1. For families of sets $M_\alpha \subset X$ and $T_\beta \subset \mathbb{R}^+$ prove $(\bigcup_\alpha M_\alpha)(\bigcup_\beta T_\beta) = \bigcup_{\alpha,\beta} M_\alpha T_\beta$, $(\bigcap_\alpha M_\alpha)t \subset \bigcap_\alpha (M_\alpha t)$, $t \in \mathbb{R}^+$.

9.2. For $M_1, M_2 \subset X$ and $t \in \mathbb{R}^+$ prove $M_1 t \setminus M_2 t \subset (M_1 \setminus M_2)t$.

9.3. Show that for a dynamical system the set inclusions in Exercises 9.1 and 9.2 may be replaced by set equalities.

9.4. Suppose $\pi: X \times \mathbb{R} \to X$ maps $X \times \mathbb{R}$ onto X and satisfies the group property of Definition 2.2. Then $\pi(x,0) = x$ for all $x \in X$.

9.5. Give an example of a semidynamical system for which the set inclusion in Exercises 9.1 and 9.2 are proper.

9.6. Suppose $U \subset X$ is open. For each $x \in U$ define $\omega_x \in \mathbb{R}^+ \cup \{\infty\}$ by $\omega_x = \sup\{t \in \mathbb{R}^+: x[0,t] \subset U\}$. Show that the assignment $x \to \omega_x$ is lower semicontinuous.

9.7. Let $\phi: (-\alpha,\infty) \to X$ be a solution of (X,π) with $\alpha > 0$. Then either $\phi(t)$ has no cluster points as $t \downarrow -\alpha$ (whereupon ϕ is maximal) or $\lim_{t \downarrow -\alpha} \phi(t)$ exists and ϕ admits a proper extension to a solution of (X,π).

9.8. The set of all critical points of (X,π) is closed.

9.9. If $x \in X$ is purely periodic, then all periods of γ_x (the unique principal solution through x) are of the form nT, where $n \in \mathbb{Z}$ and T is the fundamental period of x.

9.10. If (X,π) is a discrete semidynamical system, π_x is self-intersecting if and only if $\gamma^+(x)$ is a finite set.

9.11. A solution ϕ of (X,π) is called negative if $\mathscr{D}(\phi) \subset \mathbb{R}^-$. The range of a negative solution through $x \in X$ is called a negative orbit through x. If a negative solution through x has domain $\mathbb{R}^-$ its range is called a principal negative orbit through x.

(a) $x \in X$ is critical if and only if every neighborhood of x contains either a positive orbit or a principle negative orbit.

(b) If X is locally compact and $x \in X$ is critical, then every neighborhood of x contains either a positive orbit other than $\{x\}$ or a principal negative orbit which does not contain x.

(c) If X is locally compact, $x \in X$ is critical, and U is a neighborhood of x without start points, there is $y \in U$, $y \neq x$ such that $\gamma^+(y) \subset U$.

9.12. A set $N(x)$ is a negative orbit through $x \in X$ if and only if $N(x)$ satisfies (i) $y \in N(x)$ implies $x \in \gamma^+(y)$, and (ii) $y,z \in N(x)$ implies either $y \in \gamma^+(z)$ or $z \in \gamma^+(y)$.

9.13. A negative orbit $N(x)$ through x is principal if and only if for every $t \in \mathbb{R}^+$ there exists $y \in N(x)$ so that $yt = x$.

9.14. Suppose X is locally compact, and for each $y \in X$ let $\alpha_y = \sup\{t \in \mathbb{R}^+ : y \in Xt\}$. Then $x \in X$ is a start point if and only if $\lim_{y \to x} \alpha_y = 0$.

9.15. Let $x \in X$ be purely periodic with fundamental period $T > 0$. Suppose there exists a sequence $\{x_n\} \subset X$ of periodic points with $x_n \to x$. Then for sufficiently large n each x_n is purely periodic. Moreover, if each point x_n has fundamental period $T_n > 0$ and $T_n \to T_0 \in \mathbb{R}^+$, then $T_0 = kT$ for some $k \in N$.

9.16. Show that the result in Exercise 9.15 cannot be improved, e.g., construct an example where $T_0 = 2T$. Show how to construct an example for any given $k \in \mathbb{N}$.

10. Notes and Comments

Section 2. Semidynamical systems were first introduced by Hajek [1]. They represent a generalization of a dynamical system, the first abstract definition of which was formulated independently by both Markov [1] and Whitney [1]. The underlying concept though was originally developed by Poincaré in his study of the topological properties of orbits of second order differential equations. Birkhoff [1] further developed dynamical concepts in the context of autonomous differential equations in $\mathbb{R}^+$.

It is a feature of semidynamical systems that the mapping π_x is not assumed to be differentiable. In the case of differentiable motions see the excellent survey by Markus [2]. The type of problems studied therein are fundamentally different from the ones we explore in this book.

Sections 4-6. All of the material here is from Bhatia [2,5] and Bhatia and Hajek [1]. For the origin of start points see Flugge Lots [1]. McCann [1] has shown how to embed a semidynamical system into another semidynamical system which has no start points, and in which every maximal solution is principal.

Section 8. The reparametrization theorem is the work of Carlson [1]. It generalizes Vinograd [1,2] and Ura [1,2] on the isomorphism between local solutions of autonomous ordinary differential equations and the motions of a corresponding dynamical system.

CHAPTER II

INVARIANCE, LIMIT SETS, AND STABILITY

1. Introduction

The main concerns of this chapter are two-fold. Firstly, where do positive motions go as $t \to \infty$, and secondly, what can we say about the behavior of the resulting limiting orbits?

The first question is resolved by the introduction of positive limit sets. This is dealt with in Section 3 and again in Section 8 where Lyapunov functions are used to locate positive limit sets.

The second question raised above is treated in the remaining portions of the chapter. In Section 4 we demonstrate the existence of principal orbits in positively minimal sets. An example is given though which shows nonuniqueness is possible. (The matter of uniqueness is taken up again in Section 2 of Chapter III.) Sections 5 and 6 are devoted to a physically meaningful extension of the continuity property of semidynamical systems; namely, when $\pi(\cdot,t)$ is continuous on X, uniformly in $t \in \mathbb{R}^+$. This is better known as stability. It ensures that a "small" perturbation of the state of a stable system will not affect its limiting behavior. These ideas are developed in more detail in the metric space setting of Section 7.

Finally in Section 9 we present an example of a discrete semidynamical system which illustrates how a deterministic process may give rise to seemingly stochastic asymptotic behavior. Though the system is discrete and even finite dimensional, it provides us with a clear picture of the possibilities inherent in the simplest of nonlinear systems.

We remark that most of the results in this chapter require compact positive hulls. This restriction is not excessive since almost every application in the later chapters is set in a phase space which admits only compact positive hulls.

And now for a last word on discrete semidynamical systems. Except in the case where connectedness is involved (namely Theorem 3.5, and even here $L^+(x)$ is nonempty, compact and weakly invariant), all results hold for discrete systems as well.

2. Invariance

Definition 2.1. A set $M \subset X$ is called *positively invariant* if for each $x \in M$ we have $\gamma^+(x) \subset M$. The set M is called *invariant* if both M and $X \setminus M$ are positively invariant.

Remark 2.2. $\gamma^+(x)$ is positively invariant for every $x \in X$.

The following corollary provides a useful characterization of (positively) invariant sets. The easy proof is omitted.

Corollary 2.3. A set $M \subset X$ is positively invariant if and only if $M\mathbb{R}^+ = M$. The set M is invariant if and only if for each $x \in M$ every orbit through x lies entirely in M.

Positive invariance is preserved under the following circumstances.

Lemma 2.4. The closure of a positively invariant set is positively invariant. The union and intersection of a family of positively invariant sets are positively invariant.

Proof: We prove only the first statement; the second statement is easy to verify and its proof is left to the reader. So in view of Corollary 2.3 we need only establish that $\overline{M}\,\mathbb{R}^+ = \overline{M}$ whenever M is positively invariant. Now obviously we have $\overline{M} \subset \overline{M}\,\mathbb{R}^+$. In order to obtain the reverse inclusion we need only show that $\overline{M}\,\mathbb{R}^+ \subset \overline{M\,\mathbb{R}^+}$, as the positive invariance of M insures that $\overline{M\,\mathbb{R}^+} = \overline{M}$. So suppose $z = xt \in \overline{M}\,\mathbb{R}^+$ with $x \in \overline{M}$, $t \in \mathbb{R}^+$. There exists a net $\{x_\alpha\}$ in M so that $x_\alpha \to x$. Thus $x_\alpha t \in M\,\mathbb{R}^+$, and upon passing to the limit we obtain $xt \in \overline{M\,\mathbb{R}^+}$. Consequently $\overline{M}\,\mathbb{R}^+ \subset \overline{M\,\mathbb{R}^+} = \overline{M}$. □

Definition 2.5. For $x \in X$ the set $H^+(x) \overset{\text{def}}{=} \overline{\gamma^+(x)}$ is called the *positive hull* of x. If $H^+(x)$ is compact, the corresponding positive motion π_x is called *compact*.

Corollary 2.6. For every $x \in X$ the set $H^+(x)$ is closed and positively invariant.

In between the concepts of positive invariance and invariance lies what is called weak invariance. The motivation for this comes from positive limit sets, a discussion of which is presented in Section 3.

Definition 2.7. A set $M \subset X$ is called *weakly invariant* if for each $x \in M$ there exists an orbit through x which lies entirely in M.

Unlike positive invariance, weak invariance is not preserved under closure nor under intersections. But we do have the following lemma for arbitrary unions. The proof is immediate and so is omitted.

Lemma 2.8. The union of a family of weakly invariant sets is weakly invariant.

Compact weakly invariant sets play a large role in the qualitative study of the asymptotic behavior of semidynamical systems. The next theorem provides a characterization of such sets.

Theorem 2.9. Suppose $M \subset X$ is compact without start points. Then the following are equivalent.

(i) M is weakly invariant.

(ii) Through each point $x \in M$ there is a principal orbit $\gamma(x)$ contained in M.

(iii) For every $x \in M$ and every $\tau > 0$ there exists $y \in M$ with $y\tau = x$ and $\gamma^+(y) \subset M$.

Proof: First we show that (i) implies (ii). Let $x \in M$ and M be weakly invariant. There is an orbit $\gamma(x) \subset M$. As $\overline{\gamma(x)} \subset \overline{M} = M$ is compact and M contains no start points, then $\gamma(x)$ must be principal according to Corollary 6.15 of Chapter I.

That (ii) implies (iii) is readily seen from the definition of a principal orbit.

It remains to establish (iii) implies (i). So let $x \in M$ and set $y_0 = x$. According to (iii) we may choose $y_1 \in M$ so that $y_1 1 = x$ and $\gamma^+(y_1) \subset M$. Continue inductively to define the sequence $\{y_n\}$ in M with $y_n 1 = y_{n-1}$,

$\gamma^+(y_n) \subset M$ for each $n \in \mathbb{N}$. Define $\phi_n: [-n,\infty) \to M$ by $\phi_n(t) = y_n(t+n)$ for every $n \in \mathbb{N}$. Each ϕ_n is a solution through x and is an extension of ϕ_{n-1}, $n \in \mathbb{N}$. Take ϕ_0 to be π_x. For any $t \in \mathbb{R}$ define $\gamma_x(t)$ to be $\phi_n(t)$ where $n \in \mathbb{N}$ is chosen so that $t+n \in \mathbb{R}^+$. Then $\gamma_x(t)$ is well defined for every $t \in \mathbb{R}$, and therefore γ_x is a principal solution through x whose orbit $\gamma(x)$ lies entirely in M. Consequently M is weakly invariant. □

3. Limit Sets: The Generalized Invariance Principle

We turn to the question of where do positive motions go as $t \to \infty$?

Definition 3.1. For every $x \in X$ the set

$$L^+(x) = \{y \in X : xt_\alpha \to y \text{ for some net } \{t_\alpha\} \subset \mathbb{R}^+,\ t_\alpha \to \infty\}$$

is called the *positive limit set* of x.

The following characterization of positive limit sets is useful.

Lemma 3.2. For every $x \in X$ we have

$$L^+(x) = \bigcap_{k\in \mathbb{Z}^+} H^+(xk) = \bigcap_{t\in \mathbb{R}^+} H^+(xt).$$

Proof: We will only prove the first equality; the second one is left as an exercise. Now suppose $y \in L^+(x)$. There exists a net $\{t_\alpha\}$ in $\mathbb{R}^+$, $t_\alpha \to \infty$, with $xt_\alpha \to y$. For any $k \in \mathbb{Z}^+$ we must have $t_\alpha \geq k$ for sufficiently large α. Thus the net $\{xt_\alpha\}$ eventually is in $\gamma^+(xk)$, so $y \in H^+(xk)$. Therefore $L^+(x) \subset \bigcap\{H^+(xk): k \in \mathbb{Z}^+\}$. On the other hand suppose $y \in H^+(xk)$ for every $k \in \mathbb{Z}^+$. For each such k there exists a net $\{y_\alpha^k\} \subset \gamma^+(xk)$ with $y_\alpha^k \to y$. Let U be a neighborhood of y, and for each $k \in \mathbb{Z}^+$ we may choose α_k so that $y_{\alpha_k}^k \in U$. As $y_{\alpha_k}^k \in \gamma^+(xk)$ there must exist $t_k \geq k$ such that $y_{\alpha_k}^k = xt_k$. As $\{y_{\alpha_k}^k\}$ must converge to y and $t_k \to \infty$ we have that $y \in L^+(x)$. Therefore $\cap\{H^+(xk): k \in \mathbb{Z}^+\} \subset L^+(x)$. □

Example 3.3. We describe the positive limit sets in Example 5.11 of Chapter I. For $z = re^{i\theta}$ with $r \geq 1$, $L^+(z) = \{\xi \in \mathbb{C}: |\xi| = 1\}$. For $z = re^{i\theta}$ with $0 \leq r < 1$, $L^+(z) = \{\xi \in \mathbb{C}: |\xi| = |z|\}$.

The positive limit sets possess some desirable properties as the next theorem shows. Note that it is possible for such sets to be empty.

Theorem 3.4. For every $x \in X$ we have

(i) $L^+(x)$ is closed and positively invariant,

(ii) $L^+(xt) = L^+(x)$ for all $t \in \mathbb{R}^+$, and

(iii) $H^+(x) = \gamma^+(x) \cup L^+(x)$.

Proof: That $L^+(x)$ is closed follows from the characterization of $L^+(x)$ in Lemma 3.2. To see that $L^+(x)$ is positively invariant we can assume that $L^+(x) \neq \emptyset$, for otherwise $L^+(x)$ is trivially positively invariant. But again we see from Lemma 3.2 that $L^+(x)$ is the intersection of a family of positively invariant sets. Therefore $L^+(x)$ itself is positively invariant, and so (i) is established. This same lemma implies directly that $L^+(xt) = L^+(x)$ for every $t \in \mathbb{R}^+$ thus establishing (ii).

In order to prove (iii) we see immediately that $L^+(x) \subset H^+(x)$ from Lemma 3.2. Therefore $\gamma^+(x) \cup L^+(x) \subset H^+(x)$. To obtain the reverse inclusion let $y \in H^+(x)$, and suppose $\{y_\alpha\}$ is a net in $\gamma^+(x)$ with $y_\alpha \to y$. For each y_α there exists $t_\alpha \in \mathbb{R}^+$ so that $y_\alpha = xt_\alpha$. If the net $\{t_\alpha\}$ is bounded, there is a subnet which converges to some $t \in \mathbb{R}^+$. In this case we may assume $y_\alpha = xt_\alpha \to xt$ so that $y \in \gamma^+(x)$. If $\{t_\alpha\}$ is unbounded, there is a subnet which converges to ∞. In this case we may also assume $y_\alpha = xt_\alpha \to y \in L^+(x)$. In either case we obtain $H^+(x) \subset \gamma^+(x) \cup L^+(x)$. □

As can be seen from Example 5.11 of Chapter I and Example 3.3 above, the unit circle is positively invariant but not invariant. A closer examination though shows it to be weakly invariant. The next theorem establishes this as a general result in case the positive orbit is relatively compact. Though this appears to be a rather strong assumption, in almost every useful application we endeavor to find a topology for X so that indeed, the positive orbits are relatively compact. The phase spaces for the example in the later chapters illustrate this point. It is for this reason that the next theorem is so crucial.

<u>Theorem 3.5</u>. Suppose $\gamma^+(x)$ is relatively compact for some $x \in X$. Then $L^+(x)$ is nonempty, compact, connected, and weakly invariant. Moreover $L^+(x)$ contains no start points, and every orbit lying in $L^+(x)$ is principal.

<u>Proof</u>: The hypothesis assures us that $H^+(xk)$ is nonempty and compact for every $k \in \mathbb{N}$. As $H^+(xn) \subset H^+(xk)$ for all positive integers $n > k$, we conclude from Lemma 3.2 that $L^+(x)$ is the intersection of a family of compact sets, hence compact. Now suppose $L^+(x)$ were not connected. We may write $L^+(x) = F \cup G$, where F and G are nonempty, closed

disjoint subsets of X. As $L^+(x)$ is compact, so are F and G. They may be separated by open sets V and W; that is, $F \subset V$, $G \subset W$ with $V \cap W = \emptyset$. Set $U = V \cup W$. Then U is an open set containing $L^+(x)$. From Lemma 3.2 there is some $t_0 \in \mathbb{R}^+$ so that $H^+(xt) \subset U$ for all $t \geq t_0$. But $\gamma^+(xt_0)$ is connected as it is the continuous image of $\mathbb{R}^+$. It follows that $H^+(xt_0)$ is also connected. So either $H^+(xt_0) \subset V$ or $H^+(xt_0) \subset W$. Consequently, either $L^+(x) \subset V$ or $L^+(x) \subset W$. So either $V = \emptyset$ or $W = \emptyset$, a contradiction. Therefore $L^+(x)$ must be connected.

Next we demonstrate that $L^+(x)$ is without start points. Let $y \in L^+(x)$. There is a net $\{t_\alpha\}$ in $\mathbb{R}^+$, $t_\alpha \to \infty$, so that $xt_\alpha \to y$. Choose any $\tau > 0$ and consider the net $\{y_\alpha\}$ given by $y_\alpha = x(t_\alpha - \tau)$. This net is eventually defined, hence $\{y_\alpha\}$ lies in the compact set $H^+(x)$. $\{y_\alpha\}$ must have a cluster point $z \in H^+(x)$, hence $\{y_\alpha\tau\}$ has a cluster point $z\tau$. As $xt_\alpha = x(t_\alpha - \tau)\tau = y_\alpha\tau$, then $\{xt_\alpha\}$ also has cluster point $z\tau$. Consequently $y = z\tau$ so y cannot be a start point. Positive invariance of $L^+(x)$ ensures that $\gamma^+(z) \subset L^+(x)$. According to Theorem 2.9 (iii) we may conclude that $L^+(x)$ is weakly invariant. Finally Corollary 6.15 of Chapter I shows that every orbit in $L^+(x)$ is principal. □

Example 5.11 of Chapter I illustrates the next corollary.

Corollary 3.6. If for some $x \in X$, $\gamma^+(x)$ is a self-intersecting positive orbit, then $L^+(x)$ is the positive orbit of a periodic point.

Proof: Suppose $0 < t_1 < t_2$ with $xt_1 = xt_2$. Then $(xt_1)t = (xt_2)t = (xt_2)(t_1+t-t_1) = (xt_1)(t+t_2-t_1)$ for every $t \in \mathbb{R}^+$. Therefore xt_1 is a periodic point with period t_2-t_1. First note that $xt_1 \in L^+(x)$. Indeed, $xt_1 = (xt_1)(n(t_2-t_1))$ for every $n \in \mathbb{N}$. Thus xt_1 is the limit of the sequence $\{xt_n\}$ where $t_n = t_1+n(t_2-t_1) \to \infty$. This implies $H^+(xt_1) \subset L^+(x) = L^+(xt_1) \subset H^+(xt_1) = \overline{(xt_1)[0,t_2-t_1]} = (xt_1)[0,t_2-t_1]$. □

In the event X is a complete metric space we obtain a useful characterization of periodic points. It will be required in Chapter III. First the following lemma is needed.

<u>Lemma 3.7</u>. Suppose X is a complete metric space. If x is not a periodic point of X, then $L^+(x) \smallsetminus \gamma^+(x)$ is dense in $L^+(x)$.

<u>Proof</u>: If $x \notin L^+(x)$, then $\gamma^+(x) \cap L^+(x) = \emptyset$ and so the result holds trivially. Therefore assume $x \in L^+(x)$. As $L^+(x)$ is closed, then $L^+(x) \smallsetminus x[0,n]$ is open in $L^+(x)$ for every $n \in \mathbb{Z}^+$. Moreover, $L^+(x) \smallsetminus x[0,n]$ is dense in $L^+(x)$. To see this let $y \in L^+(x)$. For every $\varepsilon > 0$ there exists $\tau \in \mathbb{R}^+$ such that $x\tau \in B_\varepsilon(y)$. But $x\tau \notin x[0,n]$, otherwise x would be a periodic point. Consequently $x\tau \in L^+(x) \smallsetminus x[0,n]$ for every $n \in \mathbb{Z}^+$, so $L^+(x) \smallsetminus x[0,n]$ is dense in $L^+(x)$. Now according to the Baire category theorem

$$\bigcap_{n=0}^{\infty} \{L^+(x) \smallsetminus x[0,n]\} = L^+(x) \smallsetminus \bigcup_{n=0}^{\infty} x[0,n] = L^+(x) \smallsetminus \gamma^+(x)$$

must be dense in $L^+(x)$. □

<u>Theorem 3.8</u>. Suppose X is a complete metric space. If for some point $x \in X$ we have $x \in L^+(x)$, then the positive motion π_x is periodic if and only if $\gamma^+(x) = L^+(x)$.

<u>Proof</u>: If π_x is periodic, then $\gamma^+(x) = L^+(x)$ in view of Corollary 3.6. Conversely, if $\gamma^+(x) = L^+(x)$, then $L^+(x) \smallsetminus \gamma^+(x) = \emptyset$ is not dense in $L^+(x)$. According to Lemma

3.7, π_x must be periodic. □

Now we can answer the question, where do the positive motions go as $t \to \infty$?

<u>Definition 3.9</u>. Let $x \in X$ and M be a closed subset of X. We say that x is *attracted* to M if for each neighborhood V of M there exists $\tau > 0$ such that $xt \in V$ whenever $t \geq \tau$. In this case we write $xt \to M$ as $t \to \infty$.

<u>Theorem 3.10</u>. Suppose $H^+(x)$ is compact for some $x \in X$. Then x is attracted to $L^+(x)$.

<u>Proof</u>: Let V be a neighborhood of $L^+(x)$. Since $L^+(x)$ is compact and is the intersection of the nested family $\{H^+(xt): t \in \mathbb{R}^+\}$, there exists $t_0 \in \mathbb{R}^+$ so that $H^+(xt) \subset V$ for all $t \geq t_0$. Thus $xt \in V$ for every $t \geq t_0$. □

In most applications we cannot actually determine the positive limit sets, but we can often narrow down our search to some set E towards which the solution goes. (Such a set E can be found by recourse to an appropriate Lyapunov function - a method to be explored in later chapters.) Naturally E must contain the given limit set. The theorem of LaSalle which follows is one of the most important results in the application of semidynamical systems.

<u>Theorem 3.11</u>. (Generalized Invariance Principle) Suppose there exists a closed set $E \subset X$ and a set $H \subset X$ with the following property: for each $x \in H$, $xt \to E$ as $t \to \infty$. If $\gamma^+(x)$ is relatively compact, then $xt \to M$ where M is the largest weakly invariant subset of E.

<u>Proof</u>: In view of Theorem 3.10 we see that $L^+(x) \subset E$ for every $x \in H$. If M is the largest weakly invariant subset

of E, then M contains the nonempty weakly invariant set $\cup\{L^+(y): y \in H\}$. Consequently for each $x \in H$ we have $xt \to M$ as $t \to \infty$. □

4. Minimality

Now that we have discovered that positive motions go to positive limit sets, the next problem we face is to describe the structure of the positive limit sets. We have already made a start in that direction; namely, Corollary 3.6 which says if $\gamma^+(x)$ is compact, then $L^+(x)$ is the positive orbit of a periodic point. Clearly though, we can expect much less regularity than this when $\gamma^+(x)$ is noncompact. (We point out that highly irregular - indeed chaotic - behavior ensues in the limit sets of the simplest of semidynamical systems. In Section 9 we consider such a system with broad applications to biology, economics, and the social sciences.) For now we will look for conditions on positive limit sets so that positive motions therein behave in some "regular" manner. In as much as these motions might represent some "steady state" or "equilibrium" behavior, it is meaningful to ask if negative uniqueness obtains. That is, through each point of the positive limit set is there a unique principal orbit which lies therein? And if so, does the restriction of the semidynamical system to the positive limit set become a dynamical system? These questions will be partially treated here and again in Chapter III.

Definition 4.1. A set $M \in X$ is called *positively minimal* if it is closed and positively invariant but contains no nonempty proper subset with these two properties. M is called *minimal* if it is closed and invariant but contains no nonempty proper subset with these two properties.

Theorem 4.2. The following are equivalent for a set $M \subset X$.

(i) M is positively minimal.

(ii) $M = H^+(x)$ for every $x \in M$.

(iii) $M = L^+(x)$ for every $x \in M$.

Proof: If $M = \emptyset$ there is nothing to prove. So assume $M \neq \emptyset$, and let M be positively minimal. If $x \in M$, then $H^+(x) \subset M$. If $H^+(x) \subsetneq M$, then M would properly contain a nonempty, closed, positively invariant subset $H^+(x)$, contradicting the positive minimality of M. Thus (i) implies (ii). Now assume $M = H^+(x)$ for every $x \in M$. For any such x we have $L^+(x) \subset H^+(x) = M$. Choose any $y \in L^+(x)$. As $L^+(x)$ is closed and positively invariant with $y \in M$, we must have $M = H^+(y) \subset L^+(x) \subset H^+(x) = M$. This shows that $M = L^+(x)$, so (ii) implies (iii). Finally suppose $M = L^+(x)$ for every $x \in M$. Clearly M is closed and positively invariant. If M is not positively minimal then it properly contains a nonempty, closed, positively invariant subset, say M^*. For each $x \in M^*$ we get $M = L^+(x) \subset H^+(x) \subset M^* \subsetneq M$, a contradiction. Therefore, (iii) implies (i), and the proof of the theorem is concluded. □

The existence of positively minimal sets is guaranteed by the next theorem.

Theorem 4.3. Every nonempty, compact, positively invariant subset of X contains a nonempty positively minimal set.

Proof: Suppose M is a nonempty, compact, positively invariant subset of X. The collection of all nonempty, closed, positively invariant subsets of M is partially ordered by set inclusion. Now suppose $\{M_\alpha\}_{\alpha\in\Lambda}$ is a linearly ordered family of nonempty, closed, positively invariant subsets of M. In view of the compactness of M, this family has a lower bound, namely $\cap\{M_\alpha: \alpha \in \Lambda\}$. Then by an equivalent version of Zorn's Lemma we obtain that M contains a positively minimal set. □

We now have a sufficient condition for a positive motion to extend to a principal solution.

<u>Theorem 4.4</u>. If M is a compact, positively minimal subset of X, then through each $x \in M$ there is a principal orbit $\gamma(x)$ contained in M.

<u>Proof</u>: Theorem 4.2 shows that M is a positive limit set. As M is compact we may use Theorem 3.5 to complete the proof. □

Now we turn to one of the questions posed at the beginning of this section: if M is a compact, positively minimal set in X, must the principal orbit guaranteed by Theorem 4.4 be unique? And if the answer to this question is yes, does the mapping $\hat{\pi}: M \times \mathbb{R} \to M$ given by $\hat{\pi}(x,t) = \gamma_x(t)$ define a dynamical system $(M,\hat{\pi})$, where γ_x is the unique solution through x referred to above? The answer to the first question is in general, negative, as the next example demonstrates. In view of Corollary 3.5 we must search for a case where M is not the positive orbit through a periodic point. To keep matters simple, the example will be a discrete semidynamical system. This example can be extended to obtain a (continuous) semidynamical system by the method of Nemytskii and Stepanov [1, p. 381].

Example 4.5. Denote by S^1 the circle of circumference one, and let each point on S^1 be assigned an angular coordinate θ, $0 \leq \theta < 1$. Let $[a,b]$ and (a,b) respectively denote the closed and open intervals from a to b in S^1, where a precedes b in the cyclic order established by means of the coordinate θ. Suppose there is given a Cantor set C in S^1; that is, a perfect nowhere dense subset of S^1. Then C admits the representation

$$C = S^1 \setminus \bigcup_{n=0}^{\infty} (a_n, b_n),$$

where $[a_i, b_i] \cap [a_j, b_j] = \emptyset$ for $i \neq j$. We will produce a homeomorphism $h: S^1 \to S^1$ which satisfies

(i) $h^n(x) \neq x$ for every $x \in S^1$ and $n \in \mathbb{Z}^+$,

(ii) $h([a_i, b_i]) = [a_j, b_j]$ for every i and some $j \neq i$ in $\mathbb{N}$,

(iii) h preserves the cyclic order on S^1.

On another circle Γ of circumference one fix a reference point P_0, and choose a point P_1 on Γ whose distance from P_0 along Γ measured counter-clockwise is irrational. Denote this distance by γ. We can assume that $\gamma \in [0,1)$. For each $k \in \mathbb{Z}$ let P_k denote the point on Γ whose distance from P_0 along Γ measured counter-clockwise is $k\gamma$. Then the sequence $\{P_k\}_{k \in \mathbb{Z}}$ is dense in Γ. We proceed to establish a one-to-one correspondence between the set of intervals $\{(a_n, b_n)\}_{n \in \mathbb{Z}^+}$ and the sequence $\{P_k\}_{k \in \mathbb{Z}}$. To the point P_0 let there correspond the interval $I_0 \overset{\text{def}}{=} (a_0, b_0)$. To the point P_1 let there correspond the interval $I_1 \overset{\text{def}}{=} (a_1, b_1)$. To the point P_{-1} let there correspond the interval $I_{-1} \overset{\text{def}}{=} (a_n, b_n)$ with least subscript $n \in \mathbb{Z}^+$ which lies on one of the two arcs between the intervals I_0 and I_1 such that I_0, I_1, I_{-1} should have the same cyclic order on S^1 as P_0, P_1, P_{-1} do on the circle Γ. Continue in this fashion to obtain the desired correspondence $P_k \leftrightarrow I_k$, $k \in \mathbb{Z}$. The special choice of interval (a_n, b_n) with least $n \in \mathbb{N}$ ensures that all the intervals $\{(a_i, b_i)\}_{i \in \mathbb{Z}^+}$ are accounted for. The point in producing this correspondence is to relabel these intervals so that the sought after map h may be defined. Indeed, for each $k \in \mathbb{Z}$ we will demonstrate how to map I_k affinely onto I_{k+1} and extend this to all of S^1.

So for each $k \in \mathbb{Z}$ let the interval I_k have the representation $(a^{(k)}, b^{(k)})$. For $\theta \in [a^{(k)}, b^{(k)}]$ set

$$h(\theta) = a^{(k+1)} + (\theta - a^{(k)})\left[\frac{b^{(k+1)} - a^{(k+1)}}{b^{(k)} - a^{(k)}}\right].$$

In view of its affine form h is a one-to-one continuous map of $\bar{I}_k$ onto $\bar{I}_{k+1}$. As the $\{\bar{I}_k\}_{k \in \mathbb{Z}}$ are nonoverlapping, and S^1 is compact, then h is a homeomorphism of $D \overset{\text{def}}{=} \cup\{\bar{I}_k : k \in \mathbb{Z}\}$ onto itself. Because D is dense in S^1 we need only show that h extends continuously to all of S^1. Such an extension will be unique and must be a one-to-one mapping of S^1 onto S^1. This will produce the desired homeomorphism with the properties (i), (ii), and (iii) announced earlier. Now, a sufficient condition for h to admit a continuous extension to S^1 is that h is uniformly continuous on D. So for θ, $\theta + \Delta \in I_k$ we may write

$$h(\theta+\Delta) - h(\theta) = c_k \Delta, \quad c_k = \left[\frac{b^{(k+1)} - a^{(k+1)}}{b^{(k)} - a^{(k)}}\right].$$

If the sequence $\{c_k\}$ is bounded then it is clear that h is uniformly continuous. But suppose $\{c_k\}$ were not bounded. Without loss of generality we may assume (by choosing a subsequence if necessary) that $c_k \geq k$, $k \in \mathbb{Z}^+$. Then

$$b^{(k+1)}-a^{(k+1)} \geq k(b^{(k)}-a^{(k)}) \geq \ldots \geq k!(b^{(0)}-a^{(0)}).$$

For large enough k this says that $b^{(k)}-a^{(k)} > 1$, an impossibility, as S^1 has circumference one.

Define the equivalence relation $\sim$ on $S^1 \times S^1$ as follows: $\theta \sim \psi$ if and only if $\theta,\psi \in \bar{I}_k$ for some $k \in \mathbb{Z}^+$. Set $\tilde{S}^1 = S^1/\sim$, the set of equivalence classes of S^1 with the quotient topology. Denote by $\tilde{C}$ the subset of $\tilde{S}^1$ corresponding to the cantor set C. Let $\tilde{\theta} \in \tilde{S}^1$ denote the equivalence class containing $\theta \in S^1$. Thus $\tilde{S}$ is obtained from S^1 by identifying the closed intervals $\bar{I}_0, \bar{I}_1, \bar{I}_2, \ldots,$ as points. (Note, the intervals I_{-k}, $k \in \mathbb{N}$ are not included in this identification.) Corresponding to h we have the map $\tilde{h}: \tilde{S}^1 \to \tilde{S}^1$ given by $\tilde{h}(\tilde{\theta}) = \widetilde{h(\theta)}$.

We are finally ready to define the promised discrete semidynamical system. Let the phase space X be $\tilde{S}^1$. The phase map $\tilde{\pi}: \tilde{S}^1 \times \mathbb{Z}^+ \to \tilde{S}^1$ is defined by

$$\tilde{\pi}(\tilde{\theta},n) = \tilde{h}^n(\tilde{\theta}).$$

It is easy to verify that $(\tilde{S}^1,\tilde{\pi})$ is indeed a discrete semidynamical system and that $\tilde{C}$ is a compact positively minimal set. However, there are uncountably many principal solutions $\gamma_{\tilde{\theta}}$ through $\tilde{\theta} = I_0$ with orbit $\gamma(\tilde{\theta}) \subset \tilde{C}$. This is so because $h(I_{-1}) = I_0$. Consequently in view of property (i) of h, $h^n(I_0) \neq I_{-1}$ for any $n \in \mathbb{Z}^+$. Thus for each $\psi \in \bar{I}_{-1}$ there is a principal solution $\gamma_{\tilde{\theta}}$ through $\tilde{\theta} = \bar{I}_0$ with

$\gamma_{\tilde{\theta}}(-1) = \psi$. As the points of the interval $\overline{I}_{-1}$ are not included in the identification, this provides an uncountably infinite number of distinct principal solutions through the point $\tilde{\theta} = I_0$. This concludes the example.

Another question posed at the beginning of this section concerned the "regularity" of the behavior of principal motions in a positive limit set. The following theorem demonstrates that the positive motions in a compact, positively minimal set exhibit a kind of recurrent property. That is to say, the positive motion through any point x "regularly" returns close to x. We will take up this idea again in Section 3 of Chapter III.

<u>Theorem 4.6</u>. Let $M \subset X$ be a compact, positively minimal set. For every $y \in M$ and each open set V containing y there is some $T \in \mathbb{R}^+$ so that $y[t,t+T] \cap V \neq \emptyset$ for every $t \in \mathbb{R}^+$.

<u>Proof</u>: Suppose the contrary were true. Let there exist $y \in M$ and an open set V containing y so that for any $n \in \mathbb{N}$, there exists $t_n \in \mathbb{R}^+$ with $y[t_n,t_n+n] \cap V = \emptyset$. As M is compact we may assume the sequence $\{yt_n\}$ has a cluster point $x \in M$. This implies $\gamma^+(x) \cap M = \emptyset$. Otherwise, $xt \in V$ for some $t \in \mathbb{R}^+$. By continuity of π there must exist open sets U containing x and W containing t so that $UW \subset V$. Let $n_0 \in \mathbb{N}$ so that $t \in [0,n_0]$. As x is a cluster point of $\{yt_n\}$ there is $n \geq n_0$ with $yt_n \in U$. Therefore $(yt_n)t \in UW \subset V$ and so $y[t_n,t_n+n] \cap V = (yt_n)[0,n] \cap V \neq \emptyset$. This is a contradiction, so $\gamma^+(x) \cap M = \emptyset$. But this implies $y \notin H^+(x)$ which contradicts the positive minimality of M. □

5. Prolongations and Stability of Compact Sets

Definition 5.1. For every $x \in X$ the set

$$D^+(x) = \{y \in X : x_\alpha t_\alpha \to y \text{ for some nets } \{x_\alpha\} \subset X \text{ and } \{t_\alpha\} \subset \mathbb{R}^+ \text{ with } x_\alpha \to x\}$$

is called the *positive prolongation* of x.

Certainly $D^+(x) \supset H^+(x)$. But we can say even more about $D^+(x)$. Indeed, positive motions through points arbitrarily close to x tend to $D^+(x)$. Thus the proof of the next lemma is obvious. (See for example, the proofs of Lemma 3.2 and Theorem 3.4.)

Lemma 5.2. For every $x \in X$ we have

(i) $D^+(x) = \cap\{\overline{W\,\mathbb{R}^+} : W$ is a neighborhood of $x\}$,

(ii) $D^+(x)$ is closed and positively invariant, and

(iii) $H^+(x) \subset D^+(x)$.

Definition 5.3. A subset $M \subset X$ is called *stable* if every neighborhood of M contains a positively invariant neighborhood of M.

It is immediately obvious that the following characterizations of stability hold. They will be useful later on.

Proposition 5.4. The following are equivalent for a subset $M \subset X$.

(i) M is stable.

(ii) For every neighborhood U of M there is a neighborhood V of M such that $V\,\mathbb{R}^+ \subset U$.

(iii) For every net $\{x_\alpha\} \subset X$ and $\{t_\alpha\} \subset \mathbb{R}^+$ we have $x_\alpha t_\alpha \to M$ whenever $x_\alpha \to M$.

Theorem 5.5. If a subset $M \subset X$ is stable, then it is positively invariant.

Proof: Suppose $M\mathbb{R}^+ \not\subset M$, so there is an $x \in M\mathbb{R}^+ \setminus M$. Then $U = X\setminus\{x\}$ is a neighborhood of M which contains no positively invariant neighborhood of M. This contradicts the stability of M though. □

The next lemma, seemingly technical, provides an important property of stable sets.

Lemma 5.6. If $M \subset X$ is a compact stable set, then for each $x \in M$, $D^+(x)$ is a compact connected subset of M. In particular $D^+(M) = M$, where $D^+(M) = \cup\{D^+(x): x \in M\}$.

Proof: Let $x \in M$, and suppose $y \in D^+(x)\setminus M$. For any net $\{x_\alpha\} \subset M$ with $x_\alpha \to x$, we have $x_\alpha \to M$. In view of Proposition 5.4, $x_\alpha t_\alpha \to M$ for every net $\{t_\alpha\} \subset \mathbb{R}^+$. Consequently $x_\alpha t_\alpha \to y$ cannot occur, thereby contradicting $y \in D^+(x)$. Therefore $D^+(x) \subset M$. In particular $D^+(x)$ is compact. As $x \in D^+(x)$ we obtain $M \subset \cup\{D^+(x): x \in M\} \subset M$, whence $D^+(M) = M$.

Now suppose there exists $x \in M$ so that $D^+(x)$ is not connected. Let A,B be nonempty disjoint compact sets with $D^+(x) = A\cup B$. There are separating neighborhoods U_A and U_B of A and B, respectively. We may assume that $x \in A$. Choose $y \in B$. Then there exists nets $\{x_\alpha\} \subset X$ and $\{t_\alpha\} \subset \mathbb{R}^+$ with $x_\alpha \to x$ and $x_\alpha t_\alpha \to y$. We can further assume that $x_\alpha \in U_A$ and $x_\alpha t_\alpha \in U_B$. Since the sets $x_\alpha[0,t_\alpha]$ are connected and intersect U_A and its compliment, there exist $\tau_\alpha \in [0,t_\alpha]$ such that $x_\alpha\tau_\alpha \in \partial U_A$. As $x_\alpha \to M$, the stability of M ensures that $x_\alpha\tau_\alpha \to M$. In view of the compactness of M we may assume that the net $\{x_\alpha\tau_\alpha\}$ has a cluster point

$z \in M$. Clearly, $z \in \partial U_A \cap D^+(x)$. However, this yields a contradiction as $\partial U_A \cap D^+(x) = \emptyset$. Thus $D^+(x)$ must be connected. □

Corollary 5.7. If $H^+(x)$ is a compact stable set for some $x \in X$, then $D^+(x) = H^+(x)$.

Proof: Since $x \in H^+(x)$, then $D^+(x) \subset H^+(x)$. But $D^+(x)$ is positively invariant and closed, so that $H^+(x) \subset D^+(x) \subset H^+(x)$. □

Theorem 5.8. If each component of a set $M \subset X$ is stable, then so is M. Conversely, if M is compact and stable, then so is each of its components.

Proof: The proof of the first part is trivial and is omitted. For the converse part let M be stable and N be a component of M. For every $x \in N$, $D^+(x)$ is a compact connected subset of N. If N were not stable there would be a neighborhood U of N and a net $\{x_\alpha\} \subset X$ with $x_\alpha \to N$ and $x_\alpha t_\alpha \notin U$ for some net $\{t_\alpha\} \subset \mathbb{R}^+$. Therefore there exists $\tau_\alpha \in [0,t_\alpha]$ with $x_\alpha \tau_\alpha \in \partial U$. In particular, $x_\alpha \to M$, so that $x_\alpha \tau_\alpha \to M$ from the stability of M. Since M is compact we may assume (by choosing a subnet if necessary) that $x_\alpha \tau_\alpha \to y \in M$. Thus $y \in D^+(x) \cap \partial U$, which contradicts the fact that $D^+(x) \subset N$. Consequently N is stable. □

We now show how stability is related to positive minimality. First, we require a definition.

Definition 5.9. The semidynamical system (X,π) is said to be *stable* if $H^+(x)$ is stable for every $x \in X$.

Proposition 5.10. If (X,π) is both stable and Lagrange stable, then $L^+(x)$ is positively minimal and stable for every $x \in X$.

Proof: Fix $x \in X$, and let $y \in L^+(x)$. First we will show that $L^+(x) \subset D^+(y)$. Suppose $z \in L^+(x)$. There exist nets $\{t_\alpha\}$, $\{\tau_\alpha\}$ in $\mathbb{R}^+$ with $t_\alpha \to \infty$, $\tau_\alpha \to \infty$ so that $xt_\alpha \to y$ and $x\tau_\alpha \to z$. We can assume by choosing a subnet if necessary that $\tau_\alpha - t_\alpha \in \mathbb{R}^+$. As $xt_\alpha(\tau_\alpha - t_\alpha) = x\tau_\alpha \to z$, then $z \in D^+(y)$. Consequently $H^+(y) \subset L^+(x) \subset D^+(y) = H^+(y)$ by stability of $H^+(y)$. Therefore $L^+(x)$ is positively minimal. Clearly, $L^+(x)$ is also stable. □

This last proposition admits a converse.

Proposition 5.11. Suppose X is a regular topological space. If $H^+(x)$ is compact for some $x \in X$, then $H^+(x)$ is stable whenever $L^+(x)$ is stable.

Proof: We will show that for every neighborhood U of $H^+(x)$ there is a neighborhood V of $H^+(x)$ such that $V\mathbb{R}^+ \subset U$. Now, as $L^+(x)$ is stable and U is also a neighborhood of $L^+(x)$, there is a neighborhood W_0 of $L^+(x)$ such that $W_0\mathbb{R}^+ \subset U$. Also there must exist $\tau \in \mathbb{R}^+$ with $x[\tau,\infty) \subset W_0$. We claim there exists a neighborhood N of $x[0,\tau]$ with $N\mathbb{R}^+ \subset U$. Otherwise there would be nets $\{z_\alpha\} \subset X$ and $\{t_\alpha\} \subset \mathbb{R}^+$ such that $z_\alpha \to x[0,\tau]$, $z_\alpha t_\alpha \notin U$. We may assume $z_\alpha \to xt_0$ for some $t_0 \in [0,\tau]$. If $\limsup_\alpha t_\alpha < \infty$, we may also assume (by taking a subnet if necessary) that $t_\alpha \to t < \infty$. Therefore $z_\alpha t_\alpha \to (xt_0)t \in \gamma^+(x)$. But this implies $z_\alpha t_\alpha \in U$, a contradiction. Thus suppose $t_\alpha \to \infty$. Since $xt_0(\tau - t_0) = x\tau \in W_0$, then in view of the continuity of π there is a neighborhood N_0 of xt_0 with $N_0(\tau - t_0) \subset W$. This implies $N_0[\tau - t_0, \infty) \subset W_0\mathbb{R}^+ \subset U$. So eventually we must have $z_\alpha \in N_0$,

$t_\alpha > \tau - t_0$, and $z_\alpha t_\alpha \in U$. Again this is impossible. Therefore there is a neighborhood N of $x[0,\tau]$ with $N\mathbb{R}^+ \subset U$.

Observe that $\gamma^+(x\tau) \subset H^+(x) \subset W_0$. Choose W so that $H^+(x\tau) \subset W \subset \overline{W} \subset W_0$. Then $W\mathbb{R}^+ \subset U$. Finally, set $V = N \cup W$. We get $V\mathbb{R}^+ \subset N\mathbb{R}^+ \cup W\mathbb{R}^+ \subset U$. This completes the proof. □

The last two propositions are summarized in the next theorem. The characterization of stability of (X,π) provided by the theorem reduces the problem to one of determining the stability of positive limit sets. This may be especially advantageous when there is only a single limit set in (X,π).

Theorem 5.12. Suppose X is a regular topological space for the Lagrange stable semidynamical system (X,π). Then (X,π) is stable if and only if every positive limit set is stable.

6. Attraction: Asymptotic Stability of Compact Sets

We turn again to the matter of where do positive motions go as $t \to \infty$. Some new concepts are introduced first.

Definition 6.1. Define the sets

$A_w^+(M) = \{x \in X :$ for each neighborhood U of M there exists a net $\{t_\alpha\} \subset \mathbb{R}^+$, $t_\alpha \to \infty$, such that $xt_\alpha \in U\}$,

$A^+(M) = \{x \in X :$ for each neighborhood U of M there exists $\tau \in \mathbb{R}^+$ such that $x[\tau,\infty) \subset U\}$.

The sets $A_w^+(M)$ and $A^+(M)$ are called the *region of weak attraction* and the *region of attraction* of M, respectively. If $x \in A_w^+(M)$ or $x \in A^+(M)$, we say that x is *weakly attracted* or *attracted* to M, respectively (c.f. Definition 3.7 and Theorem 3.8).

The proof of the following lemma is immediate.

Lemma 6.2. For every set $M \subset X$

(i) $A^+(M) \subset A_w^+(M)$,

(ii) $A^+(M)$, $A_w^+(M)$ are invariant.

We characterize points in $A_w^+(M)$ according to the next lemma.

Lemma 6.3. For any $M \subset X$ and $x \in X$ the following are equivalent:

(i) $x \in A_w^+(M)$.

(ii) There is a net $\{t_\alpha\}$ in $\mathbb{R}^+$, $t_\alpha \to \infty$, so that either $xt_\alpha \in M$ or $M \cap L^+(x) \neq \emptyset$.

(iii) $M \cap H^+(xt) \neq \emptyset$ for all $t \in \mathbb{R}^+$.

Proof: (i) implies (ii). Let $x \in A_w^+(M)$. Suppose there is no net $\{t_\alpha\}$ in $\mathbb{R}^+$, $t_\alpha \to \infty$ with $xt_\alpha \in M$. This means there exists $\tau \in \mathbb{R}^+$ with $\gamma^+(x\tau) \subset X \setminus M$. If $M \cap L^+(x) = \emptyset$, we must have

$$H^+(x\tau) = \gamma^+(x\tau) \cup L^+(x\tau) = \gamma^+(x\tau) \cup L^+(x) \subset X \setminus M.$$

As $H^+(x\tau)$ is closed, then $X \setminus H^+(x\tau)$ is a neighborhood of M. Consequently we cannot have $x\tau \in A_w^+(M)$ nor $x \in A_w^+(M)$, a contradiction.

(ii) implies (i). Suppose $x \notin A_w^+(M)$. There is an open neighborhood U of M and $\tau \in \mathbb{R}^+$ so that $x[\tau,\infty) \subset X \setminus U$. As $X \setminus U$ is closed, then $H^+(x\tau) \subset X \setminus U$. This implies $M \cap L^+(x) = \emptyset$ and for each net $t_\alpha \to \infty$, $xt_\alpha \notin M$.

We have just established the equivalence of (i) and (ii). It is obvious though, that (iii) is also equivalent to (ii). □

Definition 6.4. A set $M \subset X$ is called a *weak attractor* or an *attractor* whenever, respectively, $A_w^+(M)$ or $A^+(M)$ is a neighborhood of M.

Proposition 6.5. If $M \subset X$ is an attractor or a weak attractor, then $A^+(M)$ or $A_w^+(M)$, respectively, are open neighborhoods of M.

Proof: We prove the case for $A^+(M)$; the proof for $A_w^+(M)$ is similar. Suppose $x \in A^+(M)$. As $A^+(M)$ is a neighborhood of M there exists an open set U with $M \subset U \subset A^+(M)$. Also there exists $\tau \in \mathbb{R}^+$ so that $x[\tau,\infty) \subset U$. Then $V = \pi^{-1}(U,\tau)$ is an open neighborhood of x. Since $y\tau \in U$ for every $y \in V$, we conclude that $V \subset A^+(M)$. Therefore $A^+(M)$ is open. □

Corollary 6.6. If M is an attractor, then $A^+(M) = A_w^+(M)$.

Proof: We need only demonstrate $A_w^+(M) \subset A^+(M)$. So let $x \in A_w^+(M)$. As $A^+(M)$ is an open neighborhood of M, there is $\tau \in \mathbb{R}^+$ so that $x\tau \in A^+(M)$. But $A^+(M)$ is invariant so $x \in A^+(M)$. □

It is clear that an attractor is also a weak attractor. The converse need not hold. In the event of a stable weak attractor though, we have the converse.

Theorem 6.7. If $M \subset X$ is a stable weak attractor, then it is an attractor.

Proof: It will be sufficient to show $A_w^+(M) \subset A^+(M)$. So let $x \in A_w^+(M)$. If U is any neighborhood of M we may find a positively invariant open neighborhood V of M with $V \subset U$ and $V \subset A_w^+(M)$. Choose $\tau \in \mathbb{R}^+$ so that $x\tau \in V$. As V is positively invariant, $x[\tau,\infty) \subset V \subset U$. This establishes $x \in A^+(M)$. □

Definition 6.8. A set $M \subset X$ is called *asymptotically stable* if it is a stable weak attractor.

Asymptotic stability is a very important concept. Most of the examples and applications which are developed in the later chapters are shown to possess asymptotically stable orbits. Such orbits are desirable, indeed, required in view of the behavior of the physical (biological, economic, etc.) system which is being modeled.

In view of Theorem 6.7 we have the following customary characterization of asymptotic stability.

Corollary 6.9. A set $M \subset X$ is asymptotically stable if and only if it is a stable attractor.

We establish the analog of Theorem 5.8 for compact asymptotically stable sets.

Theorem 6.10. Suppose X is locally connected. A nonempty compact set $M \subset X$ is asymptotically stable if and only if M has a finite number of components, each of which is asymptotically stable.

The proof of the theorem proceeds by a sequence of lemmas, some of which are of independent value.

Lemma 6.11. Suppose X is a locally connected space, and $M \subset X$ is an attractor. If A_1 is a component of $A^+(M)$, then $M_1 \overset{\text{def}}{=} A_1 \cap M$ is a nonempty attractor with $A_1 = A^+(M_1)$.

Proof: Since X is locally connected, each component of $A^+(M)$ is open (e.g. Kelley [1], p. 61). Also the sets A_1 and $A_2 \overset{\text{def}}{=} A^+(M) \setminus A_1$ are separated open sets; that is $\bar{A}_1 \cap A_2 = A_1 \cap \bar{A}_2 = \emptyset$. Set $M_2 \overset{\text{def}}{=} A_2 \cap M$ so that $M = M_1 \cup M_2$. If M_1 were empty, then A_2 would be an open neighborhood of M. Since $A^+(M)$ is invariant, then so are A_1 and A_2. Thus if $x \in A_1$, we would have $H^+(x) \subset \bar{A}_1$, whereby $H^+(x) \cap A_2 = \emptyset$, which contradicts the fact that x is attracted to M. Thus $M_1 \neq \emptyset$. By the same reasoning we obtain that $M_2 \neq \emptyset$. It follows that each point of A_1 is attracted to M_1. Indeed, if some point $x \in A_1$ were attracted to M_2, there would exist $\tau \in \mathbb{R}^+$ so that $(x\tau)\mathbb{R}^+ \subset A_2$. But this contradicts the positive invariance of A_1, so we conclude that $A_1 \subset A^+(M_1)$. As A_1 is an open invariant set containing M_1 and $A^+(M_1) \subset A^+(M) = A_1 \cup A_2$, we have $A_1 \supset A^+(M_1)$. Consequently, $A_1 = A^+(M_1)$, and M_2 is an attractor.

□

Lemma 6.12. Let M_1 and M_2 be separated by neighborhoods. If $M_1 \cup M_2$ is asymptotically stable, then so are M_1 and M_2. Moreover, $A^+(M_1)$ and $A^+(M_2)$ are disjoint.

Proof: For $k = 1,2$ let U_k be disjoint open neighborhoods of M_k. Then $U_1 \cup U_2$ is an open neighborhood of $M \overset{\text{def}}{=} M_1 \cup M_2$, which since M is stable, contains a positively invariant neighborhood V of M. But $V_k \overset{\text{def}}{=} V \cap U_k$ is a positively invariant neighborhood of M_k with $V_k \subset U_k$, $k = 1,2$ and $V_1 \cap V_2 = \emptyset$. Thus each M_k is stable. As each V_k may be chosen to be a subset of $A^+(M)$, we see that $V_k \subset A^+(M_k)$. Therefore each M_k is an attractor, and consequently is asymptotically stable. Finally, $A^+(M_1) \cap A^+(M_2) = \emptyset$. Otherwise there would be a point $x \in X$ and a $\tau \in \mathbb{R}^+$ so that $x\tau \in U_1$ and $x\tau \in U_2$. This is impossible since $U_1 \cap U_2 = \emptyset$. Therefore $A^+(M_1)$ and $A^+(M_2)$ are disjoint.

□

We now turn to the proof of Theorem 6.10.

Proof: The "if" part is trivial. So suppose M is a nonempty compact asymptotically stable subset of X. As X is locally connected, the components of $A^+(M)$ form an open cover of M. According to Lemma 6.11 each component of $A^+(M)$ intersects M in a nonempty subset. By compactness of M there are only a finite number of components of $A^+(M)$, say $A_1, A_2, \ldots, A_m$. Define $M_k = A_k \cap M$ for $k = 1,2,\ldots,m$. Then each M_k is a nonempty attractor with $A_k = A^+(M_k)$. Accordingly, $M_1, M_2, \ldots, M_m$ are (the finite number of) components of M from Lemma 6.12. Moreover, each M_k is compact and asymptotically stable since M is. □

Corollary 6.13. If $M \subset X$ is a compact, asymptotically stable set with connected region of attraction, then M is connected. The result also holds for closed sets in normal spaces.

The following example shows how attractor properties need not be carried over to components, even in the case of a compact set.

Example 6.14. Consider the semidynamical system $(\mathbb{R}, \pi)$ with the following properties (see illustration). The points

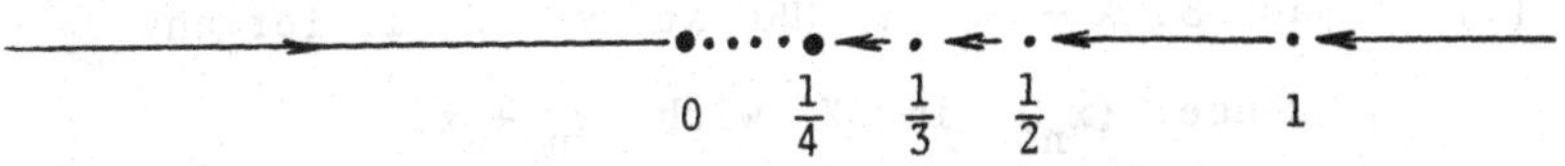

$$0 \quad \frac{1}{4} \quad \frac{1}{3} \quad \frac{1}{2} \quad 1$$

$0, \frac{1}{2}, \frac{1}{3}, \ldots$ are the only critical points and $\pi(x,t) < x$ if x is noncritical and either $x > 0$ and $t > 0$, or $x < 0$ and $t < 0$. It is easily verified that for any such semidynamical system the compact set $M = \{0, \frac{1}{2}, \frac{1}{3}, \ldots\}$ is an attractor with $A^+(M) = \mathbb{R}^+$ but none of its components is an attractor. The reason for this is that M is not stable.

7. Continuity of the Hull and Limit Set Maps in Metric Spaces

We investigate the upper and lower semicontinuity of the set valued maps $x \to H^+(x)$ and $x \to L^+(x)$ in the case X is a metric space with metric d and $H^+(x)$ is compact for every $x \in X$. This will lead to a new characterization of stability for the sets $H^+(x)$ and $L^+(x)$. Denote by $\mathscr{K}_X$ the collection of nonempty compact subsets of X.

Definition 7.1. Suppose X is a metric space. A map $S: X \to \mathscr{K}_X$ is called *upper semicontinuous* (USC) at $x \in X$ if for every $\varepsilon > 0$ there is $\delta > 0$ such that

$$S(y) \subset B_\varepsilon(S(x)) \quad \text{for every} \quad y \in B_\delta(x).$$

A map $S: X \to \mathscr{K}_X$ is called *lower semicontinuous* (LSC) at $x \in X$ if for every $\varepsilon > 0$ there is $\delta > 0$ such that

$$S(x) \subset B_\varepsilon(S(y)) \quad \text{for every} \quad y \in B_\delta(x).$$

As X has metric d we obtain the following useful characterization of semicontinuity of S at x. The proof is immediate so it is omitted.

Lemma 7.2.

(i) A map $S: X \to \mathscr{K}_X$ is USC at $x \in X$ if for any sequence $\{x_n\}$ in X with $x_n \to x$,

$$\sup\{d(y,S(x)) : y \in S(x_n)\} \to 0 \quad \text{as} \quad n \to \infty.$$

(ii) A map $S: X \to \mathscr{K}_X$ is LSC at $x \in X$ if for any sequence $\{x_n\}$ in X with $x_n \to x$,

$$\sup\{d(y,S(x_n)) : y \in S(x)\} \to 0 \quad \text{as} \quad n \to \infty.$$

<u>Definition 7.3</u>. Let X be metric. A map $S: X \to \mathscr{K}_X$ is *continuous* at $x \in X$ if it is both USC and LSC at $x \in X$.

<u>Remark 7.4</u>. Given a metric space X we may define a metric on the collection $\mathscr{K}_X$. Indeed, if $A,B \in \mathscr{K}_X$, set

$$h(A,B) = \max \{\sup_{x\in A} d(x,B),\ \sup_{y\in B} d(A,y)\}$$

Then it may be shown (e.g. Kelley [1]) that h is a metric. Moreover, a map $S: X \to \mathscr{K}_X$ is continuous at $x \in X$ (with respect to the metric topologies of X and $\mathscr{K}_X$) if and only if it is continuous in the sense of Definition 7.3.

If we assume $H^+(x)$ is compact for every $x \in X$, then we may view the sets $H^+(x)$ and $L^+(x)$ as the images of x under mappings from X to $\mathscr{K}_X$. The appropriate symbols for these maps are H^+ and L^+ respectively. Henceforth we adopt the

COMPACTNESS ASSUMPTION: (X,π) is Lagrange stable.

In view of the comments in the paragraph preceeding Theorem 3.5 of this chapter, the compactness assumption is hardly restrictive.

Definition 7.1 suggests that upper semicontinuity of H^+ is related to the stability of $H^+(x)$.

<u>Theorem 7.5</u>. H^+ is USC on $H^+(x)$ if and only if $H^+(x)$ is stable.

<u>Proof</u>: Suppose H^+ is USC on $H^+(x)$ for some $x \in X$. Let U be a neighborhood of $H^+(x)$. As $H^+(x)$ is compact we may choose $\varepsilon > 0$ so that $B_\varepsilon(H^+(x)) \subset U$. Then for every $z \in H^+(x)$ there is $\delta_z > 0$ such that $H^+(y) \subset B_{\frac{1}{2}\varepsilon}(H^+(z))$

whenever $y \in B_{\delta_z}(z)$. Set $V = \cup\{B_{\delta_z}(z): z \in H^+(x)\}$. Then V is an open neighborhood of $H^+(x)$. It follows that

$$V\,\mathbb{R}^+ \subset \{B_{\frac{1}{2}\varepsilon}(H^+(z)): z \in H^+(x)\} \subset B_\varepsilon(H^+(x)) \subset U$$

whence $H^+(x)$ is stable.

Conversely, suppose $H^+(x)$ is stable for some $x \in X$. For any $\varepsilon > 0$ there exists $\delta > 0$ so that $B_\delta(H^+(x))\,\mathbb{R}^+ \subset B_{\frac{1}{2}\varepsilon}(H^+(x))$. In particular if $y \in B_\delta(x)$, then

$$H^+(y) \subset \overline{B_\delta(x)\mathbb{R}^+} \subset \overline{B_\delta(H^+(x))\mathbb{R}^+} \subset B_{\frac{1}{2}\varepsilon}(H^+(x)) \subset B_\varepsilon(H^+(x)).$$

Consequently, H^+ is USC at x. □

Theorem 7.5 requires H^+ to be USC on all of $H^+(x)$ for stability of $H^+(x)$. In particular, USC at the point x itself is not sufficient for stability. But we can obtain a related, though weaker result, when H^+ is merely USC at x.

<u>Proposition 7.6</u>. If H^+ is USC at x, then $H^+(x) = D^+(x)$.

<u>Proof</u>: Let $y \in D^+(x)$. There exists sequences $\{x_n\}$ in X and $\{t_n\}$ in $\mathbb{R}^+$ with $x_n t_n \to y$. As $x_n t_n \in H^+(x_n)$ for every $n \in \mathbb{N}$ and H^+ is USC at x, then

$$d(x_n t_n, H^+(x)) \to 0 \quad \text{as} \quad n \to \infty.$$

But $H^+(x)$ is closed so $y \in H^+(x)$. Hence $D^+(x) \subset H^+(x)$. As $H^+(x) \subset D^+(x)$ always, then $H^+(x) = D^+(x)$. □

Next we show that H^+ is always LSC on X. But first we obtain a new characterization of $H^+(x)$ and $L^+(x)$.

<u>Definition 7.7</u>. For each $x \in X$ let

$Q^+(x) = \{y \in X$: for each sequence $\{x_n\} \subset X$ with $x_n \to x$ there exists a sequence $\{t_n\} \subset \mathbb{R}^+$ such that $x_n t_n \to y\}$,

$W^+(x) = \{y \in X$: for each sequence $\{x_n\} \subset X$ with $x_n \to x$ there exists a sequence $\{t_n\} \subset \mathbb{R}^+$ with $t_n \to \infty$ such that $x_n t_n \to y\}$.

Lemma 7.8. $Q^+(x) = H^+(x)$, $W^+(x) = L^+(x)$ for every $x \in X$.

Proof: It is easily seen that $Q^+(x) \subset H^+(x)$ and $W^+(x) \subset L^+(x)$ by choosing the sequence $x_n = x$ for every $n \in \mathbb{N}$. We show the reverse conclusions. For any $t \in \mathbb{R}^+$ we claim that $xt \in Q^+(x)$. In fact let $\{x_n\}$ be a sequence in X with $x_n \to x$ and choose $t_n = t$ for every $n \in \mathbb{N}$. Then $x_n t_n = x_n t \to xt \in Q^+(x)$. Thus $Q^+(x)$ is positively invariant. It follows from the definition of $Q^+(x)$ that $Q^+(x)$ is closed. Consequently, $\gamma^+(x) \subset H^+(x) \subset Q^+(x) \subset H^+(x)$. Thus we have $Q^+(x) = H^+(x)$.

To prove that $L^+(x) \subset W^+(x)$ it will be sufficient to show that $W^+(x) \supset \cap\{Q^+(xt)\colon t \in \mathbb{R}^+\} = \cap\{Q^+(xn)\colon n \in \mathbb{N}\}$. So let $y \in Q^+(xn)$ for each $n \in \mathbb{N}$ and suppose $\{x_k\}$ is a sequence in X with $x_k \to x$. Then $x_k n \to xn$ for every $n \in \mathbb{N}$. For each such n there exists $t_k^n \in \mathbb{R}^+$ such that

$$(x_k n)t_k^n = x_k(n+t_k^n) \to y \quad \text{as} \quad k \to \infty$$

in view of the definition of $Q^+(xn)$. For each n choose $k_n \in \mathbb{N}$ so that

$$d((x_k n)t_k^n, y) < 1/n \quad \text{for every} \quad k \geq k_n.$$

We can assume $k_1 < k_2 < \cdots$. We now construct a sequence $\{t_k\}$ in $\mathbb{R}^+$ with $t_k \to \infty$ so that $x_k t_k \to y$. For

$1 \leq k < k_2$, set $t_k = 1 + t_k^1$. For $k_n \leq k < k_{n+1}$, $n \geq 2$, set $t_k = n + t_k^n$. Then $t_k \to \infty$ and $x_k t_k \to y$. Consequently $y \in W^+(x)$. □

Proposition 7.9. The map H^+ is LSC on X.

Proof: Assume H^+ is not LSC at $x \in X$. Then there exists $\varepsilon > 0$ and a sequence $\{x_n\} \subset X$ with $x_n \to x$ such that

$$\sup\{d(y,H^+(x_n)): y \in H(x)\} \geq \varepsilon \quad \text{for every } n \in \mathbb{N}.$$

From compactness of $H^+(x)$ there exists a sequence $\{y_n\} \subset H^+(x)$ such that $d(y_n,H^+(x_n)) \geq \varepsilon$. We may assume $y_n \to y \in H^+(x) = Q^+(x)$ for some $y \in X$. Hence there is a sequence $\{t_n\}$ in $\mathbb{R}^+$ with $x_n t_n \to y$. So $d(x_n t_n, y_n) \to 0$. As $x_n t_n \in H^+(x_n)$ for every $n \in \mathbb{N}$, then

$$d(y_n,H^+(x_n)) \leq d(y_n,x_n t_n) \to 0.$$

This is impossible so H^+ must be LSC at x. □

Combining Propositions 5.10, 5.11, 7.9 with Theorem 7.5 we obtain the following characterization of the continuity of H^+. The obvious proof is omitted.

Theorem 7.10. The following are equivalent.

(i) H^+ is continuous on X.

(ii) $L^+(x)$ is stable for every $x \in X$.

(iii) (X,π) is a stable semidynamical system.

We turn to an analysis of the map L^+. It is considerably more complicated than the case for H^+.

Lemma 7.11. Suppose $S: X \to \mathcal{K}_X$ is any map which is USC at $x \in X$. Then for any sequence $\{x_n\}$ in X with $x_n \to x$, we have $S(x) \cup \{\cup\{S(x_n): n \in \mathbb{N}\}\}$ is compact.

Proof: Suppose $\{x_n\} \subset X$ with $x_n \to x \in X$, and set $Y = S(x) \cup \{\cup\{S(x_n): n \in \mathbb{N}\}\}$. For any sequence $\{y_n\}$ in Y we may assume without loss of generality that $y_n \in S(x_n)$. Then $d(y_n, S(x)) \to 0$. As $S(x)$ is compact we may find a subsequence $\{y_{n_k}\}$ and $y \in S(x)$ so that $y_{n_k} \to y$. Therefore Y is compact. □

Proposition 7.12. If the map H^+ is USC at $x \in X$, then so is the map L^+.

Proof: If $x \in L^+(x)$, then $L^+(x) = H^+(x)$; hence the proposition is proved. So assume $x \notin L^+(x)$ and proceed by contradiction; that is, suppose L^+ is not USC at x. We will first show that there exist sequences $\{x_n\} \subset X$, $\{s_n\} \subset \mathbb{R}^+$ with $x_n \to x$, $s_n \to \infty$, and $x_n s_n \to x$.

There exists $\varepsilon > 0$ and a sequence $\{x_n\}$ in X with $x_n \to x$ so that

$$\sup\{d(w, L^+(x)): w \in L^+(x_n)\} \geq 2\varepsilon.$$

We may assume that $x \notin B_{2\varepsilon}(L^+(x))$. Let $w_n \in L^+(x_n) \setminus B_{2\varepsilon}(L^+(x))$. then for each x_n there exists $t_n > n$ with $d(x_n t_n, w_n) < \varepsilon/n$. Hence $x_n t_n \notin B_\varepsilon(L^+(x))$. As H^+ is USC at x, then according to Proposition 7.6 and Lemma 7.11, $\{x_n t_n\}$ (or some subsequence thereof) converges to some $z_1 \in D^+(x) = H^+(x)$. Since for large enough $n \in \mathbb{N}$, $x_n t_n \notin B_\varepsilon(L^+(x))$, then $z_1 \in H^+(x) \setminus L^+(x)$. Thus there exists $\tau_1 \in \mathbb{R}^+$ so that $z_1 = x\tau_1 \notin B_\varepsilon(L^+(x))$. By a similar argument to the above we may assume that the sequence $\{x_n(t_n - \tau_1)\}$ converges to some $z_2 \in D^+(x) = H^+(x)$. (The choice of t_n ensures that $t_n - \tau_1 \in \mathbb{R}^+$ for sufficiently large $n \in \mathbb{N}$, and indeed, $t_n - \tau_1 \to \infty$.) Moreover, $z_2 \in H^+(x) \setminus L^+(x)$ so there exists $\tau_2 \in \mathbb{R}^+$ with $z_2 = x\tau_2$. Consequently

$$x\tau_1 = z_1 = z_2\tau_1 = (x\tau_2)\tau_1 = x(\tau_2+\tau_1)$$

implies $\tau_2 = 0$. For otherwise x would be a purely periodic point and therefore $x \in L^+(x)$ which contradicts our assumption. We conclude from this that

$$x_n(t_n-\tau_1) \to x.$$

Now choose $\delta > 0$ and $T > 0$ so that $N_\delta(x)T \subset B_\varepsilon(L^+(x))$. Let $N \in \mathbb{N}$ so that $n \geq N$ implies

$$x_n \in B_\delta(x),\ t_n-\tau_1 \geq T, \text{ and } x_n(t_n-\tau_1) \in B_\delta(x).$$

There exists $s_n \in (0,t_n-\tau_1)$ with $x_ns_n \in \partial B_\varepsilon(L^+(x))$ for every $n \geq N$. We claim

$$\liminf_{n\to\infty} (t_n-\tau_1-s_n) > 0.$$

Otherwise since $x_n(t_n-\tau_1) \to x$, then $x \in \partial B_\varepsilon(L^+(x))$. As ε is arbitrary we would have $x \in L^+(x)$, again contradicting our assumption. Choose s so that $0 < s < \liminf_{n\to\infty} (t_n-\tau-s_n)$. As before we may assume $x_n(t_n-\tau_1-s) \to z_3$ for some $z_3 \in H^+(x) \setminus L^+(x)$. Let $z_3 = x\tau_3$ for some $\tau_3 \in \mathbb{R}^+$. Then

$$x(\tau_3+s) = z_3s = \lim_{n\to\infty} x_n(t_n-\tau_1) = x.$$

But we choose $s > 0$. This means that x is a purely periodic point and therefore $x \in L^+(x)$. As this contradicts our original assumption $x \notin L^+(x)$, we must indeed have $x \in L^+(x)$. Thus $L^+(x) = H^+(x)$ and the proposition is proved. □

The following example shows the converse of Proposition 7.12 is false.

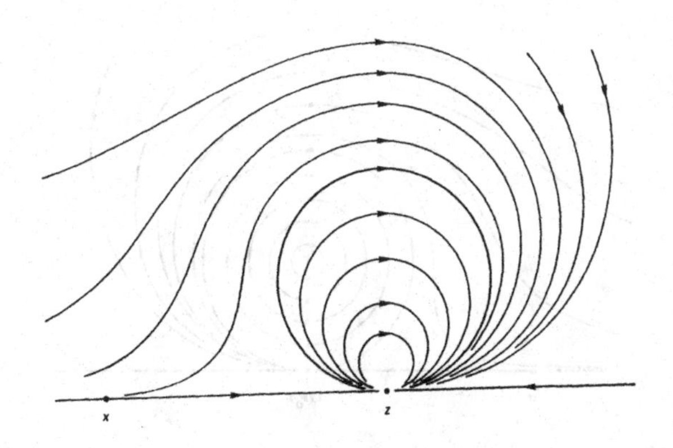

Figure 7.1

Example 7.13. In Figure 7.1 every positive orbit approaches the critical point z. The map L^+ is a constant map. However H^+ is not USC at x. The space X is assumed to have the usual topology of the plane.

As one can see in Example 7.13, $L^+(x) = \{z\}$ is not stable. This leads us to formulate a weaker form of stability in order to characterize upper semicontinuity of the map L^+.

Definition 7.14. A set $M \subset X$ is called *eventually stable* if for every neighborhood U of M there is a neighborhood V of M such that if $y \in V$ there exists $\tau = \tau(y) > 0$ so that $y[\tau,\infty) \subset U$. If τ does not depend upon $y \in V$, then M is called *uniformly eventually stable*.

Remark 7.15. Every set $M \subset X$ which is stable or an attractor is eventually stable. However, neither of these properties is necessary for eventual stability as is shown by the following example.

Example 7.16. Consider the variation of Example 7.13 as presented now in Figure 7.2. The set of critical points

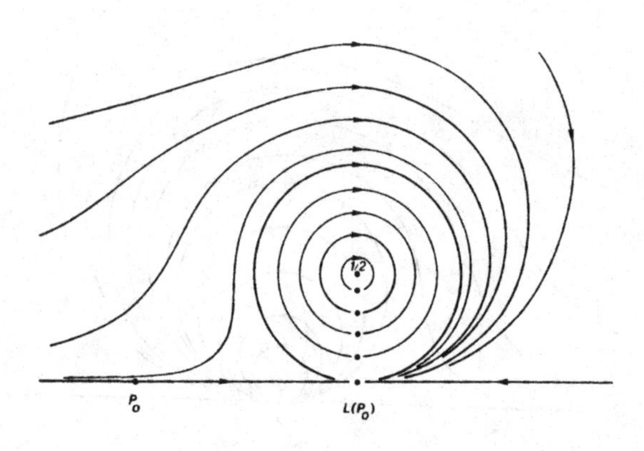

Figure 7.2

consists of $\{(x,y) : x = 0, 0 \leq y \leq \frac{1}{2}\}$. Let $P_0 = (-1,0)$. Then $L^+(P_0) = \{(0,0)\}$. This shows that $L^+(P_0)$ is eventually stable but neither stable nor an attractor.

Proposition 7.17. Given $x \in X$, the map L^+ is USC on $L^+(x)$ if and only if $L^+(x)$ is positively minimal and eventually stable.

Proof: Suppose L^+ is USC on $L^+(x)$. Let $y \in L^+(x)$. We shall prove $L^+(x) \subset L^+(y)$. Let $\{t_n\} \subset \mathbb{R}^+$ with $t_n \to \infty$ so that $xt_n \to y$. Now suppose $z \in L^+(x)$. Then $z \in L^+(xt_n) = L^+(x)$ for every $n \in \mathbb{N}$. Since

$$\sup\{d(w,L^+(y)): w \in L^+(xt_n)\} \to 0 \quad \text{as} \quad n \to \infty,$$

we have $z \in L^+(y)$. Thus $L^+(x) \subset L^+(y)$, so $L^+(x)$ is positively minimal.

We now demonstrate that $L^+(x)$ is eventually stable. Suppose $\varepsilon > 0$ and let $y \in L^+(x)$. There exists $\delta_y > 0$ so that $L^+(z) \subset B_\varepsilon(L^+(y))$ for every $z \in B_{\delta_y}(y)$. Therefore $L^+(z) \subset B_\varepsilon(L^+(x))$ as well for every $z \in B_{\delta_y}(y)$. By compactness of $L^+(x)$ there is a finite set of points

$\{y_1, y_2, \ldots, y_n\} \subset L^+(x)$ such that

$$V \overset{\text{def}}{=} \bigcup_{i=1}^{n} B_{\delta_{y_i}}(y_i) \supset L^+(x).$$

Then $L^+(z) \subset B_\varepsilon(L^+(x))$ for every $z \in V$. This shows that $L^+(x)$ is eventually stable.

Conversely, suppose $L^+(x)$ is positively minimal and eventually stable. Let $y \in L^+(x)$. For each $\varepsilon > 0$ there exists $\delta > 0$ such that for every $z \in B_\delta(L^+(x))$ there is $\tau \in \mathbb{R}^+$ with $\gamma^+(y\tau) \subset B_{\frac{1}{2}\varepsilon}(L^+(x))$. So if $z \in B_\delta(y)$, then

$$L^+(z) \subset \overline{B_{\frac{1}{2}\varepsilon}(L^+(x))} \subset B_\varepsilon(L^+(x)) = B_\varepsilon(L^+(y)).$$

This demonstrates L^+ is USC at $y \in L^+(x)$. □

The next two lemmas are required in order to characterize upper semicontinuity of the map L^+. The first of these establishes global upper semicontinuity.

<u>Lemma 7.18</u>. The map L^+ is USC on $L^+(x)$ if and only if L^+ is USC on $A^+(L^+(x))$.

<u>Proof</u>: As $L^+(x) \subset A^+(L^+(x))$, we need only show that L^+ is USC on $A^+(L^+(x))$ whenever L^+ is USC on $L^+(x)$. So let $y \in A^+(L^+(x))$. Accordingly $L^+(y) \subset L^+(x)$. As $L^+(x)$ is positively minimal from Proposition 7.17, we have $L^+(y) = L^+(x)$. Thus it will be sufficient to show L^+ is USC at $x \in A^+(L^+(x))$. Again, according to Proposition 7.17, $L^+(x)$ is eventually stable. Thus for every $\varepsilon > 0$ there exists $\eta > 0$ such that whenever $z \in B_\eta(L^+(x))$, there is $\tau = \tau(z) \in \mathbb{R}^+$ so that $\gamma^+(z\tau) \subset B_{\frac{1}{2}\varepsilon}(L^+(x))$. Choose $s \in \mathbb{R}^+$ and $\delta > 0$ such that

$$xs \in B_\eta(L^+(x)) \quad \text{and} \quad B_\delta(x)s \subset B_\eta(L^+(x)).$$

If $y \in B_\delta(x)$ we have

$$L^+(y) = L^+(ys) \subset \overline{B_{\frac{1}{2}\varepsilon}(L^+(x))} \subset B_\varepsilon(L^+(x)).$$

This shows L^+ is USC at x. □

Lemma 7.19. If $L^+(y)$ is eventually stable for each $y \in L^+(x)$ then $L^+(x)$ is positively minimal.

Proof: Suppose $L^+(x)$ is not positively minimal. There exists a positively minimal subset M in $L^+(x)$ according to Theorem 4.3. Obviously $x \notin M \subsetneq L^+(x)$. Suppose $y \in L^+(x) \setminus M$, $0 < \varepsilon < d(y,M)$, and $z \in M$. Then $L^+(z) = M$. By assumption there is a $\delta > 0$ so that for every $y' \in B_\delta(L^+(z))$, there exists $\tau = \tau(y') \in \mathbb{R}^+$ such that

$$\gamma^+(y'\tau) \subset B_{\frac{1}{2}\varepsilon}(M) = B_{\frac{1}{2}\varepsilon}(L^+(z)).$$

Since $z \in M \subset L^+(x)$, there is $s \in \mathbb{R}^+$ with $xs \in B_\delta(L^+(z))$. Thus there is $T = T(xs) \in \mathbb{R}^+$ so that

$$xs[T,\infty) = x[s+T,\infty) \subset B_{\frac{1}{2}\varepsilon}(L^+(z)).$$

Then

$$L^+(x) \subset \overline{B_{\frac{1}{2}\varepsilon}(L^+(z))} \subset B_\varepsilon(L^+(z)).$$

This is impossible as $y \in L^+(x)$, yet $y \notin B_\varepsilon(L^+(z))$. Consequently $L^+(x)$ must be positively minimal. □

Combining Proposition 7.17 and Lemmas 7.18, 7.19 and noting that $X = \cup\{A^+(L^+(x)): x \in X\}$ we have

Theorem 7.20. The map L^+ is USC on X if and only if $L^+(x)$ is eventually stable for each $x \in X$.

We turn to the matter of lower semicontinuity of the map L^+.

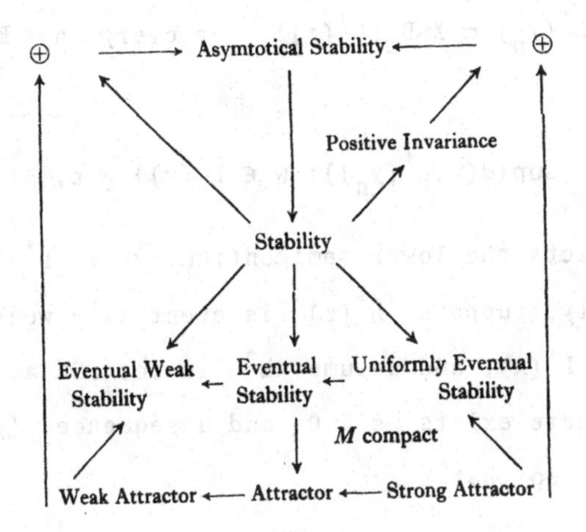

Figure 7.3

Definition 7.21. A set $M \subset X$ is called *eventually weakly stable* if for every neighborhood W of M there exists a neighborhood V of M such that for every $y \in V$ there is a sequence $\{t_n\} \subset \mathbb{R}^+$ with $t_n \to \infty$ such that $yt_n \in W$.

The relationships amongst the various stability notions which have been introduced in Sections 4,5, and 6 are diagrammed in Figure 7.3.

Proposition 7.22. Suppose $L^+(y)$ is positively minimal for each $y \in L^+(x)$. Then the map L^+ is LSC on $L^+(x)$ if and only if $L^+(z)$ is eventually weakly stable for every $z \in L^+(x)$.

Proof: First let L^+ be LSC on $L^+(x)$. Suppose $L^+(z)$ is not eventually stable for some $z \in L^+(x)$. There exists $\varepsilon > 0$ and a sequence $\{y_n\} \subset X$ which converges to $y \in L^+(z)$ so that $y_n t$ is ultimately in $X \setminus B_\varepsilon(L^+(z))$ for every $t \in \mathbb{R}^+$. Accordingly

$$L^+(y_n) \subset X \setminus B_\varepsilon(L^+(z)) \quad \text{for every} \quad n \in \mathbb{N}.$$

Therefore

$$\sup\{d(w, L^+(y_n)) : w \in L^+(y)\} \geq \varepsilon,$$

which contradicts the lower semicontinuity of L^+ at y.

Conversely, suppose $L^+(z)$ is eventually weakly stable for each $z \in L^+(x)$, and assume L^+ is not LSC at some $y \in L^+(x)$. There exists $\varepsilon > 0$ and a sequence $\{y_n\}$ converging to y so that

$$\sup\{d(w, L^+(y_n)) : w \in L^+(y)\} \geq \varepsilon \quad \text{for every} \quad n \in \mathbb{N}.$$

Compactness of $L^+(y)$ ensures that there exist $N_1 \in \mathbb{N}$ and $v \in L^+(y)$ with

$$d(v, L^+(y_n)) \geq \frac{1}{2}\varepsilon \quad \text{for every} \quad n \geq N_1.$$

Since $L^+(y)$ is positively minimal, then for every $z \in L^+(y)$ there exists $\tau = \tau(z) \in \mathbb{R}^+$ such that $z\tau \in B_{\frac{1}{4}\varepsilon}(v)$. There is a finite cover

$$\bigcup_{i=1}^{k} B_{r_i}(z) \supset L^+(y), \; z_i \in L^+(y), \; \tau_i \in \mathbb{R}^+$$

with $B_{r_i}(z_i)\tau_i \subset B_{\frac{1}{4}\varepsilon}(v)$ for every $i = 1,2,\ldots,k$. Choose $\delta > 0$ so that

$$B_\delta(L^+(y)) \subset \bigcup_{i=1}^{k} B_{r_i}(z_i).$$

According to Lemma 7.8 there exists a sequence $\{t_n\}$ in $\mathbb{R}^+$ with $t_n \to \infty$ so that $y_n t_n \to v$.

By eventual weak stability of $L^+(y)$ there is $\alpha > 0$ so that for every $z' \in B_\alpha(L^+(y))$ there exists a sequence in $\mathbb{R}^+$, $\{t'_m\}$, $t'_m \to \infty$ such that $z't'_m \in B_\delta(L^+(y))$. Choose an integer $N_2 \geq N_1$ so that $y_n t_n \in B_\alpha(L^+(y))$ for each $n \geq N_2$. Fix some $n \geq N_2$. Then there exists a sequence $\{t_k^n\}$ in $\mathbb{R}^+$ with $t_k^n \to \infty$ as $k \to \infty$ so that $(y_n t_n)t_k^n \in B_\delta(L^+(y))$. Compactness of $H^+(y_n)$ ensures $L^+(y_n t_n) \cap B_\delta(L^+(y)) = L^+(y_n) \cap B_\delta(L^+(y)) \neq \emptyset$. Let $w_n \in L^+(y_n) \cap B_\delta(L^+(y))$. Then $d(w_n, L^+(y)) < \delta$, and there is z_i with $w_n \in B_{r_i}(z_i)$ and $w_n \tau_i \in B_{\frac{1}{4}\varepsilon}(v)$ for some $i = 1,2,\ldots,k$. Thus $L^+(y_n) \cap B_{\frac{1}{4}\varepsilon}(v) \neq \emptyset$ whenever $n \geq N_2$. Hence $d(v, L^+(y_n)) < \frac{1}{4}\varepsilon$, an impossibility. Consequently L^+ is LSC on $L^+(x)$. □

Corollary 7.23. If the map L^+ is USC on $L^+(x)$, then L^+ is LSC on $L^+(x)$.

Proof: The proof is immediate upon noting that eventual stability and upper semicontinuity of L^+ on $L^+(x)$ implies positive minimality of $L^+(x)$. □

In the proof of Proposition 7.22 we notice that the positive minimality of $L^+(y)$ for each $y \in L^+(x)$ is needed in order to prove the "if part." It can be shown that even if $L^+(z)$ is eventually weakly stable for every $z \in L^+(x)$, L^+ need not be LSC on $L^+(x)$.

The main result is now at hand.

Theorem 7.24.

(i) The map L^+ is continuous on $L^+(x)$ if and only if $L^+(x)$ is positively minimal and eventually stable.

(ii) The map L^+ is continuous on X if and only if $L^+(x)$ is eventually stable for each $x \in X$.

Proof: The proof of (i) is immediate from Proposition 7.17 and Corollary 7.23. As for (ii) we see that the proof of

Lemma 7.18 also shows L^+ is LSC on $A^+(L^+(x))$ provided L^+ is LSC on $L^+(x)$ and $L^+(x)$ is positively minimal. Indeed if every $L^+(x)$ is eventually stable, then by Lemma 7.19 every $L^+(x)$ is positively minimal. Moreover, we can then infer from Proposition 7.22 that L^+ is LSC on every $L^+(x)$. Now proceed as in the proof of Lemma 7.18. It will be sufficient to show that L^+ is LSC at $x \in A^+(L^+(x))$. So let $\varepsilon > 0$ and $y \in L^+(x)$. There exists $\eta > 0$ so that $L^+(x) = L^+(y) \subset B_\varepsilon(L^+(z))$ for every $z \in B_\eta(y)$. Choose $s \in \mathbb{R}^+$ and $\delta > 0$ so that

$$xs \in B_\eta(y) \quad \text{and} \quad B_\delta(x)s \subset B_\eta(y).$$

Then $z \in B_\delta(x)$ implies $zs \in B_\eta(y)$, which in turn implies

$$L^+(x) = L^+(y) \subset B_\varepsilon(L^+(zs)) = B_\varepsilon(L^+(z)).$$

Consequently, L^+ is LSC at x. □

We may summarize the global continuity results of this section in the following diagram.

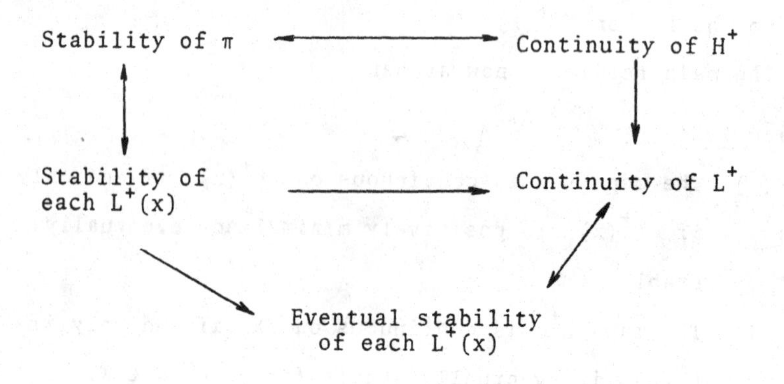

Figure 7.4

8. Lyapunov Functions: The Invariance Principle

In most applications we cannot actually determine the positive limit sets of a given semidynamical system, but we can often narrow down our search to some set E to which an orbit is attracted. Such a set E may be found by recourse to an appropriate Lyapunov function - an idea which we now explore. Naturally E must contain the given limit set. This also provides us with a tool for establishing stability of compact sets.

Definition 8.1. A function $V: X \to \mathbb{R}$ is called a *Lyapunov function* on $G \subset X$ if

(i) V is continuous on $\overline{G}$

(ii) $V(xt) \leq V(x)$ for every $t \in \mathbb{R}^+$ whenever $\gamma^+(x) \subset G$.

We begin with a lemma which locates the positive limit set for a relatively compact orbit.

Lemma 8.2. Suppose $V: G \subset X \to \mathbb{R}$ is a Lyapunov function on G. Denote by Q the largest weakly invariant subset of G. If $x \in G$ has a relatively compact positive orbit $\gamma^+(x) \subset G$, there exists $c = c(x) \in \mathbb{R}$ so that $L^+(x) \subset Q \cap V^{-1}(c)$.

Proof: V must be bounded from below on the compact set $H^+(x)$. As $V(xt)$ is nonincreasing in $t \in \mathbb{R}^+$, there must be some $c \in \mathbb{R}$ such that $\lim_{t\to\infty} V(xt) = c$. Now suppose $y \in L^+(x)$. There is a net $\{t_\alpha\}$ in $\mathbb{R}^+$, $t_\alpha \to \infty$ with $xt_\alpha \to y$. Consequently $V(y) = \lim_{t_\alpha\to\infty} V(xt_\alpha) = c$, so $L^+(x) \subset V^{-1}(c)$. But $L^+(x)$ is weakly invariant so $L^+(x) \subset Q \cap V^{-1}(c)$. □

We have the following computable criteria for a Lyapunov function. The proof can be found in Royden [1, Thm 2, p. 96].

Lemma 8.3. Suppose $V: G \subset X \to \mathbb{R}$. If

(i) V is continuous on $\overline{G}$, and

(ii) $V'(x) \leq 0$ for every $x \in G$ where

$$V'(x) = \limsup_{t \downarrow 0} \frac{V(xt) - V(x)}{t},$$

then V is a Lyapunov function for G. Moreover, $V(xt)$ is differentiable a.e. in $t \in \mathbb{R}^+$.

We can now state the LaSalle Invariance Principle. The proof is obvious in view of the preceeding lemmas. We point out here that the LaSalle Invariance Principle is a major tool in the stability analysis of all kinds of evolutionary equations. The importance of LaSalle's contribution is underscored in this book by the many applications presented in Chapters IV through VIII. Its major drawback though is the difficulty in finding a suitable function V for the system at hand.

Theorem 8.4. (LaSalle Invariance Principle). Let $V: G \subset X \to \mathbb{R}$ satisfy the following:

(i) V is continuous on G,

(ii) $V'(x) \leq 0$ for every $x \in G$, and

(iii) G is positively invariant.

Define

$$E = \{y \in \overline{G}: V'(y) = 0\}$$

and let

$$M = \text{largest weakly invariant subset of } E. \tag{8.1}$$

If $x \in G$ has a relatively compact positive orbit, there exists $c \in \mathbb{R}$ so that $xt \to M \cap V^{-1}(c)$.

Numerous examples in later chapters illustrate the use of the LaSalle Invariance Principle. In many of these instances the set M is determined to be a critical point of the semidynamical system. Moreover, we can usually choose V so that V' is continuous on $\overline{G}$. In that way E will be closed.

Corollary 8.5. Suppose $V: X \to \mathbb{R}$ satisfies the following:

(i) G is a component of $\{x \in X; V(x) < a\}$,

(ii) $\gamma^+(x)$ is relatively compact for every $x \in G$,

(iii) V satisfies the conditions of Lemma 8.3, and

(iv) V' is continuous on $\overline{G}$.

Then $\overline{M}$ is an attractor with $A^+(\overline{M}) \supset G$.

Proof: Clearly, $xt \to M$ for each $x \in G$ according to Theorem 8.4. As E is closed, then $\overline{M} \subset E \subset G$. Thus G is a neighborhood of $\overline{M}$, whereby $\overline{M}$ is an attractor with $A^+(\overline{M}) \supset G$. □

We conclude with a criteria for asymptotic stability. The following definition is required first.

Definition 8.6. A function $W: G \to \mathbb{R}^+$ is called *positive definite* with respect to a compact set $M \subset G$ if $W(x) = 0$ for $x \in M$, and if corresponding to each neighborhood U of M, there exists $\delta > 0$ so that $W(x) \geq \delta$ whenever $x \in G \setminus U$.

Theorem 8.7. Suppose $V: X \to \mathbb{R}$ satisfies the following:

(i) G is a component of $\{x \in X: V(x) < a\}$,

(ii) $\gamma^+(x)$ is relatively compact for every $x \in G$,

(iii) V, V' are continuous on $\overline{G}$,

(iv) $V'(x) \leq 0$ for every $x \in G$,

(v) W is positive definite with respect to M, and

(vi) $W(x) \leq V(x)$ for every $x \in G$.

Then M is asymptotically stable.

Proof: Let U be a neighborhood of M, and set $m = \inf\{V(x): x \in \partial U\}$. Then $m > 0$ according to Definition 8.6. Let $\hat{U} = \{x \in G: V(x) < m\}$. Then $\hat{U} \subset U$ is also a neighborhood of M but is also positively invariant in view of $V'(x) \leq 0$ in G. This means that M is stable. Corollary 8.5 shows that $M = \overline{M}$ is an attractor; therefore M is asymptotically stable. □

Remark 8.8. In most applications the space X is metric, and very often a Banach space. Thus we may take for W the function $W(x) = d(x,M)$, where d is the given metric for X.

Corollary 8.9. Suppose $V: G \subset X \to \mathbb{R}$ satisfies the following:

(i) G is positively invariant,

(ii) $\gamma^+(x)$ is relatively compact for every $x \in G$,

(iii) V, V' are continuous on $\overline{G}$,

(iv) $V'(x) < 0$ for every $x \in G \setminus M$,

(v) M is compact,

(vi) $V(x) = 0$ for every $x \in M$, and

(vii) $V(x) > 0$ for every $x \in G \setminus M$.

Then M is asymptotically stable. If $G = X$, then $A^+(M) = X$ so M is globally asymptotically stable.

9. From Stability to Chaos: A Simple Example

Population growth in some biological situations (e.g., human populations) is a continuous process and generations overlap. Appropriate mathematical models involve nonlinear functional differential equations. There are certain biological situations through (e.g., temperate zone insects) where population growth can be modeled in discrete time, and generations do not overlap. If the variable N denotes the population size of a given generation of a single species, then we take the size of the next generation to be $F(N)$ for some (continuous) function F. The appropriate mathematical model can be viewed as a discrete semidynamical system. In particular, the phase space is $\mathbb{R}^+$ and the phase map is given by

$$\pi(N,k) = F^k(N), \quad (N,k) \in \mathbb{R}^+ \times \mathbb{Z}^+. \tag{9.1}$$

$F^k(N)$ is the k-th iterate of $F(N)$; that is, $F^0(N) = N$, $F^{k+1}(N) = F(F^k(N))$. We write N_k in place of $F^k(N)$ and adopt the convention $N_{k+1} = F(N_k)$. We abandon the usual convention here and adopt the notation $(\mathbb{R}^+,F)$ to indicate the corresponding discrete semidynamical system.

For many biological populations one can expect the variable N to increase from one generation to the next when it is small, and for it to decrease when it is large. In view of this the function F will be required to satisfy the following properties: $F(0) = 0$; there is some $\alpha > 0$ so that F is increasing on $(0,\alpha)$ and decreasing beyond α; $F(N) = Nf(N)$ for some function f. This last condition insures that if the population vanishes at some generation, it will remain zero thereafter. Finally F will usually depend (continuously) on some parameters which affect the shape and steepness of the "hump" in the graph of F. These parameters typically have some biological significance.

The most elementary function F which exhibits such properties is

(9.2) $$F(N) = N[1 + r(1-N/K)],$$

where r is the usual growth rate and K is the carrying capacity of the population. By replacing N with $xK(1+r)/r$ the relationship $N_{k+1} = F(N_k)$ becomes

(9.3) $$x_{k+1} = ax_k(1-x_k)$$

where we have taken $a = 1+r$. Equation (9.3) defines a function

(9.4) $$F_a(x) = ax(1-x).$$

It is clear that x must remain in the unit interval $J = [0,1]$ in order that $0 \leq N \leq K$. Therefore we require $a \in [0,4]$. Consequently, $F_a(J) \subset J$. We denote the corresponding discrete semidynamical system by (J,F_a).

Observe that a critical point of (J,F_a) is a fixed point of F. It is readily seen that (J,F_a) has critical points $x = 0$ and $x_a = 1-a^{-1}$. We investigate the stability of these points for any continuous $F: J \to J$.

Lemma 9.1. Suppose $F: J \to J$ is continuous. If F is differentiable at the fixed point $x^* \in J$, then $\{x^*\}$ is an asymptotically stable point of (J,F) provided

(9.5) $$\left|\frac{dF(x^*)}{dx}\right| < 1.$$

On the other hand, $\{x^*\}$ is unstable whenever

(9.6) $$\left|\frac{dF(x^*)}{dx}\right| > 1.$$

Proof: Clearly x^* is critical. Set $\mu = \left|\frac{dF(x^*)}{dx}\right|$. Suppose first that $\mu < 1$ and choose $0 < \varepsilon < 1-\mu$. There exists a neighborhood V of x^* such that whenever $x \in V$,

$$F(x) = x^* + (x-x^*)\frac{dF(x^*)}{dx} + o(x-x^*),$$

where $|o(x-x^*)| < \varepsilon|x-x^*|$. Set $\beta = \mu + \varepsilon < 1$. Then we obtain

$$(9.7) \qquad |F(x) - x^*| \leq \beta|x-x^*|, \quad x \in V.$$

Suppose U is any neighborhood of x^*. Choose V as established above so that $x^* \in V \subset U$. Finally choose $\delta > 0$ so that $B_\delta(x^*) \subset V$. Then $B_\delta(x^*)$ is a positively invariant neighborhood of x^*, so we have shown x^* to be stable. To obtain asymptotic stability we need only observe that

$$|F^k(x) - x^*| \leq \beta^k|x-x^*|, \quad x \in B_\delta(x^*).$$

Thus as $\beta^k \to 0$, we have $V \subset A^+(\{x^*\})$ and therefore x^* is an attractor.

Now suppose $\mu > 1$ and choose $0 < \varepsilon < \mu-1$. There exists a neighborhood V of x^* such that whenever $x \in V$ we have as before

$$(x-x^*)\frac{dF(x^*)}{dx} = F(x) - x^* + o(x-x^*).$$

Then

$$|x-x^*|\mu \leq |F(x)-x^*| + \varepsilon|x-x^*|.$$

Set $\beta = \mu - \varepsilon > 1$. We get

$$|F(x)-x^*| \geq \beta|x-x^*|.$$

Consequently V cannot contain a positively invariant neighborhood of x^*, whence x^* is unstable. □

We turn to consider (J,F_a). According to Lemma 9.1, $x^* = 0$ is an asymptotically stable critical point whenever $a \in [0,1)$. This case will be omitted in view of the require-

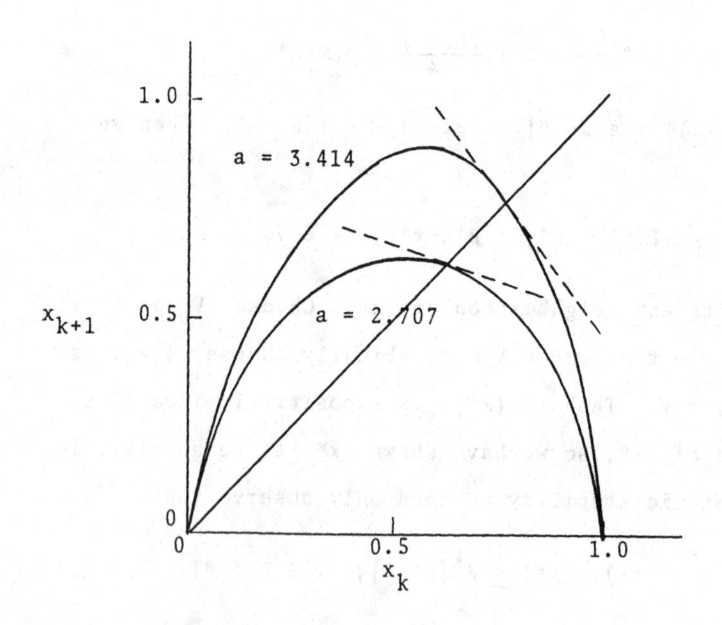

Figure 9.1

ment that $a = 1 + r \geq 1$ (r is the "growth" rate). Since $dF_a(x_a)/dx = 2-a$, the point $x^* = x_a$ is an asymptotically stable critical point whenever $a \in (1,3)$. Indeed, it is shown by May [2] that $A^+(\{x_a\}) = (0,1)$ for every $a \in (1,3)$; that is, x_a is a "global" attractor.

Figure 9.1 illustrates the behavior as a increases over $(1,4]$. The intersections of the graphs of F_a with the dashed line indicate the fixed points of F_a. It is clear that the slopes at these points decrease to less than -1 as a increases beyond 3.

Beyond $a = 3$ the asymptotically stable critical point x_a bifurcates to an asymptotically stable periodic orbit of period 2. To see this we consider the discrete semidynamical system (J,F_a^2). The critical points of this system are the solutions of $F_a^2(x) = x$ in J. This yields a cubic

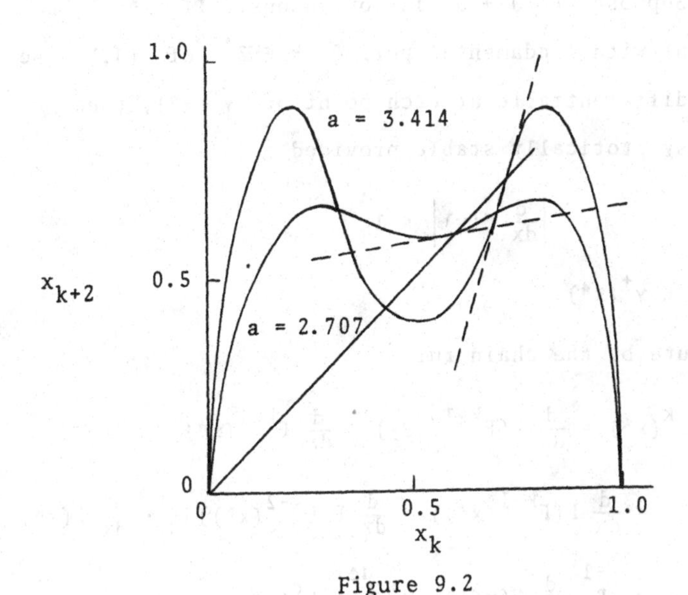

Figure 9.2

equation which may be readily factored since $x_a = 1-a^{-1}$ must also be a critical point of (J,F_a^2). We obtain

$$(9.8) \qquad (x-1+a^{-1})[a^2x^2 - (a^2+a)x + (a+1)] = 0.$$

It is readily seen that the second factor admits distinct real roots if and only if $a > 3$. The behavior of the graphs of F_a^2 corresponding to the cases presented in Fig. 9.1 are illustrated in Fig. 9.2. We see how the two additional critical points of (J,F_a^2) appear as a increases beyond 3. Clearly these points correspond to periodic points of (J,F_a), each of period 2. Denote by $x_a^{(2)}$ either one of these two points so that we obtain a periodic orbit of (J,F_a), namely $\{x_a^{(2)}, F_a(x_a^{(2)})\}$. In order to investigate the stability of this periodic orbit we need another lemma for any continuous $F: J \to J$.

Lemma 9.2. Suppose $F: J \to J$ is continuous. If x^* is a periodic point with fundamental period $k \in \mathbb{Z}^+$ of (J,F) so that F is differentiable at each point of $\gamma^+(x^*)$, then $\gamma^+(x^*)$ is asymptotically stable provided

$$\left|\frac{d}{dx} F(y)\right| < 1 \tag{9.9}$$

for every $y \in \gamma^+(x^*)$.

Proof: Compute by the chain rule

$$\begin{aligned} \frac{d}{dx} F^k(x^*) &= \frac{d}{dx} F(F^{k-1}(x^*)) \cdot \frac{d}{dx} F^{k-1}(x^*) \\ &= \frac{d}{dx} F(F^{k-1}(x^*)) \cdot \frac{d}{dx} F(F^{k-2}(x^*)) \cdots \frac{d}{dx} F(x^*) \qquad (9.10) \\ &= \prod_{r=0}^{k-1} \frac{d}{dx} F(x_r^*), \quad x_r^* \overset{\text{def}}{=} F^r(x^*). \end{aligned}$$

As each $\left|\frac{d}{dx} F(x_r^*)\right| < 1$, $r = 0,1,\ldots,k-1$, then $\left|\frac{d}{dx} F^k(x^*)\right| < 1$. Let U be a neighborhood of $\gamma^+(x^*)$. Apply Lemma 9.1 to the function F^k to obtain $\delta_0 > 0$ and $\beta_0 < 1$ so that $B_{\delta_0}(x^*) \subset U$ and

$$|F^k(x)-x^*| < \beta_0|x-x^*|$$

whenever $x \in B_{\delta_0}(x^*)$. Consequently for such x

$$\lim_{q\to\infty} F^{qk}(x) = x^*.$$

Similarly, we may prove that for each integer $r \in [0,k)$ there exists $\delta_r > 0$ and $\beta_r < 1$ so that $B_{\delta_r}(x_r^*) \subset U$ and

$$|F^k(x)-x_r^*| \le \beta_r|x-x_r^*|$$

whenever $x \in B_{\delta_r}(x_r^*)$. Consequently for such x

$$\lim_{q\to\infty} F^{qk}(x) = x_r^*.$$

Set $\delta = \min\{\delta_0,\ldots,\delta_{k-1}\}$, $\beta = \max\{\beta_0,\ldots,\beta_{k-1}\}$. Then it is easy to verify that

$$W \overset{\text{def}}{=} \bigcup_{r=0}^{k-1} B_\delta(x_r^*)$$

is a positively invariant neighborhood of $\gamma^+(x^*)$. Clearly $W \subset A^+(\gamma^+(x^*))$, so $\gamma^+(x^*)$ is a stable attractor; i.e., asymptotically stable. □

Corollary 9.3. If the Inequality (8.9) is replaced by

$$\left|\frac{d}{dx} F^k(y)\right| < 1 \tag{9.11}$$

for every $y \in \gamma^+(x^*)$, then $\gamma^+(x^*)$ is asymptotically stable. On the other hand if for some $y \in \gamma^+(x^*)$ we have that

$$\left|\frac{d}{dx} F^k(y)\right| > 1,$$

then $\gamma^+(x^*)$ is unstable. In either case, $dF^k(y)/dx$ has the same value at each $y \in \gamma^+(x^*)$.

Proof: The first part follows immediately from the proof of Lemma 9.2. The second part is established like that done in Lemma 9.1. Finally Equation (9.10) reveals that the slope of F^k at each $y \in \gamma^+(x^*)$ is the same. □

Now we return to consideration of the 2-periodic orbit of (J,F_a), namely $\{x_a^{(2)}, F_a(x_a^{(2)})\}$. In view of Corollary 9.3 this orbit is asymptotically stable for $3 < a < 1+\sqrt{6}$. The number $1+\sqrt{6} \simeq 3.449$ can be obtained by computing the value for a at which the slope of F_a^2 equals 1. This may be illustrated graphically in Figure 9.2. We see that as a increases through 3, two new fixed points of F_a^2 are born.

At this point the slope of F_a^2 at x_a becomes steeper than ± 1; therefore x_a becomes unstable. The transition from $a = 2.707$ to $a = 3.414$ in Figure 9.2 suggests that the slope of F_a^2 at $x_a^{(2)}$ has value 1 at the birth of the 2-periodic orbit, and then decreases to -1 as the hump in F_a steepens.

We may continue this analysis to show that beyond $a = 1 + \sqrt{6}$ the 2-periodic orbit will become unstable and bifurcate to an (initially) asymptotically stable 4-periodic orbit. This, in turn, gives way to an 8-periodic orbit, and onwards to a sequence of (initially) asymptotically stable 2^n-periodic orbits. In each case as an asymptotically stable 2^n-periodic orbit becomes stable, it bifurcates to produce a new and initially asymptotically periodic 2^{n+1}-periodic orbit. It is shown by May [2] that $n \to \infty$ as a approaches some critical value $a_c = 3.57$. Indeed, each $a \in (1,a_c)$ gives rise to a unique asymptotically stable 2^n-periodic orbit for some $n = n(a)$. According to May [2], these are the only periodic orbits possible for $a \in (1,a_c)$. The value of a for which the 2^n-periodic orbit first appears will be denoted by a_{2^n}. May [2] also proves (by recourse to an appropriate Lyapunov function) that each of these orbits, when asymptotically stable, has the interval $(0,1)$ as its region of attraction. It follows that each such orbit is the unique attracting positive limit set of the system (J,F_a). For each $a \in (1,a_c)$ we denote the corresponding (attracting) positive limit set by L_a. Figure 9.3 illustrates the bifurcation phenomena as a increases to a_c.

Beyond a_c there ensues what is called by Li and Yorke [1] a region of "chaos". For each $a > a_c$ there are periodic

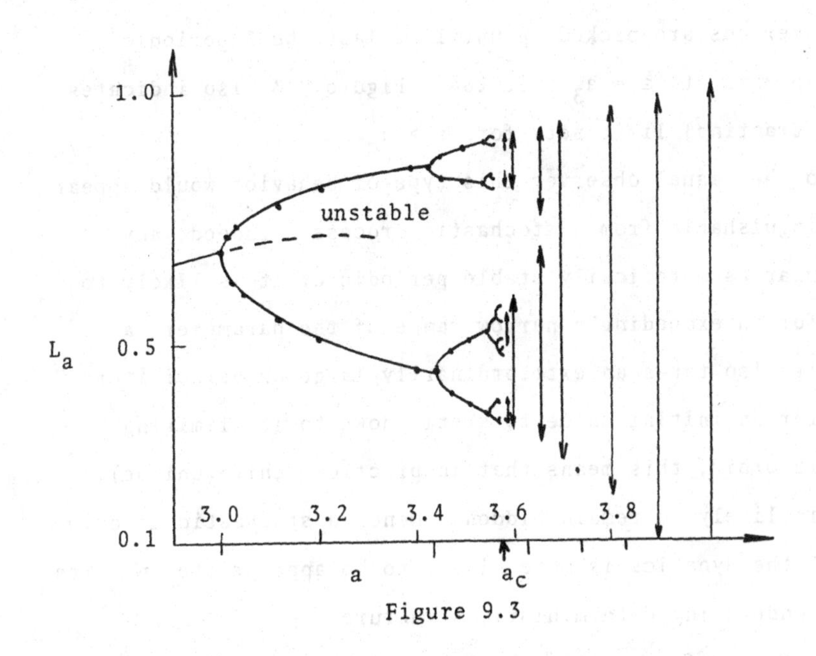

Figure 9.3

points for every fundamental period $k \in \mathbb{N}$, yet an uncountable number of initial points which are not attracted to some periodic orbit. This is in distinction to the case for $a < a_c$ where every point in $(0,1)$ is attracted to L_a. Also, though there are periodic orbits of every period for each $a > a_c$ only one of these orbits is asymptotically stable (Henon [1]).

As a passes a_c, May and Oster [1] have shown that at first, all the periodic points referred to above have even periods. Moreover, x_k oscillates about x_a. These points can have very large fundamental periods (5,726) for values of a just slightly greater than a_c. As a increases further, the first odd periodic point appears at $a \approx 3.6786$. Computer studies have indicated that these odd periodic points can have very large fundamental periods also. But as a

continues to increase, periodic points with smaller odd fundamental periods are picked up until at last the 3-periodic point appears at $a = a_3 \approx 3.8284$. Figure 9.3 also indicates the (attracting) limit sets for $a > a_c$.

To the casual observer this type of behavior would appear indistinguishable from a stochastic process. Indeed, any particular asymptotically stable periodic orbit is likely to exist for an exceedingly narrow range of the parameter a. Since it also takes an extraordinarily large number of iterations for an initial value to settle down to its limiting periodic orbit, this means that in practice, this (unique) orbit is likely to remain hidden. Hence a stochastic description of the dynamics is more likely to be appropriate in spite of the underlying deterministic structure.

In terms of population biology which suggested the function F_a in the first place, a large growth rate has disturbing implications. For one, it says that it may be impossible to distinguish data that have been generated by a "simple" deterministic process, from that of true stochastic noise or experimental error in sampling or measurement. Thus erratic fluctuations in census data may not at all imply an unpredictable environment or sampling errors. Secondly, long term prediction may be impossible as proximate initial values may eventually lead to diverging behavior.

The model provided by (J,F_a) also offers insight into the turbulent behavior of fluids. The Navier-Stokes equations which model fluid dynamics contain a parameter, the Reynolds number R, whose role is analogous to a in F_a. The onset of turbulent flow occurs as R increases through some critical value R_c. According to a model of Landau and Lifschitz [1] we may think of fluid flow past a solid body for $R < R_c$ as a critical point of an associated semidynamical system. This critical point is stable, and indeed, is an attractor. Experimental data seems to indicate that as R passes through R_c, the critical point becomes unstable, hence steady flow becomes impossible. The critical point appears to bifurcate into an asymptotically stable periodic orbit. As R is increased further, this periodic orbit becomes unstable, and in turn, bifurcates into an asymptotically stable doubly periodic (toroidal) flow. When R increases still further, more and more new periods appear in succession. The range of Reynolds numbers between successive appearances of new frequencies rapidly diminishes in size. Thus a sequence of smaller and smaller yet increasing dimensionally tori appear, each a stable attractor for successively smaller ranges of R. This is what Landau and Lifschitz refer to as turbulent flow. Recent investigations by Ruelle and Takens [1], though, indicate the Landau-Lifschitz model is incorrect and needs modification.

We close with a remarkable result of Li and Yorke [2] which guarantees "chaotic" behavior for arbitrary continuous F whenever (J,F) possesses a point of period 3. The proof is not included.

<u>Theorem 9.4</u>. Suppose $F: J \to J$ is continuous. Assume there is a point $y \in J$ for which

$$F^3(y) \leq y < F(y) < F^2(y).$$

then

(i) for every $k \in \mathbb{N}$ there is a periodic point in J of period k,

(ii) there is an uncountable set $S \subset J$ (containing no periodic points) which satisfies the following conditions:

(9.12) for every $x,y \in S$ with $x \neq y$,

$$\limsup_{k\to\infty} |F^k(x) - F^k(y)| > 0$$

$$\liminf_{k\to\infty} |F^k(x) - F^k(y)| = 0;$$

(9.13) for every $x \in S$ and periodic point $y \in J$,

$$\limsup_{k\to\infty} |F^k(x) - F^k(y)| > 0.$$

10. Exercises

10.1. Show that if $M \subset X$ is invariant, then M is positively invariant.

10.2. Show that the intersection and union of a family of positively invariant sets is positively invariant.

10.3. Prove the following: A closed set M is positively invariant if and only if for every $x \in \partial M$ there exists $\varepsilon > 0$ so that $x[0,\varepsilon) \subset M$.

10.4. Given an example for which M is positively invariant but $\operatorname{Int} M$ and ∂M are not.

10.5. For $x \in X$ set $F(x) = \{y \in X\colon yt = x$ for some $t \in \mathbb{R}^+\}$. Prove that $F(x)$ is invariant.

10.6. Prove that a closed set $M \subset X$ is weakly invariant if and only if for every nonstart point $x \in \partial M$ there exists $y \in M$ and $\varepsilon > 0$ such that $y\varepsilon = x$ and $y[0,2\varepsilon] \subset M$.

10.7. Show that the intersection of a nested family of compact weakly invariant sets without start points is weakly invariant.

10.8. Show by example that the intersection of weakly invariant sets need not be weakly invariant.

10.9. Prove that the positive orbit of a periodic point is a weakly invariant compact set.

10.10. Prove that two positively minimal subsets of X either coincide or are disjoint.

10.11. Suppose $M \subset X$ is positively minimal. If for some $x \in M$ there is a compact neighborhood W of x such that $\overline{W \cap M}$ is compact, show that M is compact.

10.12. Show that if $M \subset X$ is positively minimal, for every open set U containing $y \in M$ there exists $\tau \in \mathbb{R}^+$ such that $y\mathbb{R}^+ \subset U[0,\tau]$.

10.13. A subset $M \subset X$ is called weakly minimal if it is closed and weakly invariant but contains no nonempty proper subset with these two properties.

(a) Prove that a compact positively minimal set is weakly minimal without start points.

(b) A compact weakly minimal set is positively minimal.

10.14. For each $x \in X$ the set $J^+(x) \overset{\text{def}}{=} \{y \in X: x_\alpha t_\alpha \to y$ for some nets $x_\alpha \to x$ and $t_\alpha \to \infty\}$ is called the positive prolongational limit set of x. Prove the following:

(a) $J^+(x) = \cap\{\overline{W\mathbb{R}^+t}: t \in \mathbb{R}^+, W$ a neighborhood of $x\}$.

(b) $J^+(x)$ is closed and positively invariant.

(c) $D^+(x) = \gamma^+(x) \cup J^+(x)$.

(d) $J^+(x) \subset J^+(xt)$ for every $t \in \mathbb{R}^+$.

(e) $L^+(x) \subset J^+(x)$.

(f) $y \in L^+(x)$ implies $J^+(x) \subset J^+(y)$.

10.15. Prove that a necessary and sufficient condition for the existence of a positively minimal subset of X is the existence of some $y \in X$ for which $H^+(y) \subset A_w^+(y)$. In this case, show that $H^+(y)$ is the unique positively minimal subset of $A_w^+(y)$.

10.16. Show that $D^+(M) = M$ if and only if for each $x \notin M$ and $y \in M$ there exist neighborhoods U of x and V of y such that $U \cap V\mathbb{R}^+ = \emptyset$.

10.17. Given a set $M \subset X$ define $A_s^+(M)$ to consist of all points $x \in X$ so that if U is any neighborhood of M, there is a neighborhood V of x and $\tau > 0$ with $V[\tau,\infty) \subset U$. Show that;

(a) $A_s^+(M) \subset A^+(M) \subset A_w^+(M)$.

(b) If $A_s^+(M)$ is a neighborhood of M, then $A_s^+(M)$ is an open invariant neighborhood of M.

(c) If M is an attractor, then $A^+(M) = A_s^+(M)$.

(d) If $M \subset X$ is compact, then $A_s^+(M) \subset \{x \in X: \emptyset \neq J^+(x) \subset M\}$ (see Exercise 10.14).

(e) For any $M \subset X$, $x \in A_w^+(M)$ implies $J^+(x) \subset J^+(M)$.

(f) If $M \subset X$ is a compact positively invariant attractor, then M is stable.

10.18. Suppose $y \in X$ is a critical point which is contained in the closure of an open set U and suppose N is any neighborhood of y. Assume that:

(i) V is a Lyapunov function of $G = U \cap N$,

(ii) $M \cap G$ is either empty or equal to $\{y\}$ (M is defined by Equation (8.1)),

(iii) $V(x) < \beta$ on G for every $x \neq y$, and

(iv) $V(y) = \beta$ and $V(x) = \beta$ for every x on that part of the boundary of G within N.

Then y is unstable.

10.19. Show that the function $V(N) = (N - K)^2$ is a Lyapunov function for discrete semidynamical system, $N_{k+1} = F(N_k)$ given by Equation (9.2). In particular verify that

$$\Delta V_k = (N_{k+1}-N_k)(N_{k+1} + N_k - 2K) \leq 0$$

for all $N_k > 0$ if and only if $0 < r < 2$. Thus conclude that point $N^* = K$ is a globally asymptotically solution of (9.2) whenever $0 < r < 2$.

11. Notes and Comments

Many concepts in this chapter (and the next) were first formulated in the context of dynamical systems. For a thorough discussion of their origin, see Bhatia and Szegö [1], especially the notes and comments to their Chapter 5.

Section 2. Weak invariance seems first to have been used by Hale [1] in establishing properties of positive limit sets of functional differential equations. The problem of extending solutions of such equations backwards in time gave rise to the notion of weak invariance (Hale refers to weak invariance as just "invariance.") In the event X is locally compact, then $\overline{M}$ is weakly invariant without start points whenever M is.

Section 3. The origin of a positive limit set goes back to Birkhoff [1]. The statement and proof of Theorem 3.5 is due to Bhatia and Hajek [1]. If X is locally compact, then $L^+(x)$ is always weakly invariant without start points. Moreover in locally compact X, $L^+(x)$ is the largest weakly invariant subset of $H^+(x)$ without start points. Lemma 3.7 is from Bhatia and Hajek [2].

Section 4. Theorem 4.3 was first established by Birkhoff [1] for minimal sets. Theorem 4.4 is also true if compactness of M is replaced by local compactness of X. Furthermore it can be shown that in a locally compact X, every positively minimal set is compact. Example 4.5 is due to Bhatia and Chow [1], and Nemytskii and Stepanov [1, p. 381]. Theorem 4.6 is a precursor of recurrence which is taken up in Chapter III.

Section 5. The concept of stability is of course, long known in differential equations. In the event X is locally compact, then Lemma 5.6 admits a converse; that is, if $D^+(M) = M$ for some set M, then M is stable. Both results are from Bhatia and Hajek [1]. Theorem 5.8 is due to Bhatia [3]. Proposition 5.10 comes from the discussion of characteristic 0 by Bhatia and Hajek [1, Chap. 12]. The converse, Proposition 5.11, is due to Saperstone and Nishihama [1].

Section 6. Weak attraction was first used by Bhatia [1]. Lemma 6.3 is from Bhatia [3]. Proposition 6.5 and Theorem 6.7 are from Bhatia and Hajek [1]. Theorem 6.10 is due to Bhatia [3] as is Example 6.14. Egawa [1] shows that the usual description of a semidynamical system near a compact positively minimal set does not hold when local compactness of X is dropped.

Section 7. All of the results in this section are due to Saperstone and Nishihama [1].

Section 8. The use of Lyapunov functions as a tool for locating positive limit sets comes from LaSalle [1,2,3,4,6,9]. The results on attraction and asymptotic stability, namely Corollary 8.5 and Theorem 8.7 are also from LaSalle [5] and

LaSalle [8, p. 32]. This last reference provides an excellent survey of results on stability in discrete semidynamical systems.

Section 9. This example has been used independently by Li and Yorke [1,2] to study turbulent behavior of fluid flow and May [2] to model population dynamics. Lemma 9.1 is standard; Lemma 9.2 and Corollary 9.3 are from Li and Yorke [1]. Figure 9.3 was produced by Li and Yorke [1]. The references by May [1,3] are easily read review articles. Additional references may be found in May [2] and May and Oster [1]. Marotto [1] has shown how to extend Theorem 9.4 to a differentiable function $F: \mathbb{R}^d \to \mathbb{R}^d$.

CHAPTER III

MOTIONS IN METRIC SPACE

1. Introduction

Stability and attraction for sets as defined in Chapter II only yield information on the behavior of motions near a positively invariant set. Much detail is obscured by only considering the positive orbit or its hull. As Sell points out in [1], stability and attraction of sets are too crude to give much information about the behavior of motions within a set. A finer tool is needed here. In part, this can be accomplished by endowing the phase space with a metric or uniform structure. Also, the consideration of almost periodic motions requires completeness of the phase space. This added structure allows us to answer the question posed in the opening paragraph of Section 4 of Chapter II; namely, when does a semidynamical system extend (uniquely) to a dynamical system? Moreover, in this setting we can complete the classification of compact positively minimal sets into the closure of recurrent, uniformly recurrent, almost periodic, periodic, and critical motions.

Section 2 concerns Lyapunov stability of motions. It is shown how this relates to the stability of positively invariant sets. The major result is how a stable semidynamical system extends uniquely to a dynamical system. Section 3 develops a characterization of compact positively minimal sets in terms of recurrent motions. Almost periodicity, a natural sequel to recurrence, is taken up in Section 4. There are two major results here. One concerns the relationship between Lyapunov stability and almost periodicity. The other is a characterization of the limiting behavior of almost periodic motions. The main objective of Section 5 is a generalization of the classical Poincaré-Bendixon theorem for an autonomous differential equation in the plane. The Poincaré-Bendixson theorem says (in the language of this book) that if π_x is a bounded positive motion which "stays away" from critical points, then $L^+(x)$ is a periodic orbit. Section 6 presents an application of much of the material developed in the earlier sections. Specifically we demonstrate the existence of periodic solutions of differential equations.

We make formal now what was suggested above. Assume throughout this chapter that the phase space X is metric with metric d.

2. Lyapunov Stable Motions

Definition 2.1. Suppose F is an arbitrary nonempty subset of X. The positive motion π_x is called *Lyapunov stable relative to* F if for every $\varepsilon > 0$ there exists $\delta > 0$ such that if $y \in F \cap B_\delta(x)$, then

$$d(yt,xt) < \varepsilon \quad \text{for every} \quad t \in \mathbb{R}^+.$$

If $F = X$ above, then π_x is called just *Lyapunov stable*. The positive motion π_x is called *uniformly Lyapunov stable relative to* F if for every $\varepsilon > 0$ there exists $\delta > 0$ such that if $y \in F \cap B_\delta(x\tau)$ for $\tau \in \mathbb{R}^+$, then

$$d(yt, x(\tau+t)) < \varepsilon \quad \text{for every} \quad t \in \mathbb{R}^+.$$

The semidynamical system (X,π) is called *Lyapunov stable* if for every $x \in X$ the positive motion π_x is Lyapunov stable.

The stability of a set M is a property of M with respect to neighboring positive orbits. Indeed, this kind of stability is a geometrical concept which depends only upon the range of the motions π_x near M and not upon the motions themselves. Surely, differing positive motions can possess the same positive orbit. Thus we see that Lyapunov stability of a positive motion is distinct form stability of a set. But the concept of uniform Lyapunov stability provides a means of bridging that distinction. In fact we see that uniform Lyapunov stability of a positive motion π_x is just Lyapunov stability of every motion π_z belonging to the positive orbit $\gamma^+(x)$, where the corresponding δ does not depend on z. This seems to take us close to the meaning of stability of the set $\gamma^+(x)$. We obtain the following relationship in the case of compact motions.

Proposition 2.2. If π_x is a compact positive motion which is uniformly Lyapunov stable, then $H^+(x)$ is stable.

Proof: Let U be a neighborhood of $H^+(x)$ and choose $\varepsilon > 0$ so that $B_\varepsilon(H^+(x)) \subset U$. Let $\delta > 0$ exist in view of the uniform Lyapunov stability of π_x. Set $V = \cup\{B_\delta(x\tau):$ $\tau \in \mathbb{R}^+\}$. Then V is an open neighborhood of $H^+(x)$. For otherwise there would exist $y \in H^+(x)$ with $d(y, x\tau) \geq \delta$ for every $\tau \in \mathbb{R}^+$. This is impossible, so V is an open neighborhood of $H^+(x)$. It follows that $V\mathbb{R}^+ \subset U$; therefore $H^+(x)$ is stable. □

In the event that $x \in X$ is critical, Proposition 2.2 admits a converse. (In this case uniform Lyapunov stability is equivalent to Lyapunov stability.)

Theorem 2.3. A critical motion π_x is Lyapunov stable if and only if the set $\{x\}$ is stable.

Proof: The "only if" part follows from Proposition 2.2. Now suppose the set $\{x\}$ is stable. Then $\{x\}$ is positively invariant from Theorem 5.5 of Chapter II, and so the motion π_x must be critical. Given any $\varepsilon > 0$ there exists $\delta > 0$ such that $B_\delta(x)\,\mathbb{R}^+ \subset B_\varepsilon(x)$. As $xt = x$ for every $t \in \mathbb{R}^+$, then $y \in B_\delta(x)$ implies

$$yt \in B_\delta(x)\,\mathbb{R}^+ \subset B_\varepsilon(x) = B_\varepsilon(xt) \quad \text{for every } t \in \mathbb{R}^+.$$

This says π_x is Lyapunov stable. □

We can characterize the positively minimal sets of a Lyapunov stable semidynamical system as follows.

Theorem 2.4. If (X,π) is a Lyapunov stable semidynamical system, then a subset $M \subset X$ is positively minimal if and only if $M = L^+(x) \neq \emptyset$ for some $x \in X$.

Proof: If M is positively minimal then $M = L^+(x)$ for each $x \in M$. Conversely suppose $L^+(x) \neq \emptyset$ for some $x \in X$. Let $y \in L^+(x)$. There exists a sequence $\{t_n\} \subset \mathbb{R}^+$ with $t_n \to \infty$ so that $xt_n \to y$. Naturally, $L^+(y) \subset H^+(y) \subset L^+(x)$. We claim that $L^+(x) \subset L^+(y)$. So let $z \in L^+(x)$. There exists a sequence $\{\tau_n\} \subset \mathbb{R}^+$ with $\tau_n \to \infty$ so that $x\tau_n \to z$. We may assume (by choosing a subsequence if necessary) that

$s_n \overset{\text{def}}{=} \tau_n - t_n \to \infty$. Then $(xt_n)s_n = x\tau_n \to z$. Now let $\varepsilon > 0$. There exists $\delta > 0$ so that if $xt_n \in B_\delta(y)$, then $x\tau_n = (xt_n)s_n \in B_{\frac{1}{2}\varepsilon}(ys_n)$ for every s_n by Lyapunov stability of π_y. Since for sufficiently large n we must have $xt_n \in B_\delta(y)$ and $x\tau_n \in B_{\frac{1}{2}\varepsilon}(z)$, then

$$d(z, ys_n) \leq d(z, x\tau_n) + d(x\tau_n, ys_n) < \varepsilon.$$

Thus $ys_n \to z$ so that $z \in L^+(y)$. It follows that $L^+(y) = L^+(x)$ so that $L^+(x)$ is indeed positively minimal. □

We turn to a question first posed in Section 4 of Chapter II. That is, under what conditions (if any) does the restriction of a semidynamical system (X,π) to a closed positively invariant set M define a dynamical system $(M,\hat{\pi})$? The next theorem answers this question. Note that the set M need not be compact.

Theorem 2.5. Suppose X is a complete metric space and $M \subset X$ is positively minimal. If every positive motion in M is Lyapunov stable relative to M, then the restriction of (X,π) to M extends uniquely to a dyanamical system $(M,\hat{\pi})$.

Proof: First we will show $\pi^t: M \to M$ is one-to-one for every $t \in \mathbb{R}^+$. So let $y \in M$. Since M is positively minimal, there exists a sequence $\{\tau_n\} \subset \mathbb{R}^+$ with $\tau_n \uparrow \infty$ so that $y\tau_n \to y$. By continuity of π the sequence of maps $\{\pi^{\tau_n}\}$ converges pointwise to the identity map on $\gamma^+(y)$. We claim that the convergence is pointwise on all of M. So let $x \in M$ be arbitrary. There exists a sequence $\{y_k\} \subset \gamma^+(y)$ such that $y_k \to x$. Consider the inequality

$$d(x, x\tau_n) \leq d(x, y_k) + d(y_k, y_k\tau_n) + d(y_k\tau_n, x\tau_n). \tag{2.1}$$

For every $\varepsilon > 0$ there exists $k_1 \in \mathbb{N}$ so that $k \geq k_1$ implies

$$d(x,y_k) < \varepsilon/3 \quad \text{and} \quad d(y_k\tau_n, x\tau_n) < \varepsilon/3 \quad \text{for each} \quad n \in \mathbb{N}.$$

Also there exists $n_1 = N(k_1) \in \mathbb{N}$ so that $d(y_k, y_k\tau_n) < \varepsilon/3$ whenever $k \geq k_1$ and $n \geq n_1$. Thus in view of (2.1) we have $x\tau_n \to x$ so that $\{\pi^{\tau_n}\}$ does indeed converge to the identity map on M. Now if π^t were not one-to-one for some $t \in \mathbb{R}^+$, there would exist $x,y \in M$, $x \neq y$ so that $xt = yt$. Pick any $0 < \varepsilon < d(x,y)$. There is $n_2 \in \mathbb{N}$ such that

$$d(x,x\tau_n) < \varepsilon/2 \quad \text{and} \quad d(y,y\tau_n) < \varepsilon/2 \quad \text{for each} \quad n \geq n_2.$$

Consequently $d(x,y) < \varepsilon$, a contradiction. Thus π^t is one-to-one on M for every $t \in \mathbb{R}^+$.

Next we show that π^t maps M onto M for each $t \in \mathbb{R}^+$. Fix $y \in M$ and $t \in \mathbb{R}^+$, and consider the sequence $\{\tau_n\}$ established above with $y\tau_n \to y$. For sufficiently large $k,m,n \in \mathbb{N}$ (so that the appropriate numbers, e.g., $\tau_k - t$, are defined), we have

$$d(y(\tau_k-t),y(\tau_m-t)) \leq d(y(\tau_k-t),y(\tau_n+\tau_k-t))$$
$$+ d(y(\tau_n+\tau_k-t),y(\tau_m+\tau_k-t)) + d(y(\tau_m+\tau_k-t),y(\tau_m-t)).$$

Let $\varepsilon > 0$. In view of the Lyapunov stability of π_y there exists $N \in \mathbb{N}$ so that $k,m,n \geq N$ implies

$$d(y(\tau_k-t),y(\tau_n+\tau_k-t)) < \varepsilon/3$$
$$d(y(\tau_m-t),y(\tau_k+\tau_m-t)) < \varepsilon/3.$$

For each fixed $k \in \mathbb{N}$, $\{y(\tau_k+\tau_n-t)\}$ is Cauchy as it converges to $y(\tau_k-t)$. Consequently, $\{y(\tau_n-t)\}$ is also Cauchy and hence converges to some $z \in M$ by completeness of X. Then $zt = y$ so π^t indeed maps M onto M.

So for each $t \in \mathbb{R}^+$, $\pi^{-t} = (\pi^t)^{-1}$ is a well-defined mapping of M onto M. Now we conclude by establishing the continuity of π^{-t}. Let $\{y_k\}$ be a sequence in M with $y_k \to y$. For each y_k there exists a unique $z_k \in M$ such that $z_k t = y_k$. Also there exists a unique $z \in M$ such that $zt = y$. Then π^{-t} is continuous at y if and only if $z_k \to z$. For sufficiently large $n \in \mathbb{N}$ consider the inequality

$$d(z,z_k) \leq d(z,y(\tau_n-t)) + d(y(\tau_n-t),y_k(\tau_n-t)) + d(y_k(\tau_n-t),z_k).$$

Let $\varepsilon > 0$. There exists $k_3 \in \mathbb{N}$ so that for sufficiently large $n \in \mathbb{N}$ we have

$$d(y(\tau_n-t),y_k(\tau_n-t)) < \varepsilon/3 \quad \text{for each} \quad k \geq k_3.$$

Moreover there exists $n_3 = N(k_3) \in \mathbb{N}$ such that $n \geq n_3$ implies

$$d(z,y(\tau_n-t)) < \varepsilon/3 \quad \text{and} \quad d(y_k(\tau_n-t),z_k) < \varepsilon/3.$$

Therefore $d(z,z_k) < \varepsilon$ so we must have $z_k \to z$. This establishes π^t is a homeomorphism of M onto M. In view of Corollary 2.8, Chapter I, the restriction of (X,π) to M extends uniquely to a dynamical system $(M,\hat{\pi})$. □

Corollary 2.6. Under the hypothesis of Theorem 2.5 the dynamical system $(M,\hat{\pi})$ is bilaterally Lyapunov stable in the sense that for each $x \in M$ and $\varepsilon > 0$ there exists $\delta > 0$ so that $y \in M \cap B_\delta(x)$ implies $d(\hat{\pi}(y,t),\hat{\pi}(x,t)) < \varepsilon$ for every $t \in \mathbb{R}$.

Proof: Choose the sequence $\{\tau_n\} \subset \mathbb{R}^+$ as in the proof of Theorem 2.5 so that for any $(x,t) \in M \times \mathbb{R}^+$,

$$x(\tau_n - t) \to \pi^{-1}(x,t) = \hat{\pi}(x,-t).$$

By stability of π_x, given any $\varepsilon > 0$ there exists $\delta > 0$ so that $y \in B_\delta(x)$ implies $d(y(\tau_n - t), x(\tau_n - t)) < \varepsilon/3$ whenever $\tau_n - t \in \mathbb{R}^+$. Consider

$$d(\hat{\pi}(y,-t),\hat{\pi}(x,-t)) \le d(\hat{\pi}(y,-t),y(\tau_n - t)) + d(y(\tau_n - t),x(\tau_n - t)) + d(x(\tau_n - t),\hat{\pi}(x,-t)).$$

Both the first and third terms on the right side of the last inequality can be made less than $\varepsilon/3$ for sufficiently large $n \in \mathbb{N}$. Consequently

$$d(\hat{\pi}(y,-t),\hat{\pi}(x,-t)) < \varepsilon \quad \text{for each} \quad t \in \mathbb{R}^+$$

as the choice of δ was independent of t or τ_n. Thus $(X,\hat{\pi})$ is bilaterally Lyapunov stable. □

3. Recurrent Motions

A stated objective in Section 4 of Chapter II was to characterize the behavior of principal motions in a compact positive limit set. Under the assumption of positive minimality we noticed that the positive motion through any point x of a compact positive limit set "regularly returns" close to x. This behavior is formalized in the next definition.

Definition 3.1. A positive motion π_x is said to *recur* at $y \in X$ if for each $\varepsilon > 0$ there exists $L > 0$ so that $(xt)[0,L] \cap B_\varepsilon(y) \neq \emptyset$ for every $t \in \mathbb{R}^+$. π_x is called *recurrent* if it recurs at x. The set of all points $y \in X$ at which π_x recurs will be denoted by $\mathscr{R}^+(x)$.

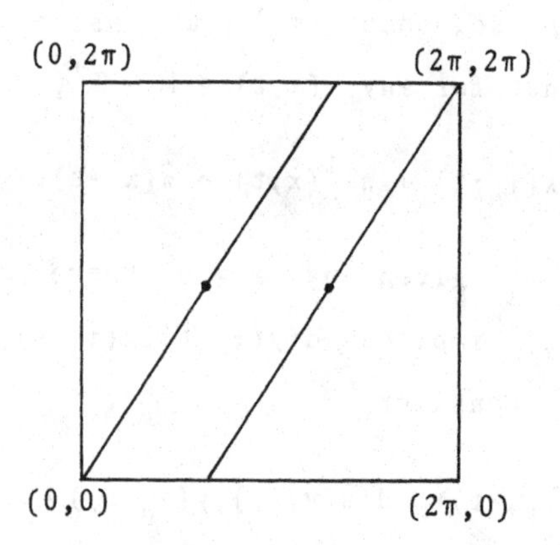

It is clear that periodic motions are recurrent. An immediate consequence of Definition 3.1 and Theorem 4.6 of Chapter II is

Corollary 3.2. If M is a compact positively minimal subset of X, then every positive motion in M is recurrent.

Example 3.3. The space $X = S^1 \times S^1$ is the torus. We represent X here by the square $\{(x_1,x_2) \in \mathbb{R}^2: 0 \leq x_i \leq 2\pi, i = 1,2\}$ with the opposite edges identified. Endow $S^1 \times S^1$ with the metric $d((x_1,x_2),(y_1,y_2)) = \sqrt{(x_1-y_1)^2+(x_2-y_2)^2}$, where the differences x_1-y_1, x_2-y_2 are taken mod 2π. We define a semidynamical system with phase space $S^1 \times S^1$ as follows: fix $\alpha \in \mathbb{R}$ to be irrational. For any $(x_1,x_2) \in S^1 \times S^1$ and $t \in \mathbb{R}^+$ set

$$\pi((x_1,x_2),t) = (x_1+t,x_2+\alpha t)$$

where the addition is mod 2π. Because α is irrational we see that no positive orbit is periodic. Indeed, for any $(x_1,x_2) \in S^1 \times S^1$, the hull $H^+(x_1,x_2) = S^1 \times S^1$. Thus $S^1 \times S^1$ itself is a compact positively minimal set and every positive motion in $S^1 \times S^1$ is recurrent.

The following properties of $\mathcal{R}^+(x)$ are found to be useful.

Lemma 3.4.

(i) $\mathcal{R}^+(x)$ is a closed, positively invariant subset of $L^+(x)$.

(ii) $\mathcal{R}^+(x) = \cap\{H^+(z): z \in H^+(x)\}$.

(iii) If $\mathcal{R}^+(x) \neq \emptyset$, it is the unique positively minimal subset of $H^+(x)$.

Proof: (i) Suppose $\{y_n\}$ is a sequence in $\mathcal{R}^+(x)$ with $y_n \to y$. Let $\varepsilon > 0$. Then $y_n \in B_\varepsilon(y)$ for all sufficiently large $n \in \mathbb{N}$. For any such y_n there must be $\delta > 0$ so that $B_\delta(y_n) \subset B_\varepsilon(y)$. As $y_n \in \mathcal{R}^+(x)$ there exists $L > 0$ so that $(xt)[0,L] \cap B_\delta(y_n) \neq \emptyset$ for every $t \in \mathbb{R}^+$. Clearly, $(xt)[0,L] \cap B_\varepsilon(y) \neq \emptyset$ for every $t \in \mathbb{R}^+$. Consequently $y \in \mathcal{R}^+(x)$ and $\mathcal{R}^+(x)$ is closed.

Now let $y \in \mathcal{R}^+(x)$, $\tau \in \mathbb{R}^+$, and suppose $\varepsilon > 0$. There exists $\delta > 0$ such that $B_\delta(y) \subset \pi^{-1}(B_\varepsilon(y\tau),\tau)$. Moreover, there exists $L > 0$ such that $(xt)[0,L] \cap B_\delta(y) \neq \emptyset$ for every $t \in \mathbb{R}^+$. Then we must have

$$(xt)[0,\tau+L] \cap B_\varepsilon(y\tau) \neq \emptyset \quad \text{for every} \quad t \in \mathbb{R}^+.$$

This means that $y\tau \in \mathcal{R}^+(x)$; that is, $\mathcal{R}^+(x)$ is positively invariant.

Finally $y \in \mathcal{R}^+(x)$ implies $y \in \cap\{H^+(z): z \in \gamma^+(x)\} = L^+(x)$ from Lemma 3.2 of Chapter II.

(ii) Set $\mathcal{P} = \cap\{H^+(z): z \in H^+(x)\}$. We claim that $\mathcal{R}^+(x) \subset \mathcal{P}$. So let $y \in \mathcal{R}^+(x)$. Given any $\varepsilon > 0$ there exists $L > 0$ so that $z[0,L] \cap B_{\frac{1}{2}\varepsilon}(y) \neq \emptyset$ for every $z \in \gamma^+(x)$.

Now suppose $\{z_n\}$ is a sequence in $\gamma^+(x)$ with $z_n \to z \in H^+(x)$. For each z_n there exists $t_n \in [0,L]$ with $z_n t_n \in B_{\frac{1}{2}\varepsilon}(y)$. We may assume (by choosing a subsequence if necessary) that $t_n \to t \in [0,L]$. Then

$$z_n t_n \to zt \in z[0,L] \cap B_{\frac{1}{2}\varepsilon}(y) \subset z[0,L] \cap B_\varepsilon(y) \neq \emptyset.$$

As this holds for every $z \in H^+(x)$ we have that

$$y \in H^+(z) \quad \text{for every} \quad z \in H^+(x).$$

Therefore $y \in \mathscr{P}$ so $\mathscr{R}^+(x) \subset \mathscr{P}$ as claimed.

Now $\mathscr{P}$ must be positively minimal as it is the intersection of all the closed, positively invariant subsets of $H^+(x)$. Since $\mathscr{R}^+(x)$ itself is a closed and positively invariant subset of $H^+(x)$, we must have $\mathscr{R}^+(x) = \mathscr{P}$.

(iii) If $H^+(x)$ contains two nonempty, disjoint positively minimal subsets, then P and hence $\mathscr{R}^+(x)$ must be empty. This contradicts our assumption though. The proof is now complete. □

We are ready to characterize recurrent positive motions.

<u>Theorem 3.5</u>. A set $M \subset X$ is compact and positively minimal if and only if $M = H^+(x)$ for some compact recurrent positive motion π_x.

<u>Proof</u>: Suppose M is a compact, positively minimal subset of X. For every $x \in M$ we must have $H^+(x) = M$. Then π_x is a compact positive motion and recurrent according to Corollary 3.2. Conversely, suppose π_x is a compact recurrent positive motion. Set $M = H^+(x)$. As π_x is recurrent, then $x \in \mathscr{R}^+(x)$. Thus $\mathscr{R}^+(x) \neq \emptyset$. In view of (iii) of Lemma 3.4 we have $M = H^+(x) \supset \mathscr{R}^+(x)$ where $\mathscr{R}^+(x)$ is

positively minimal. But $\mathscr{R}^+(x) \neq \emptyset$ implies $H^+(x) = \mathscr{R}^+(x)$. Consequently, $M = \mathscr{R}^+(x)$ is positively minimal. □

The characterization of recurrence provided in Theorem 3.5 requires compactness of π_x. This requirement may be dropped by strengthening the concept of recurrence. We begin with the following observation.

Remark 3.6. If the positive motion π_x is recurrent, then π_x recurs at every point along $\gamma^+(x)$. To see this let $\varepsilon > 0$ and $y = xs \in \gamma^+(x)$. Choose $\delta > 0$ by continuity of π so that $B_\delta(x)s \subset B_\varepsilon(xs)$. There exists $L > 0$ such that $(xt)[0,L] \cap B_\delta(x) \neq \emptyset$ for every $t \in \mathbb{R}^+$. Thus

$$\emptyset \neq \{(xt)[0,L] \cap B_\delta(x)\}s \subset (xt)[0,s+L] \cap B_\varepsilon(xs)$$

for every $t \in \mathbb{R}^+$. This shows π_x recurs at xs. We see though, that the interval $[0,s+L]$ depends upon s. If the interval can be chosen independently of s, then we can express the recurrence of π_x at every $y \in \gamma^+(x)$ as follows: for every $\varepsilon > 0$ there exists $L = L(\varepsilon) > 0$ so that

$$\gamma^+(x) \subset B_\varepsilon((xt)[0,L])$$

for every $t \in \mathbb{R}^+$. This motivates the following definition.

Definition 3.7. Let $M \subset X$. The positive motion π_x is said to *recur uniformly on the set* M if for every $\varepsilon > 0$ there exists $L = L(\varepsilon) > 0$ such that

$$M \subset B_\varepsilon((xt)[0,L]) \quad \text{for every } t \in \mathbb{R}^+.$$

When $M = \gamma^+(x)$, we say π_x is *uniformly recurrent*.

Corollary 3.8. If the positive motion π_x is uniformly recurrent, then $\gamma^+(x)$ is precompact.

Proof: Choose $t = 0$ in Definition 3.7. Then $\gamma^+(x) \subset B_\varepsilon(x[0,L])$. As the set $x[0,L]$ is compact, then $\gamma^+(x)$ must be precompact. □

Obviously, a uniformly recurrent positive motion is recurrent. The converse holds for a compact positive motion.

Lemma 3.9. If the positive motion π_x is compact and recurrent, then it is uniformly recurrent.

Proof: As $M = H^+(x)$ must be compact, given any $\varepsilon > 0$, we can find a finite set $\{x_1,x_2,\ldots,x_m\} \subset M$ with $M \subset \bigcup_{i=1}^{m} B_{\frac{1}{2}\varepsilon}(x_i)$. Moreover for each $i = 1,2,\ldots,m$ there exists $L_i > 0$ such that $x_i \in B_{\frac{1}{2}\varepsilon}((xt)[0,L_i])$ for every $t \in \mathbb{R}^+$. Set $L = \max\{L_1,L_2,\ldots,L_m\}$. Then for every $t \in \mathbb{R}^+$ we obtain

$$H^+(x) = M \subset \bigcup_{i=1}^{m} B_{\frac{1}{2}\varepsilon}(x_i) \subset B_\varepsilon((xt)[0,L]).$$

This shows that π_x is uniformly recurrent. □

In the event that X is complete, we may drop the compactness requirement on π_x to obtain the promised generalization of Theorem 3.5.

Theorem 3.10. Suppose X is a complete metric space. Then $M \subset X$ is compact, positively minimal if and only if $M = H^+(x)$ for some uniformly recurrent positive motion π_x in M.

Proof: If M is compact and positively minimal, then $M = H^+(x)$ for some recurrent positive motion π_x according to Theorem 3.5. As π_x must be compact, then it must be uniformly recurrent by Lemma 3.8.

Conversely, suppose π_x is uniformly recurrent, and set $M = H^+(x)$. As $\gamma^+(x)$ is precompact and X is complete, then

$M = \overline{\gamma^+(x)}$ is compact. The positive minimality of M follows from Theorem 3.5. □

Corollary 3.11. Suppose X is a complete metric space. The positive motion π_x uniformly recurs on $L^+(x)$ if and only if $L^+(x)$ is a compact positively minimal set.

4. Almost Periodic Motions

We examine more closely the nature of recurrent positive motions. Let us reformulate recurrence in the next lemma. Its proof is obvious.

Lemma 4.1. The positive motion π_x is recurrent if and only if for every $\varepsilon > 0$ there exists $L = L(\varepsilon) > 0$ so that every interval in $\mathbb{R}^+$ of length L contains a point τ with $d(x,x\tau) < \varepsilon$.

Now suppose not just x, but every point y in some positive orbit $\gamma^+(x)$ "regularly returns" close to y within some fixed interval of time which is independent of y. Then we would have a recurrent motion which is "nearly" or "almost" periodic. But since we literally mean this, the following definition is appropriate.

Definition 4.2. The positive motion π_x is called *almost periodic* if for every $\varepsilon > 0$ there exists $L = L(\varepsilon) > 0$ so that every interval in $\mathbb{R}^+$ of length L contains a point τ with

$$d(xt,x(t+\tau)) < \varepsilon \quad \text{for every} \quad t \in \mathbb{R}^+.$$

It is obvious that an almost periodic positive motion is recurrent. What is not so obvious is that an almost periodic positive motion is uniformly recurrent without the requirement of completeness or compactness.

Lemma 4.3. If the positive motion π_x is almost periodic, then it is uniformly recurrent.

Proof: Let $\varepsilon > 0$ and suppose $L = L(\varepsilon)$ be given by almost periodicity of π_x. Let $s,t \in \mathbb{R}^+$. If $s \leq t$, choose $\tau \in [t-s,t-s+L]$ so that $d(xs,x(s+\tau)) < \varepsilon$. Then $xs \in B_\varepsilon(xt[0,L])$. If $t < s \leq t+L$, then obviously $xs \in B_\varepsilon(xt[0,L])$. Finally, if $s > t+L$, choose $\tau \in [s-t-L, s-t]$ so that $d(xs,x(s-\tau)) = d(x(s-\tau+\tau),x(s-\tau)) < \varepsilon$. As $s-\tau \in [t,t+L]$, then $xs \in B_\varepsilon(xt[0,L])$. Consequently, $\gamma^+(x) \subset B_\varepsilon((xt)[0,L])$ for every $t \in \mathbb{R}^+$, so π_x is uniformly recurrent. □

From Lemma 4.3 and Corollary 3.8 we obtain

Corollary 4.4. If the positive motion π_x is almost periodic, then $\gamma^+(x)$ is precompact.

The converse of Lemma 4.3 is false. We will exhibit (in Example 4.12) a uniformly recurrent positive motion which is not almost periodic. First we must develop some properties of almost periodic motions. We begin by introducing the concept of a relatively dense set.

Definition 4.5. A set $D \subset \mathbb{R}^+$ is called *relatively dense* in $\mathbb{R}^+$ if there exists $L > 0$ such that $[t,t+L] \cap D \neq \emptyset$ for every $t \in \mathbb{R}^+$. We define a relatively dense set in $\mathbb{R}$ analogously.

The set $\mathbb{Z}^+$ is relatively dense; the set of prime numbers is not. In view of Definition 4.5 we obtain the following characterizations of an almost periodic positive motion. The proof is immediate.

Corollary 4.6. The positive motion π_x is almost periodic if and only if for every $\varepsilon > 0$ the set

$$D(\varepsilon) = \{\tau \in \mathbb{R}^+ : \sup_{t \in \mathbb{R}^+} d(xt, x(t+\tau)) < \varepsilon\}$$

is relatively dense in $\mathbb{R}^+$.

The relatively dense set defined in Corollary 4.6 is called the set of ε-periods of π_x. We observe that if π_x is almost periodic, then so is π_y whenever $y \in \gamma^+(x)$. Moreover, the ε-periods of π_y are also ε-periods of π_y. Periodic motions are also almost periodic.

Recurrence may be reformulated in terms of a relatively dense set of ε-periods. In particular, the positive motion π_x is recurrent if and only if for each $\varepsilon > 0$ the set $\{\tau \in \mathbb{R}^+ : d(x, x\tau) < \varepsilon\}$ is relatively dense in $\mathbb{R}^+$.

When X is complete we obtain a characterization of the positive limit set of an almost periodic motion.

Theorem 4.7. Suppose X is a complete metric space and π_x is an almost periodic positive motion for some $x \in X$. Then

(i) $H^+(x)$ is compact and positively minimal,

(ii) π_y is almost periodic for every $y \in L^+(x)$, and

(iii) π_x uniformly recurs on $L^+(x)$.

Proof: (i) $\gamma^+(x)$ is precompact according to Corollary 4.4. Since X is complete, then $H^+(x) = \overline{\gamma^+(x)}$ is compact. Also, as π_x is a compact recurrent positive motion, then $H^+(x)$ is positively minimal by Theorem 3.5. (ii) Let $\varepsilon > 0$ and choose the relatively dense set $D = D(\varepsilon)$ in $\mathbb{R}^+$ and $L = L(\varepsilon) > 0$ according to Corollary 4.5. Now suppose $y \in L^+(x)$. There is a sequence $\{t_n\} \subset \mathbb{R}^+$ with $t_n \to \infty$ so that $xt_n \to y$. For every such t_n and $\tau \in D$ we have

$$\sup_{t\in\mathbb{R}^+} d(x(t_n+t), x(t_n+t+\tau)) < \varepsilon.$$

Proceeding to the limit we obtain for every $\tau \in D$

$$\sup_{t\in\mathbb{R}^+} d(xt,y(t+\tau)) \leq \varepsilon.$$

This establishes π_y is almost periodic. (iii) This follows from Corollary 3.11. □

The proof of Theorem 4.7 has established more than just the almost periodicity of every π_y, $y \in L^+(x)$. It demonstrated that the relatively dense set of ε-periods may be chosen independently of all points y in the compact positive limit set $L^+(x)$. This leads us to a new definition.

Definition 4.8. A positively invariant subset $M \subset X$ is called *equi-almost periodic* if for every $\varepsilon > 0$, the set

$$\{\tau \in \mathbb{R}^+ : \sup_{\substack{x\in M\\ t\in\mathbb{R}^+}} d(xt,x(t+\tau)) < \varepsilon\}$$

is relatively dense in $\mathbb{R}^+$.

In view of Theorem 4.7 we obtain the following result.

Corollary 4.9. Suppose π_x is an almost periodic positive motion in the complete metric space X. Then $L^+(x)$ is a compact, positively minimal equi-almost periodic set.

We turn to the matter of determining conditions for a positive motion to be almost periodic. As might have been suspected, it is related to recurrence and Lyapunov stability of motions.

Theorem 4.10. A positive motion π_x is almost periodic if and only if it is recurrent and Lyapunov stable relative to $\gamma^+(x)$.

Proof: Suppose π_x is almost periodic. Clearly, if we set $y = x$ in Definition 4.1, we have that π_x is recurrent. Now let $\varepsilon > 0$. There exists a relatively dense set $D = D(\varepsilon/3)$ and $L = L(\varepsilon/3) > 0$ which satisfy Definition 4.5. Since $H^+(x)$ is compact from Theorem 4.7, then π is uniformly continuous on the (compact) set $H^+(x) \times [0,L]$. There exists $\delta > 0$ such that for every $s_1, s_2 \in \mathbb{R}^+$ with $d(xs_1, xs_2) < \delta$ we have $d((xs_1)s, (xs_2)s) < \varepsilon/3$ for every $s \in [0,L]$. For any $t > L$ there exists $\tau \in D$ and $s \in [0,L]$ so that $t = \tau + s$. Then

$$d((xs_1)t, (xs_2)t) \leq d((xs_1)(s+\tau), (xs_1)s)$$
$$+ d((xs_1)s, (xs_2)s) + d((xs_2)(s+\tau), (xs_2)s).$$

As $d(x(s_i+s), x(s_i+s+\tau)) < \varepsilon/3$ for s_i, $s \in \mathbb{R}^+$, $i = 1,2$, and $\tau \in D$, then $d((xs_1)t, (xs_2)t) < \varepsilon$. It follows that π_x is Lyapunov stable relative to $\gamma^+(x)$.

Conversely, suppose π_x is a recurrent positive motion which is Lyapunov stable relative to $\gamma^+(x)$. given any $\varepsilon > 0$ there exists $\delta > 0$ such that $y \in \gamma^+(x) \cap B_\delta(x)$ implies $d(yt, xt) < \varepsilon$ for every $t \in \mathbb{R}^+$. By recurrence there exists a relatively dense set $D = D(\delta)$ in $\mathbb{R}^+$ such that $d(x, x\tau) < \delta$ for all $\tau \in D$. Therefore as $x\tau \in \gamma^+(x) \cap B_\delta(x)$, we have for every $\tau \in D$,

$$d(xt, x(\tau+t)) < \varepsilon \quad \text{for every} \quad t \in \mathbb{R}^+. \qquad \square$$

Example 4.11. The semidynamical system defined in Example 3.3 is Lyapunov stable. An easy computation shows that $d(\pi((x_1,x_2),t), \pi((y_1,y_2),t)) = d((x_1,x_2),(y_1,y_2))$ for all $t \in \mathbb{R}^+$. Thus every positive motion is almost periodic.

Example 4.12. The positive motions through the set $\tilde{C}$ of Example 4.5 of Chapter II are uniformly recurrent but not almost periodic. Endow the space $\tilde{S}^1$ with the usual $\tilde{S}$ Euclidean metric of the plane. Indeed, $\tilde{C}$ is a compact positively minimal set, hence the hull of a uniformly recurrent positive motion. On the other hand, the positive motions in $\tilde{C}$ are not Lyapunov stable. To see this let the length of the longest interval I_{k_0} be denoted by d_0. No matter how close together we choose two points $\tilde{\theta}_1, \tilde{\theta}_2 \in \tilde{C}$, there is another interval I_{k_1} which is contained between θ_1 and θ_2, the preimages of $\tilde{\theta}_1$ and $\tilde{\theta}_2$ respectively, in $\tilde{S}^1$. We can find $n \in \mathbb{Z}^+$ so that $h^n(I_{k_1}) = I_{k_0}$. Consequently,

$$d(\tilde{\pi}(\tilde{\theta}_1,n),\tilde{\pi}(\tilde{\theta}_2,n)) \geq d_0,$$

which contradicts Lyapunov stability.

Corollary 4.13. Suppose π_x has a compact, positively minimal hull. If π_x is Lyapunov stable relative to $\gamma^+(x)$, then π_x is almost periodic.

The proof of Theorem 4.10 also shows that if π_x is almost periodic, then π_x is uniformly Lyapunov stable relative to $\gamma^+(x)$. It can be shown that π_x is even uniformly Lyapunov stable relative to $H^+(x)$. Before proceeding we require the following properties of almost periodic motions.

Lemma 4.14. If a positive motion is almost periodic, then it is uniformly continuous on $\mathbb{R}^+$.

Proof: Suppose π_x is almost periodic. Given $\varepsilon > 0$ let $D = D(\varepsilon/3)$ and $L = L(\varepsilon/3)$ exist in view of Corollary 4.5. Now π_x is uniformly continuous on $[0,L]$. Choose $\delta > 0$

so that if $t_1, t_2 \in [0,L]$ with $|t_1-t_2| < \delta$, then $d(xt_1,xt_2) < \varepsilon/3$. For any $s,t \in \mathbb{R}^+$ with $|s-t| < \delta$ there exists $\tau \in D$ so that $s_0 \overset{\text{def}}{=} s-\tau$, $t_0 \overset{\text{def}}{=} t-\tau \in [0,L]$. Then as

$$d(xs,xt) \leq d(x(s_0+\tau),xs_0) + d(xs_0,xt_0)$$
$$+ d(xt_0,s(t_0+\tau)) < \varepsilon,$$

we have that π_x is uniformly continuous on $\mathbb{R}^+$. □

Lemma 4.15. Suppose X is a complete metric space. If the positive motion π_x is almost periodic, then for every sequence $\{t_n\}$ in $\mathbb{R}^+$ there is a subsequence $\{t_n'\}$ of $\{t_n\}$ so that the sequence of functions $\{\pi_{xt_n'}\}$ converges uniformly on $\mathbb{R}^+$.

Proof: Let $\{t_n\}$ be a sequence in $\mathbb{R}^+$. Given $\varepsilon > 0$ let there exist a relatively dense set $D = D(\frac{1}{4}\varepsilon)$ and $L = L(\frac{1}{4}\varepsilon) > 0$ from Corollary 4.6. For each $n \in \mathbb{N}$ we may write $t_n = \tau_n + s_n$ with $\tau_n \in D$ and $s_n \in [0,L]$. Let τ be a cluster point of $\{\tau_n\}$. By uniform continuity of π_x choose $\delta > 0$ so that $d(xs,st) < \frac{1}{4}\varepsilon$ whenever $|s-t| < \delta$. Denote by $\{t_n'\}$ those members of $\{t_n\}$ which lie in $[\tau-\delta,\tau+\delta]$. Then

$$d(\pi_{xt_n'}(t),\pi_{xt_m'}(t))$$
$$\leq d(x(t_n'+t),x(\tau_n'+t)) + d(x(\tau_n'+t),xt)$$
$$+ d(xt,x(\tau_m'+t)) + d(x(\tau_m'+t),x(t_m'+t)) < \varepsilon.$$

Thus $\{\pi_{xt_n'}\}$ is uniformly Cauchy, so the sequence converges uniformly on $\mathbb{R}^+$ to some continuous mapping from $\mathbb{R}^+$ to X. □

We conclude our characterization of almost periodic motions by relating them to uniform Lyapunov stable motions. The first result is an extension of Theorem 4.10.

Theorem 4.16. Suppose X is a complete metric space. If the positive motion π_x is almost periodic, then every positive motion in $H^+(x)$ is uniformly Lyapunov stable relative to $H^+(x)$. Consequently, the restriction of (X,π) to $H^+(x)$ is itself a uniformly Lyapunov stable semidynamical system.

Proof: Suppose π_x is an almost periodic positive motion. Let $\varepsilon > 0$. The proof of Theorem 4.10 shows that π_x is uniformly Lyapunov stable relative to $\gamma^+(x)$. Indeed, there exists $\delta = \delta(\frac{1}{2}\varepsilon)$ so that for every $s,\tau \in \mathbb{R}^+$ with $d(xs,x\tau) < \delta$, we have

$$d((xs)t,(x\tau)t) < \tfrac{1}{2}\varepsilon \quad \text{for every} \quad t \in \mathbb{R}^+.$$

We will show for any $z \in H^+(x)$ that the positive motion π_z is uniformly Lyapunov stable relative to $H^+(x)$. So let $y \in H^+(x) \cap B_{\frac{1}{2}\delta}(z\tau)$ for some $\tau \in \mathbb{R}^+$. There are sequences $\{s_n\}$ and $\{t_n\}$ in $\mathbb{R}^+$ with $xs_n \to y$, $xt_n \to z$. Choose a positive integer N so that

$$d(xs_n,y) < \tfrac{1}{4}\delta, \quad d((xt_n)\tau,z\tau) < \tfrac{1}{4}\delta \quad \text{for every} \quad n \geq N.$$

Then for every $n \geq N$,

$$d(xs_n,(x\tau)t_n) \leq d(xs_n,y) + d(y,z\tau) + d(z\tau,(xt_n)\tau) < \delta.$$

Consequently,

$$d(x(s_n+t),x(\tau+t_n+t)) < \tfrac{1}{2}\varepsilon \quad \text{for all} \quad n \geq N \quad \text{and} \quad t \in \mathbb{R}^+.$$

In view of Lemma 4.15 we may assume (by choosing a subsequence if necessary) that

$$x(s_n+t) \to yt, \quad x(\tau+t_n+t) \to z(\tau+t) \text{ uniformly in } t \in \mathbb{R}^+.$$

Hence whenever $n \geq N$,

$$d(x(s_n+t),yt) < \tfrac{1}{4}\varepsilon, \; d(x(\tau+t_n+t),z(\tau+t)) < \tfrac{1}{4}\varepsilon$$

for every $t \in \mathbb{R}^+$. Finally we obtain whenever $y \in H^+(x) \cap B_{\frac{1}{2}\delta}(z\tau)$ that

$$d(yt,z(\tau+t)) < \varepsilon \text{ for every } t \in \mathbb{R}^+. \qquad \square$$

Remark 4.17. Theorem 4.16 remains true when X is not complete, but $H^+(x)$ is the positively minimal hull of an almost periodic motion.

Corollary 4.18. If π_x is an almost periodic positive motion in a complete metric space X, then the restriction of (X,π) to $H^+(x)$ extends uniquely to a dynamical system $(H^+(x),\hat{\pi})$.

Proof: $H^+(x)$ is a compact positively minimal set according to Theorem 4.7. From Theorem 4.16 we have that every positive motion in $H^+(x)$ is (uniformly) Lyapunov stable with respect to $H^+(x)$. The conclusion follows by Theorem 2.5. $\square$

Theorem 4.16 admits a converse which will be used later. Completeness of X is not required.

Theorem 4.19. Suppose π_x is a compact positive motion which is uniformly Lyapunov stable relative to $L^+(x)$. Then $L^+(x)$ is a compact positively minimal equi-almost periodic set.

Proof: We first establish that $L^+(x)$ is positively minimal. So let $y \in L^+(x)$. We must show $H^+(y) = L^+(x)$. Choose any

$z \in L^+(x)$ and $\varepsilon > 0$. By hypothesis there exists $\delta > 0$ so that $y \in L^+(x) \cap B_\delta(xs)$ for some $s \in \mathbb{R}^+$ implies

$$d(yt, x(s+t)) < \tfrac{1}{2}\varepsilon \text{ for every } t \in \mathbb{R}^+.$$

Choose sequences $\{s_n\}$, $\{t_n\}$ in $\mathbb{R}^+$, $s_n \to \infty$, $t_n \to \infty$ so that $xs_n \to y$, $xt_n \to z$. Fix an index $m \in \mathbb{N}$ so that $d(xs_m, y) < \delta$. Next choose $N \in \mathbb{N}$ satisfying $t_n - s_m \in \mathbb{R}^+$ for every $n \geq N$. Finally select an integer $k \geq N$ such that $d(xt_k, z) < \tfrac{1}{2}\varepsilon$. If we set $\tau = t_k - s_m$, then

$$d(y\tau, xt_k) = d(y\tau, x(s_m+\tau)) < \tfrac{1}{2}\varepsilon$$

as $y \in L^+(x) \cap B_\delta(xs_m)$. Consequently

$$d(y\tau, z) \leq d(y\tau, xt_k) + d(xt_k, z) < \varepsilon.$$

This proves $B_\varepsilon(z) \cap \gamma^+(y) \neq \emptyset$, so z must belong to $H^+(y)$. Therefore $L^+(x) \subset H^+(y)$. As $L^+(x) \supset H^+(y)$ always holds, we have $L^+(x) = H^+(y)$ so $L^+(x)$ is positively minimal.

Next we show that π_y is almost periodic for each $y \in L^+(x)$. As $L^+(x)$ is compact positively minimal, then π_y is recurrent according to Corollary 3.2. If we can demonstrate that π_y is stable relative to $\gamma^+(y)$, then we may conclude from Theorem 4.8 that π_y is almost periodic. So let $\varepsilon > 0$. Since π_x is uniformly stable relative to $L^+(x)$, there exists $\delta > 0$ so that $y \in L^+(x) \cap B_\delta(xs)$ for some $s \in \mathbb{R}^+$ implies $d(yt, x(s+t)) < \tfrac{1}{2}\varepsilon$ for every $t \in \mathbb{R}^+$. Now suppose $z \in L^+(x)$ with $d(y,z) < \tfrac{1}{2}\delta$. There is a $\tau \in \mathbb{R}^+$ for which $d(x\tau, y) < \tfrac{1}{2}\delta$. Consequently $d(x\tau, z) < \delta$. It follows that

$$d(yt, x(\tau+t)) < \tfrac{1}{2}\varepsilon, \quad d(zt, x(\tau+t)) < \tfrac{1}{2}\varepsilon$$

for every $t \in \mathbb{R}^+$. Then as $L^+(x) = H^+(y)$ we see that $z \in H^+(y) \cap B_{\frac{1}{2}\delta}(y)$ implies

$$d(yt,zt) \leq d(yt,x(\tau+t)) + d(zt,x(\tau+t)) < \varepsilon$$

for every $t \in \mathbb{R}^+$. This establishes π_y is stable relative to $H^+(y)$, hence stable relative to $\gamma^+(y)$.

Finally, we see from Corollary 4.9 that the motions π_y, $y \in L^+(x)$, are equi-almost periodic. Indeed, the proof of Corollary 4.9 does not require completeness of X for this result. □

5. Asymptotically Stable Motions

There is a "motion" counterpart to the concept of attraction of sets. As one might guess, the present effort will be concerned with positive motions, not positively invariant sets. We will formulate a definition for asymptotic orbital stability of a positive motion in terms of closeness of neighboring positive motions. The relationship with the concepts of Section 6 of Chapter II will be examined. Our principle result is a generalization of the Poincaré-Bendixon theorem for bounded positive motions in $\mathbb{R}^2$ (cf. Hale [1], p. 54).

Definition 5.1. A positive motion π_y is said to be *orbitally attracted* to the positive motion π_x if there exists $\tau \in \mathbb{R}^+$ such that

$$\lim_{t\to\infty} d(yt,x(\tau+t)) = 0.$$

Denote by $\mathscr{A}^+(x)$ the set of all points $y \in X$ for which the positive motion π_y is attracted to π_x. Call this set the *region of orbital attraction* of π_x. The positive motion π_x is called an *orbital attractor* if $\mathscr{A}^+(x)$ is a neighborhood of $H^+(x)$.

As in the case of stability, attraction can refer either to positive motions or positively invariant sets. It will always be made clear though whether we are dealing with attracting motions or attracting sets. The two definitions are related in the case of compact motions.

Proposition 5.2. If a compact positive motion π_x is an orbital attractor, then $H^+(x)$ and $L^+(x)$ are attractors (in the sense of Definition 6.1 of Chapter II).

Proof: Suppose $y \in \mathcal{A}^+(x)$. As $L^+(x) \neq \emptyset$, then we must have $yt \to L^+(x)$. Indeed, for any $\varepsilon > 0$ we can choose $T > 0$ such that

$$d(yt,x(\tau+t)) < \tfrac{1}{2}\varepsilon \quad \text{and} \quad d((x\tau)t,L^+(x)) < \tfrac{1}{2}\varepsilon$$

for some $\tau \in \mathbb{R}^+$ and every $t > T$. Consequently

$$d(yt,L^+(x)) < \varepsilon \quad \text{for every} \quad t > T.$$

This means that $y \in A^+(L^+(x))$. If W is an open set for which $H^+(x) \subset W \subset \mathcal{A}^+(x)$, then $H^+(x)$ and $L^+(x)$ are attractors. □

Definition 5.3. Suppose F is an arbitrary nonempty subset of X. The positive motion π_x is called *orbitally asymptotically stable relative to* F provided

(i) π_x is uniformly Lyapunov stable relative to F, and

(ii) π_x is an orbital attractor.

If $F = X$, we say π_x is just *orbitally asymptotically stable.*

Orbitally asymptotically stable motions and asymptotically stable sets are related according to the next result.

Theorem 5.4. If a compact positive motion π_x is orbitally asymptotically stable, its hull $H^+(x)$ is asymptotically stable. If π_x is a critical motion, then π_x is orbitally asymptotically stable if and only if the set $\{x\}$ is asymptotically stable.

Proof: The truth of the first statement follows directly from Propositions 2.2 and 5.2 and the fact that $H^+(x) \subset \mathscr{A}^+(x) \subset A^+(L^+(x)) \subset A^+(H^+(x))$. For the second statement we need only observe that a critical motion π_x is an orbital attractor if and only if the set $\{x\}$ is an attractor. Now apply Theorem 2.3 to complete the proof. □

Our final result is the aforementioned generalization of the Poincaré-Bendixon Theorem. An application is provided in the next section.

Theorem 5.5. Suppose X is a complete metric space. If π_x is a compact positive motion which is orbitally asymptotically stable relative to $L^+(x)$, then $L^+(x)$ is a periodic orbit.

Proof: In view of Theorem 4.19, $L^+(x)$ is a compact positively minimal set, all of whose positive motions are equi-almost periodic. We will show there exists $y \in L^+(x)$ such that $\gamma^+(y) = L^+(x) = L^+(y)$, whereby y is a periodic point by Theorem 3.8 of Chapter II.

First note that $L^+(x) \subset \mathscr{A}^+(x)$. Indeed, as π_x is an orbital attractor, there exists $\delta > 0$ such that

$$L^+(x) \subset B_\delta(H^+(x)) \subset \mathscr{A}^+(x).$$

Next observe that $y \in L^+(x)$ implies $L^+(x) \subset \mathscr{A}^+(y)$. To see this let $z \in L^+(x)$. There must exist $s,\tau \in \mathbb{R}^+$ so that

$$\lim_{t\to\infty} d(yt,x(s+t)) = 0, \quad \lim_{t\to\infty} d(zt,x(\tau+t)) = 0.$$

We may assume $\tau - s \in \mathbb{R}^+$. Then

$$\lim_{t\to\infty} d(y(\tau-s+t), x(\tau+t)) = 0$$

so

$$\lim_{t\to\infty} d(zt, y(\tau-s+t)) = 0.$$

This shows that $z \in \mathscr{A}^+(y)$.

According to Theorem 4.16 every positive motion π_y in $L^+(x)$ is uniformly Lyapunov stable relative to $H^+(y) = L^+(x)$. Thus Corollary 4.18 shows that the restriction of (X,π) to $L^+(x)$ extends uniquely to a dynamical system $(L^+(x), \hat{\pi})$. Now suppose there is no point $\hat{y} \in L^+(x)$ for which $\gamma^+(y) = L^+(x)$. Fix any point $y_0 \in L^+(x)$ and choose $z \in L^+(x) \setminus \gamma^+(y_0)$. Then $z \neq y_0(\tau-s)$. Set $\varepsilon = d(z, y_0(\tau-s))$. Choose $\delta > 0$ in view of bilaterial stability of the (principal) motion through y_0 from Corollary 2.6. For sufficiently large $t_0 \in \mathbb{R}^+$ we must have

$$d(zt_0, y_0(\tau-s+t_0)) < \delta,$$

since $L^+(x) \subset \mathscr{A}^+(y_0)$. Then Corollary 2.6 shows that

$$\varepsilon = d(z,y(\tau-s)) = d((zt_0)(-t_0), (y_0(\tau-s+t_0))(t_0)) < \varepsilon.$$

This is impossible, so we must have $\gamma^+(y) = L^+(x)$ for some $y \in L^+(x)$. □

Corollary 5.6. Suppose X is a complete metric space. If π_x is a compact positive motion which is orbitally asymptotically stable, then $L^+(x)$ is a periodic orbit whose every positive motion is orbitally asymptotically stable.

A simple example suffices to demonstrate that π_x itself need not be periodic.

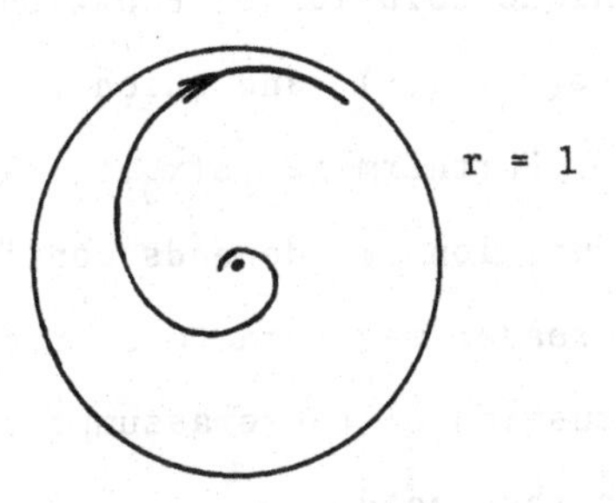

Figure 5.1

Example 5.7. Consider the 2-dimensional semidynamical system with phase portrait given in Figure 5.1. The space X is the closed unit ball in $\mathbb{R}^2$, which in polar coordinates is $\{(r,\theta): 0 \leq r \leq 1, 0 \leq \theta<2\pi\}$. The phase map is

$$\pi((r,\theta),t) = \begin{cases} (0,0), & r=0,\ 0\leq\theta<2\pi,\ t\geq 0 \\ ([1-(1-r^{-1})e^{-t}]^{-1}, \theta+t), & 0<r\leq 1,\ 0\leq\theta<2\pi,\ t\geq 0. \end{cases}$$

The origin $(0,0)$ is a critical point; the orbit $r = 1$ is 2 -periodic. The motion through every (r,θ) for $0 < r < 1$ is orbitally asymptotically stable. Yet these motions are not periodic.

6. Periodic Solutions of an Ordinary Differential Equation

In this section we develop a semidynamical system to deal with solutions of the differential equation

$$\frac{dx}{dt} = f(x,t) \tag{6.1}$$

where $f: \mathbb{R}^d \times \mathbb{R}^+ \to \mathbb{R}^d$ is continuous. By a solution of Equation (6.1) we mean a function $\psi: [a,b) \to \mathbb{R}^d$ for some $0 \leq a < b \leq +\infty$ so that $\dot{\psi}(t) = f(\psi(t),t)$ for every $t \in [a,b)$. Henceforth it is assumed that for every point $(x_0,t_0) \in \mathbb{R}^d \times \mathbb{R}^+$

there exists a unique solution of Equation (6.1) which we shall denote by $\phi(x_0,t_0;\cdot)$ and which satisfies $\phi(x_0,t_0;t_0) = x_0$. Furthermore $\phi(x_0,t_0;t)$ is defined for all $t \geq t_0$, and the function ϕ depends continuously on (x_0,t_0,t). (The reader may turn to Section 1 of Chapter IV for a fuller discussion of these assumptions and some conditions under which they hold.)

We shall establish the existence of a periodic solution of Equation (6.1) under hypotheses to be set fourth later. The major tool here will be Theorem 5.5. To begin with we must assume f is periodic; that is, there is some $\omega > 0$ for which

$$f(x,t+\omega) = f(x,t) \quad \text{for every} \quad (x,t) \in \mathbb{R}^d \times \mathbb{R}^+.$$

(We do not assume the period ω is minimal.) A solution ψ of Equation (6.1) is called *harmonic* if ψ is a periodic function with period $k\omega$ for some $k \in \mathbb{N}$. Without loss of generality we will take $\omega = 2\pi$.

Next we define an appropriate semidynamical system for this problem. Though the phase space we construct is finite dimensional, it has the advantage of simplicity. Most of the remaining applications in the book deal with infinite dimensional phase spaces. So identify the hyperspaces

$$H_0 = \{(x,t) \in \mathbb{R}^d \times \mathbb{R}^+ : t = 0\}$$

and

$$H_{2\pi} = \{(x,t) \in \mathbb{R}^d \times \mathbb{R}^+ : t = 2\pi\}.$$

Then we can map the space $\mathbb{R}^d \times \mathbb{R}^+$ onto the open torus $\mathbb{R}^d \times S^1$. The mapping goes like $(x,t) \to (x,e^{it})$, where $i = \sqrt{-1}$. Points in S^1 will be represented by e^{it} for some $t \in \mathbb{R}^+$ (mod 2π). The space $\mathbb{R}^d \times S^1$ admits the metric

$$d((x,e^{it}),(y,e^{is})) = |x-y| + \sqrt{1-\cos(t-s)},$$
$$(x,t),\ (y,s) \in \mathbb{R}^d \times S^1.$$

Now define the mapping $\pi\colon \mathbb{R}^d \times S^1 \times \mathbb{R}^+ \to \mathbb{R}^d \times S^1$ by

$$\pi((x_0,e^{it_0}),t) = (\phi(x_0,t_0;t_0+t),e^{i(t_0+t)}).$$

Proposition 6.1. The pair $(\mathbb{R}^d \times S^1,\pi)$ is a semidynamical system.

Proof: We must verify Definition 2.1 of Chapter I. The initial value property (i) follows directly from the definition of ϕ. The semigroup property (ii) holds as a result of the uniqueness of solutions of Equation (6.1) through each $(x_0,t_0) \in \mathbb{R}^d \times \mathbb{R}^+$. Indeed, fix $s \in \mathbb{R}^+$ and set for every $t \in \mathbb{R}^+$

$$x(t) = \phi(x_0,t_0;t_0+s+t)$$
$$y(t) = \phi(\phi(x_0,t_0;t_0+s),t_0+s;t_0+s+t).$$

Both $x(\cdot)$ and $y(\cdot)$ are solutions of Equation (6.1) for which $x(0) = y(0)$. By uniqueness they must agree at every $t \in \mathbb{R}^+$. Therefore

$$\begin{aligned}\pi(\pi((x_0,e^{it_0}),s),t) &= \pi((\phi(x_0,t_0;t_0+s),e^{i(t_0+s)}),t)\\ &= (\phi(\phi(x_0,t_0;t_0+s),t_0+s;t_0+s+t),e^{i(t_0+s+t)})\\ &= (\phi(x_0,t_0;t_0+s+t),e^{i(t_0+s+t)})\\ &= \pi((x_0,e^{it_0}),s+t).\end{aligned}$$

Finally, π is continuous (property (iii) of Definition 2.1 of Chapter I) in view of the continuity assumption made earlier about ϕ. □

The following definitions concerning the behavior of solutions of Equation (6.1) are required. They will be treated again in Chapter IV.

Definition 6.2. A solution ψ of Equation (6.1) is called *uniformly stable* if for every $\varepsilon > 0$ there exists $\delta = \delta(\varepsilon) > 0$ such that if $t_0 \in \mathbb{R}^+$ and $|x_0 - \psi(t_0)| < \delta$, then

$$|\phi(x_0,t_0;t) - \psi(t)| < \varepsilon \quad \text{for all} \quad t \geq t_0.$$

Definition 6.3. A solution ψ of Equation (6.1) is called *asymptotically stable* if it is uniformly stable and there is $\beta > 0$ such that if $t_0 \in \mathbb{R}^+$ and $|x_0 - \psi(t_0)| < \beta$, then

$$\lim_{t \to \infty} |\phi(x_0,t_0;t) - \psi(t)| = 0.$$

The following theorem is the chief result of this section. We assume f in Equation (6.1) is periodic.

Theorem 6.4. Suppose Equation (6.1) has a bounded solution which is asymptotically stable. Then Equation (6.1) has a harmonic solution which is asymptotically stable.

Proof: Denote by $\phi(x_0,t_0;\cdot)$ a bounded asymptotically stable solution of Equation (6.1). The corresponding positive motion $\pi_{(x_0,e^{it_0})}$ must be compact. We will first show this positive motion is uniformly Lyapunov stable. So let $\varepsilon > 0$ and choose the corresponding $\delta = \delta(\frac{1}{4}\varepsilon) < \frac{1}{2}\varepsilon$ by uniform stability of $\phi(x_0,t_0;\cdot)$. Set

$$B = \sup_{t \geq t_0} |f(\phi(x_0,t_0;t),t)|.$$

As $\phi(x_0,t_0;t)$ is bounded on $[t_0,\infty)$ and f is continuous, then $B < \infty$. Fix $\tau \in \mathbb{R}^+$ and let $(x,e^{ir}) \in \mathbb{R}^d \times S^1$ so that

$$|x-\phi(x_0,t_0;t_0+\tau)| + \sqrt{1-\cos(t_0+\tau-r)} < \tfrac{1}{2}\delta.$$

Choose $\eta < \frac{1}{2}B^{-1}\delta$ such that if $r_0 \equiv r \pmod{2\pi}$, then $|t_0+\tau-r_0| < \eta$ implies

$$\sqrt{1-\cos(t_0+\tau-r_0)} < \tfrac{1}{2}\delta, \quad |\phi(x_0,t_0;r_0)-\phi(x_0,t_0;t_0+\tau)| < \tfrac{1}{2}\delta.$$

There is no loss in generality in replacing r by r_0. So

$$\begin{aligned}|x-\phi(x_0,t_0;r_0)| &\le |x-\phi(x_0,t_0;t_0+\tau)| + |\phi(x_0,t_0;t_0+\tau) - \phi(x_0,t_0;r_0)| \\ &< \tfrac{1}{2}\delta + \tfrac{1}{2}\delta = \delta.\end{aligned}$$

In view of the uniform stability of $\phi(x_0,t_0;\cdot)$ we have

$$|\phi(x,r_0;r_0+t)-\phi(x_0,t_0;r_0+t)| < \tfrac{1}{4}\varepsilon \quad \text{for all} \quad t \in \mathbb{R}^+.$$

Now consider the estimate for all $t \in \mathbb{R}^+$

$$\begin{aligned}|\phi(x_0,t_0;r_0+t) - \phi(x_0,t_0;t_0+\tau+t)| &\le \int_{r_0+t}^{t_0+\tau+t} |f(\phi(x_0,t_0;s),s)|ds \le B|t_0+\tau-r_0| \\ &< B\eta < \tfrac{1}{2}\delta < \tfrac{1}{4}\varepsilon.\end{aligned}$$

Then

$$\begin{aligned}&|\phi(x_0,r_0;r_0+t) - \phi(x_0,t_0;t_0+\tau+t)| \\ &\quad\le |\phi(x,r_0;r_0+t) - \phi(x_0,t_0;r_0+t)| + |\phi(x_0,t_0;r_0+t) \\ &\qquad\qquad - \phi(x_0,t_0;t_0+\tau+t)| \\ &\quad< \tfrac{1}{4}\varepsilon + \tfrac{1}{4}\varepsilon = \tfrac{1}{2}\varepsilon \quad \text{for all} \quad t \in \mathbb{R}^+.\end{aligned}$$

Thus $|x-\phi(x_0,t_0;t_0+\tau)| + \sqrt{1-\cos(t_0+\tau-r_0)} < \frac{1}{2}\delta$ implies

$$|\phi(x,r_0;r_0+t) - \phi(x_0,t_0;t_0+\tau+t)| + \sqrt{1-\cos(t_0+\tau-r_0)} < \varepsilon,$$

and so in view of Definition 2.1, the mapping π and the metric defined above on $\mathbb{R}^d \times S^1$, the positive motion $\pi_{(x_0,e^{it_0})}$ is uniformly Lyapunov stable. An almost identical analysis, this time using the fact that $\phi(x_0,t_0;\cdot)$ is asymptotically stable, shows that the positive motion $\pi_{(x_0,e^{it_0})}$ is orbitally attracting. Now by Corollary 5.6 $L^+(x_0,e^{it_0})$ is a periodic orbit whose every positive motion is asymptotically orbitally stable. Choose any $(y,e^{is}) \in L^+(x_0,e^{it_0})$. Then $\phi(y,s;\cdot)$ must be a periodic solution of Equation (6.1). It is clear that the period of $\phi(y,s;\cdot)$ must be an integral multiple of 2π; that is, $\phi(y,s;\cdot)$ is harmonic. Moreover, $\phi(y,s;\cdot)$ must also be asymptotically stable as the corresponding positive motion $\pi_{(y,e^{is})}$ is asymptotically orbitally stable. This is evident from the definitions of π and the metric on $\mathbb{R}^d \times S^1$. □

Theorem 6.4 has a surprising consequence in the event f in Equation (6.1) does not depend upon t. Then Equation (6.1) can be written

$$\frac{dx}{dt} = f(x). \tag{6.2}$$

We call Equation (6.2) an autonomous differential equation.

Theorem 6.5. Suppose Equation (6.2) has a periodic solution ψ which is asymptotically stable. Then for every $t \in \mathbb{R}^+$, $\psi(t)$ is an equilibrium point of Equation (6.2); that is, $\psi(t) \equiv c$ for some constant $c \in \mathbb{R}^d$.

Proof: If $\psi(t)$ were not an equilibrium point of Equation (6.2), then ψ would have a minimum period $\omega_0 > 0$. However, $f(x)$ is periodic in t of any period $\omega > 0$. Choose ω so that and ω_0 are incommensurable; that is, ω/ω_0 is irrational. This requires points in $\mathbb{R}^d \times S^1$ to be represented by $(x, e^{2\pi it/\omega})$. The motion in $\mathbb{R}^d \times S^1$ through $(x_0, 1) = (\psi(0), 1)$ corresponding to the ω_0-periodic solution ψ can be represented by $\pi_{(x_0,1)}(t) = (\psi(t), e^{2\pi it/\omega})$. But the motion $\pi_{(x_0,1)}$ is almost periodic, not periodic. Consequently $\pi_{(x_0,1)}$ is a compact positive motion which (as demonstrated in the proof of Theorem 6.4) is asymptotically orbitally stable. According to Corollary 5.6, $L^+(x_0,1)$ is a periodic orbit. But we also see from Theorem 4.7 that $H^+(x_0,1)$ is positively minimal. Hence $H^+(x_0,1) = L^+(x_0,1)$ which contradicts the almost periodicity of $\pi_{(x_0,1)}$. We must conclude that $\psi(t)$ is an equilibrium point of Equation (6.2). □

7. Exercises

Unless otherwise stated, assume all motions are in a given semidynamical system (X,π), where X is a metric space with metric d.

7.1. Show that if π_x is Lyapunov stable relative to F, then for any $\tau \in \mathbb{R}^+$, $\pi_{x\tau}$ is Lyapunov stable relative to F.

7.2. Prove that if π_x is Lyapunov stable, then $D^+(x) = H^+(x)$.

7.3. Suppose (X,π) is a Lyapunov stable semidynamical system.

(i) Prove the set $\Omega \overset{\text{def}}{=} \cup\{L^+(x) : x \in X\}$ is closed.

(ii) Prove that $y \notin \Omega$ if and only if there exists a neighborhood W of y such that $W \cap Wt = \emptyset$ for every $t \in \mathbb{R}^+$.

7.4. Prove that if π_x is recurrent then so is $\pi_{x\tau}$ for every $\tau \in \mathbb{R}^+$.

7.5. Show the following are equivalent

(i) π_x is recurrent.

(ii) $x \in \mathscr{R}^+(x)$.

(iii) $H^+(x) = \mathscr{R}^+(x)$.

7.6. Prove that if π_x is periodic, then it is recurrent and and $\mathscr{R}^+(x) = \gamma^+(x)$.

7.7. Show that if π_x is uniformly recurrent, then so is $\pi_{x\tau}$ for every $\tau \in \mathbb{R}^+$.

7.8. Show that π_x is uniformly recurrent if and only if $x \in \mathscr{R}^+(x)$ and $\mathscr{R}^+(x)$ is precompact.

7.9. Prove that for every $y \in L^+(x)$ we have $\mathscr{R}^+(x) \subset \mathscr{R}^+(y)$.

7.10. Prove that if π_x is uniformly recurrent, then every point in $L^+(x)$ is uniformly recurrent. [Hint: use Exercises 7.8 and 7.9.]

7.11. Show that a positive motion π_x recurs uniformly on the set M if and only if M is a precompact subset of $\mathscr{R}^+(x)$.

7.12. Show that if π_x is almost periodic, then so is $\pi_{x\tau}$ for every $\tau \in \mathbb{R}^+$.

7.13. Prove that every periodic motion is almost periodic.

7.14. The semidynamical system (X,π) is called distal if whenever $x,y \in X$, there exists $\delta > 0$ such that $d(xt,yt) \geq \delta$ for every $t \in \mathbb{R}^+$. Establish each of the following.

(i) If (X,π) is distal, then $xt_n \to z$, $yt_n \to z$ implies $x = y$ for any sequence $\{t_n\} \subset \mathbb{R}^+$ and points $x,y,z \in X$.

(ii) The converse of (i) holds provided (X,π) is Lagrange stable.

(iii) If π_x is an almost periodic motion in a complete metric space, the restriction of (X,π) to $H^+(x)$ is distal.

(iv) If (X,π) is Lagrange stable and distal, then $H^+(x)$ is positively minimal for every $x \in X$.

(v) If (X,π) is Lagrange stable and distal, then (X,π) extends uniquely to a dynamical system.

7.15. Give an example of a compact motion π_x with asymptotically stable hull $H^+(x)$, yet π_x is not asymptotically orbitally stable.

7.16. Suppose $L^+(x) \neq \emptyset$ for some $x \in X$ and let $y \in \mathscr{A}^+(x)$. Show that $L^+(y) \subset L^+(x)$.

7.17. Suppose ψ is an asymptotically stable solution of Equation (6.2). Prove that if ψ is bounded, then $\lim_{t\to\infty} \psi(t)$ exists and is an asymptotically stable equilibrium point of Equation (6.2).

7.18. Suppose ψ is an almost periodic solution of Equation (6.2) that is asymptotically stable. Show that $\psi(t)$ is an equilbrium point of Equation (6.2); that is, $\psi(t) \equiv c$, a constant.

8. Notes and Comments

We have selected only those properties of positive motions which will be useful to us in the applications to follow in later chapters. Indeed, the subject of almost periodic motions has been treated in detail far beyond our needs here. See Corduneanu [1] for additional references.

Section 2. The definition of a Lyapunov stable motion originates from Lyapunov [1]. Lyapunov stability of a semidynamical system (X,π) is also referred to as equicontinuity.

Uniform Lyapunov stability as employed here comes from Sell [1]. Theorem 2.4 is standard; it may be found in Della Riccia [1]. Theorem 2.5 and Corollary 2.6 are also due to Della Riccia [1]. Further properties of Lyapunov stable systems are developed there. For additional characterizations of Lyapunov stability in terms of Lyapunov functions, see Bhatia and Hajek [2].

Section 3. Recurrent motions were introduced by Birkhoff [1]. Actually, recurrence to Birkhoff is what we call uniform recurrence. The set $\mathscr{R}^+(x)$ was studied by Bhatia and Hajek [3]. Lemma 3.4 is from Bhatia and Chow [1]. Theorem 3.5 is referred to as the "Birkhoff Recurrence Theorem." The notion of uniform recurrence on a set M is due to Bhatia and Chow [1]. Theorem 3.10 and Corollary 3.11 are from them also.

Section 4. Many of the results in this section can be found in Sell [3]. We use the so-called Bohr definition of almost periodicity here. Lemma 4.15 is due to Bochner [1]. Bochner's definition of an almost periodic function is that precisely given by Lemma 4.15. The definitions are, in fact, equivalent. Theorem 4.16 is from Sell [3]. Bhatia and Chow [1] obtained a similar result. Theorem 4.19 is due to Deysach and Sell [1].

Section 5. The definition of orbital asymptotic stability is in one sense stronger, and in another sense weaker than the customary definition of asymptotic stability of positive motions. On the one hand, the customary definition requires only Lyapunov stability -- not uniform Lyapunov stability. On the other hand, orbital attraction is weaker than the usual kind of attraction; namely, $d(xt,yt) \to 0$ as $t \to \infty$. Indeed, the orbit $r = 1$ of Example 5.7 is not asymptotically stable in the customary sense. This is because any two distinct points on the orbit $r = 1$ remain the same distance apart for all times. But a suitable time translation of one of these points to put it in phase with the other will provide us with the usual approach to zero. Definition 5.3 is especially suitable for periodic orbits. See Coddington and Levinson [1] for further motivation on this. Finally, Theorem 5.5 is due to Sell [1]. It is an extension of Theorem 4.19.

Section 6. These results are all due to Sell [1]. One must be careful in dealing with autonomous differential equations, however. Since the solutions of Equation (6.2) themselves are the positive motions of a semidynamical system with phase space $\mathbb{R}^d$ (see Section 1, Chapter IV), we might mistakenly substitute Definition 5.3 for the asymptotic stability called for in Theorem 6.5. Indeed, as the motions in Example 5.7 are solutions to the system $\dot{r} = r(1-r)$, $\dot{\theta} = 1$, one could hastily conclude that the periodic orbit $r = 1$ of Example 5.7 is critical. One can check that the orbit $r = 1$ is not asymptotically stable in the sense of Definition 6.3 though. Sell [3] has also treated Theorem 6.4 in an infinite dimensional setting. We will use this approach in Chapter IV. Finally we point out that Theorem 6.4 and 6.5 are by no means the sharpest results possible for this problem. The point is to demonstrate the usefulness of the semidynamical system framework in order to obtain Theorem 6.4. By putting the ordinary differential equaton in a more general setting and using the semidynamical system machinery, we must sacrifice using specialized ordinary differential equation techniques. Thus Yoshizawa [4], for example, is able to obtain even stronger results without recourse to the semidynamical framework.

CHAPTER IV

NONAUTONOMOUS ORDINARY DIFFERENTIAL EQUATIONS

1. Introduction

The solutions of the autonomous ordinary differential equation

$$(1.1) \qquad \dot{x} = f(x)$$

(where $\dot{x}$ stands for $\frac{dx}{dt}$) give rise to a semidynamical (even dynamical) system on $\mathbb{R}^d$ provided $f: W \to \mathbb{R}^d$ is continuous on the open subset $W \subset \mathbb{R}^d$ and the solutions of Equation (1.1) through any point $(x_0,t_0) \in W \times \mathbb{R}^+$ are uniquely defined and remain in W for all time. In fact, if $\phi(x_0;t)$ denotes the solution of Equation (1.1) through $(x_0,0)$ evaluated at time $t \in \mathbb{R}^+$, it can be verified that (W,ϕ) is a semidynamical system.

The situation for the nonautonomous ordinary differential equation

$$(1.2) \qquad \dot{x} = f(x,t)$$

is not as nice. Suppose $f: W \times \mathbb{R}^+ \to \mathbb{R}^d$ is continuous, where W is open in $\mathbb{R}^d$, and for any $(x_0,t_0) \in W \times \mathbb{R}^+$ there is one and only one solution $\phi(x_0,t_0;\cdot)$ of Equation (1.2) for which

$\phi(x_0,t_0;t_0) = x_0$ and $\phi(x_0,t_0;t) \in W$ for all $t \geq t_0$. Then these solutions do not even define a (local) semidynamical system with phase space W. We can, though, transform Equation (1.2) into an autonomous differential equation by incorporating the time variable into the phase space. Set $\hat{x} = (x,t) \in W \times \mathbb{R}^+$ and $\hat{f} = (f,1)$, where 1 denotes the constant scalar function with value one. Consequently a semidynamical system does obtain on the phase space $\hat{W} = W \times \mathbb{R}^+$. Namely, it again can be verified that $(\hat{W},\hat{\pi})$ is a semidynamical system, where $\hat{\pi}: \hat{W} \times \mathbb{R}^+ \to \hat{W}$ is given by $\hat{\pi}(x_0,t_0,t) = (\phi(x_0,t_0;t_0+t),t_0+t)$. (This resembles the semidynamical system of Section 6, Chapter III.)

This system, though, is of little interest. It possesses no critical points, no periodic, no recurrent, nor even bounded trajectories. Stability must be examined with regard to noncompact sets. The purpose of this chapter, therefore, is to demonstrate how to generate a semidynamical system from Equation (1.2) and its solutions. Also some very interesting properties of the asymptotic behavior of the solutions of Equation (1.2) can be obtained via the theory of semidynamical systems.

Before proceeding to the formulation of the desired semidynamical system, we review for the reader some basic terminology associated with solutions of the differential equation (2.1). Let W be an open set in $\mathbb{R}^d$ and consider a function $f: W \times \mathbb{R}^+ \to \mathbb{R}^d$. Assume that f is continuous in $x \in W$ for each $t \in \mathbb{R}^+$ and measurable in $t \in \mathbb{R}^+$ for each $x \in W$. By a *solution* of Equation (1.2) we mean an absolutely continuous function ϕ and an interval $J \subset \mathbb{R}^+$ so that $\phi: J \to W$

and satisfies $\dot{\phi}(t) = f(\phi(t),t)$ almost everywhere (a.e.) in $t \in J$. The solution ϕ is called *noncontinuable* if there does not exist an extension of ϕ to a solution of Equation (1.2) on some interval which properly contains J. In this event J is called the *maximal interval of definition* of the solution ϕ. Given a point $(x_0,t_0) \in W \times \mathbb{R}^+$, an *initial value problem for Equation (1.2)* consists of finding a noncontinuable solution ϕ of Equation (1.2) which satisfies $\phi(t_0) = x_0$. In this case we refer to ϕ as a *solution through* (x_0,t_0). If a noncontinuable solution ϕ of Equation (1.2) through (x_0,t_0) has $[t_0,\infty)$ as its maximal interval of definition, then ϕ is called *global*. (We do not concern ourselves at this point with solutions defined to the left of t_0.) Henceforth we make the following hypothesis concerning solutions of Equation (1.2):

H_1: for every $(x_0,t_0) \in W \times \mathbb{R}^+$ there exists a unique noncontinuable solution ϕ of Equation (1.2) through (x_0,t_0). This solution also depends continuously upon the initial value (x_0,t_0), uniformly in t belonging to bounded subsets of $[t_0,\infty)$.

H_2: every noncontinuable solution of Equation (1.2) is global.

In the next section we will provide sufficient conditions for H_1 to be satisfied (see Remark 2.3). In regard to H_2 there is little loss of generality in making this assumption. In fact, if only local existence (of noncontinuable solutions of Equation (1.2)) is supposed, it is possible to construct a local semidynamical system according to the same method that the semidynamical system is constructed in Section 2. The only difference is that the (global) interval of definition $[t_0,\infty)$ is replaced by some (maximal) interval of definition $[t_0,\omega_\phi)$, where ω_ϕ depends upon the solution ϕ of Equation (1.2). The resulting local semidynamical system may then be reparametrized in view of Corollary 8.7 of Chapter I to obtain a (global) semidynamical system. Thus, it is sufficient to begin with global solutions of Equation (1.2) in the first place. In either case the (global) semidynamical systems are isomorphic in the sense of Definition 8.4 of Chapter I.

2. Construction of the Skew Product Semidynamical System

The basic reason why solutions to the autonomous differential Equation (1.1) define a semidynamical system is that time translations of Equation (1.1) are still solutions of Equation (1.1). But this is not true in the case of the nonautonomous Equation (1.2). If $\phi(t)$ is a solution of $\dot{x} = f(x,t)$ evaluated at time t, then $\phi(t+s)$ is a solution of $\dot{x} = f(x,t+s)$ evaluated at time $t+s$. Thus the time translation of a solution of Equation (1.2) is a solution of a similar time translation of Equation (1.2). If we can somehow incorporate this time translation of f into the state of the system, then we will be on our way towards the desired semidynamical system.

Definition 2.1. For any function $g\colon W \times \mathbb{R}^+ \to \mathbb{R}^d$ and any $s \in \mathbb{R}^+$ we call g_s the s-*translate* of g where $g_s(x,t) = g(x,t+s)$.

Denote by $\phi(f,x_0;t)$ the solution of Equation (1.2) through $(x_0,0)$ evaluated at time t. In accordance with our notation convention, the solution may be represented by

$\phi(f,x_0;\cdot)$. Therefore $\phi(f,x_0;s+\cdot)$ is the solution of $\dot{x} = f_s(x,t)$ through $(\phi(f,x_0;s),0)$. But this solution can also be represented by $\phi(f_s,\phi(f,x_0;s);\cdot)$. In view of the uniqueness property of solutions of Equation (1.2) we must have $\phi(f,x_0;t+s) = \phi(f_s,\phi(f,x_0;s);t)$. This suggests that if you allow f and all of its translates f_s, $s \in \mathbb{R}^+$, to be members of some appropriately defined function space, $\mathcal{F}$, then the mapping defined by

$$\pi((f,x_0),t) = (f_t,\phi(f,x_0;t)) \tag{2.1}$$

is an appropriate candidate for the phase map of a semidynamical system with phase space $\mathcal{F} \times W$. The remainder of this section is devoted to demonstrating this. The resulting semidynamical system $(\mathcal{F} \times W,\pi)$ is called *skew product*. It is clear that $\mathcal{F}$ must be equipped with a topology in which the mapping from $\mathcal{F} \times R^+$ to $\mathcal{F}$ given by $(f,t) \to f_t$ is continuous.

We begin with the specification of the function f corresponding to the differential equation (1.2). Let $W \subset \mathbb{R}^d$ be open and fix a function $f\colon W \times \mathbb{R}^+ \to \mathbb{R}^d$ which is continuous in $x \in W$ for each $t \in \mathbb{R}^+$ and is Lebesgue measurable in $t \in \mathbb{R}^+$ for each $x \in W$. (Unless otherwise specified, all measurable sets, measurable functions, and definite integrals will be of Lebesgue type.) Also suppose for each compact set $K \subset W$ there exist two locally integrable functions m_K and l_K so that

$$|f(x,t)| \le m_K(t) \quad \text{for every} \quad (x,t) \in K \times \mathbb{R}^+, \tag{2.2}$$

$$|f(x,t)-f(y,t)| \le l_K(t)|x-y| \quad \text{for every} \quad (x,t),(y,t) \in K \times \mathbb{R}^+. \tag{2.3}$$

Moreover, the functions m_K and l_K must satisfy

$\mathcal{F}_1$: for every $\varepsilon > 0$ there exists $\delta = \delta_K(\varepsilon) > 0$ such that if E is a measurable set in $\mathbb{R}^+$, contained in an interval $[s,s+1]$ and with measure less than δ, then $\int_E m_K(\tau)d\tau < \varepsilon$,

$\mathcal{F}_2$: there exists a constant $L_K > 0$ so that $\int_s^{s+1} l_K(\tau)d\tau \leq L_K$ for every $s \in \mathbb{R}^+$.

Definition 2.2. A function $g: W \times \mathbb{R}^+ \to \mathbb{R}^d$ which is continuous in $x \in W$ for each $t \in \mathbb{R}^+$, measurable in $t \in \mathbb{R}^+$ for each $x \in W$, and for which there exists a locally integrable function m_K satisfying the Inequality (2.2) for each compact set $K \subset W$, is said to fulfill the *Carathéodory conditions*. If there exists a locally integrable function l_K satisfying the Inequality (2.3) for each compact set $K \subset W$, then g is called *locally Lipschitz* in x with Lipschitz "constant" $l_K(t)$.

Remark 2.3. It can be shown (cf. Hale [4], pp. 28-30) that if f fulfills the Carathéodory conditions, there exists a noncontinuable solution of Equation (1.2) through each point of $W \times \mathbb{R}^+$. If, in addition, f is locally Lipschitz in x, then there is only one noncontinuable solution of Equation (1.2) through each point (x_0,t_0) of $W \times \mathbb{R}^+$. Moreover, the solution depends continuously in (x_0,t_0,t). Thus the hypothesis H_1 of Section 1 is well supported.

Remark 2.4. Conditions $\mathcal{F}_1$ and $\mathcal{F}_2$ deserve some comment. If m_K is integrable on all of $\mathbb{R}^+$, then $\nu(E) = \int_E m_K(\tau)d\tau$ defines a (countably additive positive) measure on the Lebesgue measurable subsets of $\mathbb{R}^+$ and is absolutely continuous with respect to Lebesgue measure. This means that for every $\varepsilon > 0$ there exists $\delta > 0$ so that if $E \subset \mathbb{R}^+$ has Lebesgue measure less than δ, then $\int_E m_K(\tau)d\tau < \varepsilon$. But since m_K is only locally integrable, we must take into account the local nature of the absolute continuity property. Thus, for $E \subset [s,s+1]$, we must have that δ depends upon s (as well as K and ε). As condition $\mathscr{F}_1$ asserts δ is independent of $s \in \mathbb{R}^+$, we see that condition $\mathscr{F}_1$ is a strengthening of the absolute continuity property of $\int_E m_K(\tau)d\tau$ to be uniform with respect to each bounded interval of integration for m_K. Condition $\mathscr{F}_2$ says that for each K, the family of $L^1([0,1]; \mathbb{R}^+)$ functions $\{\ell_s : s \in \mathbb{R}^+\}$, defined by $\ell_s(t) = l_K(s+t)$ is bounded in $L^1([0,1]; \mathbb{R}^+)$ by L_K.

Example 2.5. We see immediately that if m_K is bounded on $\mathbb{R}^+$, then condition $\mathscr{F}_1$ is automatically satisfied. So consider the scalar differential equation defined by

$$f(x,t) = -x^2 \ln t, \quad (x,t) \in \mathbb{R} \times (0,1],$$

and extend f to $\mathbb{R} \times \mathbb{R}^+$ so that for each fixed $x \in \mathbb{R}$ $f(x,\cdot)$ is periodic with period 1. Observe that f fulfills the Carathéodory conditions by taking

$$m_K(t) = K^2 |\ln \tilde{t}|, \quad |x| \leq K, \quad t \in \mathbb{R}^+,$$

where $\tilde{t} = t \pmod 1$. Though m_K is unbounded, it is locally integrable and satisfies condition $\mathscr{F}_1$. Moreover, f is locally Lipschitz in x with

$$l_K(t) = 2K|\ln \tilde{t}|, \quad |x| \leq K, \quad t \in \mathbb{R}^+.$$

As $\int_0^1 |\ln (\tau)| d\tau = 1$, we see by symmetry considerations that

$\int_s^{s+1} 1_K(\tau)d\tau = 2K$ for every $s \in \mathbb{R}^+$. Thus condition $\mathscr{F}_2$ is fulfilled by taking $L_K = 2K$.

We now turn to the definition of the space $\mathscr{F}$. $\mathscr{F}$ consists of all functions $g: W \times \mathbb{R}^+ \to \mathbb{R}^d$ which satisfy the Carathéodory conditions and are locally Lipschitz in x. Furthermore the corresponding locally integrable functions $m_{K,g}$ and $1_{K,g}$ (whose dependence on g is made explicit) must satisfy

$\mathscr{F}_1'$: for every $\varepsilon > 0$ and every measurable subset $E \subset [s,s+1]$ with measure less than $\delta_K(\varepsilon)$, $\int_E m_{K,g}(\tau)d\tau < \varepsilon$,

$\mathscr{F}_2'$: $\int_s^{s+1} 1_{K,g}(\tau)d\tau < L_K$ for every $s \in \mathbb{R}^+$.

Here $\delta_K(\cdot)$ and L_K are provided by the original function f and remain fixed for all $g \in \mathscr{F}$.

Remark 2.6. Condition $\mathscr{F}_1'$ says that the absolute continuity of the measure $\int_E m_{K,g}(\tau)d\tau$ is uniform with respect to $g \in \mathscr{F}$. Compare this with Remark 2.4.

Next we endow $\mathscr{F}$ with a metric topology. The proof of the lemma is left as an exercise.

Lemma 2.7. $\mathscr{F}$ is closed under t-translations. That is, if $g \in \mathscr{F}$, then $g_t \in \mathscr{F}$ for every $t \in \mathbb{R}^+$.

The candidate for a metric ρ on $\mathscr{F}$ is as follows. Fix a dense sequence $\{x_i\} \subset W$ and a dense sequence $\{s_j\} \subset \mathbb{R}^+$. For any $g,h \in \mathscr{F}$, set

$$(2.4)\quad \rho(g,h) = \sum_{i,j=1}^{\infty} 2^{-(i+j)}\min\{1, \left|\int_0^{s_j} [g(x_i,t)-h(x_i,t)]dt\right|\}.$$

Theorem 2.8. $\mathcal{F}$ is a metric space with metric ρ.

Proof: We show here that $\rho(g,h) = 0$ implies $g = h$. The verification of the remaining axioms for a metric are straightforward so are left to the reader. For each positive integer i consider the absolutely continuous function of $s \in \mathbb{R}^+$ given by $F_i(s) = \int_0^s [f(x_i,t) - h(x_i,t)]dt$. If $\rho(f,h) = 0$, then $F_i(s) = 0$ on the dense sequence $\{s_j\} \subset \mathbb{R}^+$. By continuity of F_i it follows that $F_i(s) = 0$ for all $s \in \mathbb{R}^+$. Taking derivatives we obtain $0 = \dot{F}_i(s) = g(x_i,s) - h(x_i,s)$ for all $s \in \mathbb{R}^+$. As this holds for every x_i in the countably dense set $\{x_i\} \subset W$, then the continuity of g,h in the x-variable implies $g(x,s) = h(x,s)$ for all $(x,s) \in W \times \mathbb{R}^+$. Thus $g = h$. □

It will be useful to have a characterization of convergence in $\mathcal{F}$. The following key lemma provides that.

Lemma 2.9. (Convergence) A sequence $\{g_n\} \subset \mathcal{F}$ converges to $g_0 \in \mathcal{F}$ if and only if for each $(x,s) \in W \times \mathbb{R}^+$,

$$\lim_{n\to\infty} \int_0^s g_n(x,t)dt = \int_0^s g_0(x,t)dt. \tag{2.5}$$

Proof: First suppose $g_n \to g_0$ according to Equation (2.5). Let $\varepsilon > 0$ and choose a positive integer m so that $\sum_{i,j=m}^{\infty} 2^{-(i+j)} < \tfrac{1}{2}\varepsilon$. If $\{x_i\} \subset W$ and $\{s_j\} \subset \mathbb{R}^+$ are the sequences specified in the definition of the metric ρ, then there exists a positive integer N such that $n \geq N$ implies

$$\left|\int_0^{s_j} [g_n(x_i,t) - g_0(x_i,t)]dt\right| < \tfrac{1}{2}m^{-2}\varepsilon, \quad i,j = 0,1,\ldots,m-1.$$

Consequently $n \geq N$ implies

$$\rho(g_n,g_0) \le \sum_{i,j=0}^{m-1} 2^{-(i+j)}\tfrac{1}{2}m^{-2}\varepsilon + \sum_{i,j=m}^{\infty} 2^{-(i+j)} \le \varepsilon.$$

Conversely, suppose $\rho(g_n,g_0) \to 0$. We first show that the convergence of Lemma 2.9 is satisfied for all pairs of the form (x_i,s), $s \in \mathbb{R}^+$. So let $\varepsilon > 0$ and choose $\delta = \delta_K(\frac{1}{4}\varepsilon)$, where K is any compact set containing x_i and $\delta_K(\cdot)$ is specified in condition $\mathcal{F}_1$. Pick s_j so that $|s-s_j| < \delta$. If we set $h_n = g_n - g_0$, then

$$\left|\int_0^s h_n(x_i,t)dt\right| \le \left|\int_s^{s_j} h_n(x_i,t)dt\right| + \left|\int_0^{s_j} h_n(x_i,t)dt\right|.$$

There exists a positive integer N so that $n \ge N$ implies $\left|\int_0^{s_j} h_n(x_i,t)dt\right| < \frac{1}{2}\varepsilon$. This is so because $\rho(g_n,g_0) \to 0$ implies that the convergence specified in Lemma 2.9 must hold for every pair (x_i,s_j). Now according to condition $\mathcal{F}_1'$ there exist locally integrable functions $\{m_{K,g_n}\}_{n=0}^{\infty}$ so that

$$\left|\int_s^{s_j} h_n(x_i,t)dt\right| \le \int_s^{s_j} m_{K,g_n}(t)dt + \int_s^{s_j} m_{K,g_0}(t)dt$$

$$< \tfrac{1}{4}\varepsilon + \tfrac{1}{4}\varepsilon = \tfrac{1}{2}\varepsilon.$$

Therefore $\lim_{n\to\infty} \int_0^s g_n(x_i,t)dt = \int_0^s g_0(x_i,t)dt$ for every x_i of the specified dense sequence $\{x_i\}$. It remains to show that this last limit is satisfied for every $(x,s) \in W \times \mathbb{R}^+$. For any $\varepsilon > 0$ and closed ball B centered at x, choose x_i interior to B so that $|x-x_i| < \varepsilon/[2(s+1)L_B]$. Now

$$\left|\int_0^s [g_n(x,t) - g_0(x,t)]dt\right| \le \left|\int_0^s [g_n(x,t) - g_n(x_i,t)]dt\right|$$

$$+ \left|\int_0^s [g_n(x_i,t) - g_0(x_i,t)]dt\right| + \left|\int_0^s [g_0(x_i,t)-g_0(x,t)]dt\right|.$$

The second term on the right side of the inequality tends to zero as previously established. Consider the first term on the right side. It is bounded by $\int_0^s |g_n(x,\tau) - g_n(x_i,\tau)|d\tau$, which according to condition $\mathscr{F}_2'$ is further bounded by

$$\int_0^s 1_{B,g_n}(\tau)|x-x_i|d\tau < (s+1)L_B|x-x_i| < \tfrac{1}{2}\varepsilon.$$

The third term on the right side is bounded similarly. □

The following sequence of lemmas lead to the establishment of the semidynamical system on $\mathscr{F} \times W$.

Lemma 2.10. If $g \in \mathscr{F}$, $x \in W$, and $s,t \in \mathbb{R}^+$, then $\phi(g_t,\phi(g,x;t);s) = \phi(g,x;t+s)$.

Proof: This has already been established following Definition 2.1.

Lemma 2.11. Let $\{g_n\} \subset \mathscr{F}$ converge to $g_0 \in \mathscr{F}$. If $\{\psi_n\}$ is a sequence of continuous functions on $[0,s]$ which converges uniformly to ψ_0, then

$$\int_0^s g_n(\psi_n(\tau),\tau)d\tau \to \int_0^s g_0(\psi_0(\tau),\tau)d\tau$$

for every $s \in \mathbb{R}^+$.

Proof: Let $\varepsilon > 0$, $K \subset W$ be a compact set which contains $\{\psi_n(t): t \in [0,s],\ n \in \mathbb{Z}^+\}$. One can find a piecewise constant function $p: \mathbb{R}^+ \to \mathbb{R}^d$ so that

$$|\psi_0(t) - p(t)| < \varepsilon/[4(s+1)L_K] \quad \text{for every } t \in [0,s].$$

Because of the uniform convergence of ψ_n to ψ_0 on $[0,s]$, we may assume that for sufficiently large n,

$$|\psi_n(t) - \psi_0(t)| < \varepsilon/[4(s+1)L_K] \quad \text{for every } t \in [0,s].$$

Then using the same estimation technique found in the proof of Lemma 2.9 we obtain for sufficiently large n

$$
\begin{aligned}
\left|\int_0^s [g_n(\psi_n(t),t)-g_0(\psi_0(t),t)]dt\right| &\leq \int_0^s |g_n(\psi_n(t),t)-g_n(p(t),t)|dt \\
&+ \left|\int_0^s [g_n(p(t),t)-g_0(p(t),t)]dt\right| \\
&+ \int_0^s |g_n(p(t),t)-g_0(\psi_0(t),t)|dt \\
&\leq \tfrac{1}{2}\varepsilon + \left|\int_0^s [g_n(p(t),t)-g_0(p(t),t)]dt\right| + \tfrac{1}{4}\varepsilon.
\end{aligned}
$$

The convergence Lemma 2.9 insures that the remaining integral can be made less than $\tfrac{1}{4}\varepsilon$ for sufficiently large n by considering the (finite number of) subintervals of $[0,s]$ on which $p(t)$ is constant. □

Lemma 2.12. ϕ is continuous on $\mathcal{F} \times W \times \mathbb{R}^+$.

Proof: Let $(g_n,x_n) \to (g_0,x_0)$ in $\mathcal{F} \times W$. For simplicity use the notation $\phi_n(s) = \phi(g_n,x_n;s)$, $n \in \mathbb{Z}^+$. Then $\{\phi_n\}$ is equicontinuous on $\mathbb{R}^+$. In fact let $\varepsilon > 0$ and B denote any closed ball which contains x_0 in its interior. If $|s - s_0| < \delta = \delta_B(\varepsilon)$, with $s, s_0 \in \mathbb{R}^+$, we have

$$|\phi_n(s) - \phi_n(s_0)| \leq \int_s^s |\dot{\phi}_n(t)|dt \leq \int_{s_0}^s m_{B,g_n}(t)dt < \varepsilon.$$

Thus $\{\phi_n\}$ is equicontinuous on $\mathbb{R}^+$. If it can be demonstrated that $\phi_n(s) \to \phi_0(s)$ uniformly on intervals $[0,b] \subset \mathbb{R}^+$, then ϕ will be continuous on $\mathcal{F} \times W \times \mathbb{R}^+$.

Now the uniformly bounded and equicontinuous sequence $\{\phi_n\}$ on intervals $[0,b]$ is precompact in the space of bounded continuous functions on $\mathbb{R}^+$ into $\mathbb{R}^d$ with the topology of uniform convergence on compact subsets of $\mathbb{R}^+$. If it can be shown that each limit point of the sequence $\{\phi_n\}$ is the (unique) solution ϕ_0, then the sequence converges uniformly on all intervals $[0,b] \subset \mathbb{R}^+$ to ϕ_0.

ϕ_n is the unique solution to the integral equation

$$x(t) = x_n + \int_0^t g_n(x(s),s)ds \tag{2.6}$$

for every $n \in \mathbb{Z}^+$. Suppose ϕ is a limit point of sequence $\{\phi_n\}$. Assume (by taking a subsequence if necessary) that $\phi_n \to \phi$ uniformly on $[0,b]$. An application of Lemma 2.11 shows that $\phi(t)$ satisfies Equation (2.6) for the case $n = 0$. Uniqueness of solutions of Equation (2.6) implies that $\phi(t) = \phi(g_0,x_0;t)$ on $[0,b]$. □

Lemma 2.13. The mapping from $\mathbb{R}^+$ into $\mathcal{F}$ given by $t \to g_t$ is continuous.

Proof: Fix $(x,s) \in K \times \mathbb{R}^+$, where $K \subset W$ is compact, and let $\varepsilon > 0$. For $t \in \mathbb{R}^+$, $\tau \in \mathbb{R}$ with $t + \tau \in \mathbb{R}^+$ we have

$$\begin{aligned} &\left|\int_0^s g_{t+\tau}(x,u)du - \int_0^s g_t(x,u)du\right| \\ &= \left|\int_\tau^{\tau+s} g(x,t+u)du - \int_0^s g(x,t+u)du\right| \\ &= \left|\int_s^{s+\tau} g(x,t+u)du - \int_0^\tau g(x,t+u)du\right| \\ &\le \left|\int_s^{s+\tau} m_{K,g}(t+u)du + \int_0^\tau m_{K,g}(t+u)du\right| \\ &< \tfrac{1}{2}\varepsilon + \tfrac{1}{2}\varepsilon = \varepsilon \quad \text{if} \quad |\tau| < \delta_K(\tfrac{1}{2}\ ). \end{aligned}$$

This establishes the continuity of $t \to g_t$ in terms of the equivalent topology given by Lemma 2.9. □

Lemma 2.14. The mapping from $\mathcal{F} \times \mathbb{R}^+$ onto $\mathcal{F}$ given by $(g,t) \to g_t$ is continuous.

Proof: Let $(f_n,t_n) \to (f_0,t_0)$ in $\mathcal{F} \times \mathbb{R}^+$. Let g_n denote the t_n-translate of f_n, g_0 the t_0-translate of f_0, and h_n the t_n-translate of f_0. It will be shown that

$$\rho(g_n,g_0) \leq \rho(g_n,h_n) + \rho(h_n,g_0) \to 0.$$

Now $\rho(h_n,g_0) \to 0$ from Lemma 2.13. Consider $\rho(g_n,h_n)$. Fix $(x,s) \in K \times \mathbb{R}^+$, $K \subset W$ is compact. Letting $p_n = f_n - f_0$, we have

$$\int_0^s g_n(x,t)dt - \int_0^s h_n(x,t)dt = \int_0^{s+t_n} p_n(x,t)dt - \int_0^{t_n} p_n(x,t)dt.$$

We will show that both of these last two integrals tend to zero as $n \to \infty$. The case for the first integral is presented. The second integral is similar. Now

$$\left|\int_0^{s+t_n} p_n(x,t)dt\right| \leq \left|\int_0^{s+t_0} p_n(x,t)dt\right| + \left|\int_{s+t_0}^{s+t_n} p_n(x,t)dt\right|.$$

The first term on the right side tends to zero in view of the assumption $f_n \to f_0$ in $\mathcal{F}$. The second term is bounded by

$$\int_{s+t_0}^{s+t_n} m_{K,f_n}(t)dt + \int_{s+t_0}^{s+t_n} m_{K,f_0}(t)dt$$

which also tends to zero from condition $\mathcal{F}_1'$. Thus $(g,t) \to g_t$ is continuous. □

The proof of the following corollary is left as an exercise.

Corollary 2.15. The mapping $\pi^*: \mathcal{F} \times \mathbb{R}^+ \to \mathcal{F}$ given by $\pi^*(g,t) = g_t$ defines a semidynamical system $(\mathcal{F},\pi^*)$.

We are finally ready to establish the desired semidynamical system. Define $\pi: \mathcal{F} \times W \times \mathbb{R}^+ \to \mathcal{F} \times W$ by

$$\pi(g,x,t) = (g_t, \phi(g,x;t)).$$

Theorem 2.16. The pair $(\mathcal{F} \times W, \pi)$ is a semidynamical system and is called the *skew product* semidynamical system associated with Equation (1.2).

Proof: The proof is evident from Lemmas 2.10, 2.12, and 2.14. □

Remark 2.17. $\mathcal{F} \times W$ is a metric space with metric $\hat{\rho}$ defined by $\hat{\rho}((f,x),(g,y)) = \rho(f,g) + |x-y|$.

3. Compactness of the Space $\mathcal{F}$

As a consequence of the definition of $\mathcal{F}$ and the very weak topology established by the metric ρ, we can show that $\mathcal{F}$ is compact. This obviously desirable property lends itself to obtaining an invariance principle (Section 4) and stability results (Section 5). Of even more significance is that the phase space $\mathcal{F} \times W$ will be locally compact.

The compactness of $\mathcal{F}$ rests on the following two lemmas whose proofs are omitted. The proof of the first may be found in Dunford and Schwartz [1, IV-8-11]. The second, a purely technical result, was established by Artstein [2].

Lemma 3.1. A necessary and sufficient condition for a set $\mathcal{G} \subset L^1([a,b]; \mathbb{R}^d)$ to be weakly sequentially compact is that for every $\varepsilon > 0$ there exists $\mu(\varepsilon) > 0$ so that if E is a measurable subset of $[a,b]$ with measure less than $\mu(\varepsilon)$, then $|\int_E g(\tau)d\tau| < \varepsilon$, uniformly in $g \in \mathcal{G}$. Thus the measures $\int_E g(\tau)d\tau$ are absolutely continuous with respect to Lebesgue measure, uniformly in $g \in \mathcal{G}$.

Lemma 3.2. Let $\{b_n\}$ be a bounded sequence in $L^1([a,b]; \mathbb{R}^+)$. There exists some $b_0 \in L^1([a,b]; \mathbb{R}^+)$ with the same L^1-bound as $\{b_n\}$ so that if $\{c_n\}$ is a sequence in $L^1([a,b]; \mathbb{R}^d)$ which converges weakly to $c_0 \in L^1([a,b]; \mathbb{R}^d)$ with $|c_n(t)| \leq b_n(t)$ a.e., $n = 1,2,\ldots$, then $|c_0(t)| \leq b_0(t)$ a.e.

Remark 3.3. Lemma 3.1 is only a special case of the characterization of weak sequential compactness in L^1. The more general result in Dunford and Schwartz [1, IV-8.11] requires, in addition to the uniform absolute continuity with respect to $\mathcal{G}$, that $\mathcal{G}$ be L^1-bounded. But in our case the L^1-boundedness of $\mathcal{G}$ follows from the uniform absolute continuity property itself. For if $\mathcal{G}$ were not L^1-bounded, we could find a sequence $\{h_n\} \subset \mathcal{G}$ so that $\int_a^b |h_n(\tau)|d\tau > [n(b-a) + 1]$, $n = 1,2,\ldots$. So for each $n = 1,2,\ldots$, there must exist an interval $I_n \subset [a,b]$ of length $1/n$ so that $\int_{I_n} |h_n(\tau)|d\tau > 1$, $n = 1,2,\ldots$. But this contradicts the uniform absolute continuity with respect to $\mathcal{G}$ as expressed in Lemma 3.1. Thus, $\mathcal{G}$ must be L^1-bounded.

Theorem 3.4. $\mathcal{F}$ is compact.

Proof: Suppose $\{g_k\}$ is a sequence in $\mathcal{F}$. We shall produce a convergent subsequence. Fix $(x,s) \in W \times \mathbb{R}^+$. Observe that the sequence $\{g_k(x,\cdot)\}$ belongs to $L^1([0,s]; \mathbb{R}^d)$ as a consequence of the Carathéodory conditions. Let K be the compact set $\{x\}$, and suppose $\varepsilon > 0$. For each g_k there exists $m_k \in L^1([0,s]: \mathbb{R}^+)$ so that by condition $\mathcal{F}_1'$,

$$\left|\int_E g_k(x,\tau)d\tau\right| \leq \int_E m_K(\tau)d\tau < \varepsilon, \quad k = 1,2,\ldots$$

for every measurable set $E \subset [0,s]$ with measure less than $\delta_K(\varepsilon)$. Note that we have used the interval $[0,s]$ in place of $[s,s+1]$. This is clearly permissible in view of the condition $\mathscr{F}_1'$. According to Lemma 3.1 we may assume (by taking a subsequence if necessary) that $\{g_k(x,\cdot)\}$ converges weakly to some $g_0(x,\cdot) \in L^1([0,s]; \mathbb{R}^d)$. We will first show that this weak limit can be extended to all $(x,t) \in W \times \mathbb{R}^+$ in such a way that $g_0 \in \mathscr{F}$.

Let $\{x_i\}$ be a dense subset of W. By the standard diagonalization procedure we can find a subsequence $\{g_{k_n}\}$ of $\{g_k\}$ so that for every x_i the sequence $\{g_{k_n}(x_i,\cdot)\}$ converges weakly in $L^1([0,s]; \mathbb{R}^d)$ to $g_0(x_i,\cdot)$. Denote g_{k_n} by g_n.

In order to show that $g_0 \in \mathscr{F}$, the domain of g_0 must be extended to all of $W \times \mathbb{R}^+$. This will be accomplished by proving that g_0 is continuous on a dense subset of $K \times [0,s]$ for any compact $K \subset W$. Additionally it will be seen that g_0 satisfies conditions $\mathscr{F}_1'$ and $\mathscr{F}_2'$. The proof is concluded upon proving $\rho(g_n,g_0) \to 0$.

Let $K \subset W$ be compact, the closure of an open set, and suppose $\{l_{K,g_n}\}$ is the corresponding sequence of Lipschitz "constants" for the sequence $\{g_n\}$ in $\mathscr{F}$. Note that $\{l_{K,g_n}\} \subset L^1([0,s]; \mathbb{R}^+)$ and is uniformly L^1-bounded according to $\mathscr{F}_2'$. Denote by $\hat{l}_K$ the $L^1([0,s]; \mathbb{R}^+)$ function whose existence is postulated by Lemma 3.2. Then $\hat{l}_K$ also satisfies the condition $\mathscr{F}_2'$. For x_i, $x_j \in K$ and $t \in [0,s]$ define $h_n(t) = [g_n(x_i,t) - g_n(x_j,t)]/|x_i-x_j|$. $h_n(t)$ converges weakly in $L^1([0,s]; \mathbb{R}^d)$ to $h_0(t) = [g_0(x_i,t) - g_0(x_j,t)]/|x_i-x_j|$. As the Lipschitz assumption implies

$|h_n(t)| \leq l_{K,g_n}(t)$ a.e., then $|h_0(t)| \leq \hat{l}_K(t)$ a.e. Thus $g_0(\cdot,t)$ satisfies the Lipschitz condition on a dense subset of K. In particular, $g_0(\cdot,t)$ is continuous on a dense subset of K, so extends uniquely to K. The extension is also Lipschitz. As W can be represented by $\bigcup_{j=1}^{\infty} K_j$ for closed balls $\{K_j\}$, then g_0 is locally Lipschitz in x and condition $\mathscr{F}_2'$ is readily satisfied.

The verification of condition $\mathscr{F}_1'$ is next. Let $K \subset W$ be as above and $\{m_{K,g_n}\}$ be the corresponding sequence of "Caratheodory type bounds" for the sequence $\{g_n\}$. In view of condition $\mathscr{F}_1'$ we may assume $\{m_{K,g_n}\}$ is a weak sequentially precompact family in $L^1([0,s];\mathbb{R}^+)$. Let $\hat{m}_K$ be a weak limit of $\{m_{K,g_n}\}$. As $|g_n(x_i,t)| \leq m_{K,g_n}(t)$ on $[0,s]$, the weak convergence of $\{g_n(x_i,\cdot)\}$ to $g_0(x_i,\cdot)$ on $[0,s]$ for each x_i implies $|g_0(x_i,t)| \leq \hat{m}_K(t)$ on $[0,s]$. The continuity of g_0 in x implies $|g_0(x,t)| \leq \hat{m}_K(t)$ for all $(x,t) \in K \times [0,s]$. Since $\{m_{K,g_n}\}$ converges weakly to m_K, it follows by continuity that $|\int_E \hat{m}_K(\tau)d\tau| < \varepsilon$ whenever E has measure less than $\delta_K(\varepsilon)$.

Finally we show $\rho(g_n,g_0) \to 0$. For each fixed $x \in W$, we have $\int_0^s 1\cdot g_n(x,\tau)d\tau \to \int_0^s 1\cdot g_0(x,\tau)d\tau$ by weak convergence of $\{g_n(x,\cdot)\}$ to $g_0(x,\cdot)$ in $L^1([0,s];\mathbb{R}^d)$. Here 1 represents the constant function 1 in $L^\infty([0,s];\mathbb{R}^+)$. But this convergence is precisely that established by Lemma 2.9. Thus, $\rho(g_n,g_0) \to 0$. □

Corollary 3.5. The phase space $\mathscr{F} \times W$ is locally compact.

4. The Invariance Principle for Ordinary Differential Equations

Where do solutions of Equation (1.2) go as $t \to \infty$? As you might expect, they tend to positive limit sets. The problem is though, that these limit sets need not be even positively invariant with respect to solutions of Equation (1.2). So fix an open set $W \subset \mathbb{R}^d$ and consider the ordinary differential equation

$$\dot{x} = f(x,t) \tag{4.1}$$

for some $f \in \mathcal{F}$. In particular, f satisfies the Carathéodory conditions and is locally Lipschitz in $x \in W$. Also the locally integrable functions m_K and l_K satisfy properties $\mathcal{F}_1$ and $\mathcal{F}_2$. Henceforth we will assume f to be fixed.

Consider the semidynamical system $(\mathcal{F},\pi^*)$ established in Corollary 2.15, and let $H^+(f)$ be the positive hull of f in this system. $H^+(f)$ is compact in view of Theorem 3.4. It is of interest to know how for any $g \in H^+(f)$ the locally integrable functions $m_{K,g}$ and $l_{K,g}$ depend upon m_K and l_K, the locally integrable functions corresponding to f.

<u>Proposition 4.1</u>. Suppose the functions m_K and l_K are constant for each compact set $K \subset W$. Then each $g \in H^+(f)$ admits the very same constant functions; that is, we may take $m_{K,g} = m_K$ and $l_{K,g} = l_K$.

<u>Proof</u>: Let $K \subset W$ be compact and consider any $g \in H^+(f)$. There exists a sequence $\{t_n\} \subset \mathbb{R}^+$ so that $f_{t_n} \to g$. Fix $x,y \in K$ and observe that the hypothesis implies

$$|f(x,t_n+t) - f(y,t_n+t)| \leq l_K|x-y| \quad \text{for all } n \in \mathbb{N},\ t \in \mathbb{R}^+.$$

Set $h_n(t) = [f_{t_n}(x,t) - f_{t_n}(y,t)]/|x-y|$, and consider the sequence $\{h_n\} \subset L^1([0,s];\mathbb{R}^d)$ for $s > 0$. It follows from Lemma 3.1 that $\{h_n\}$ converges weakly to h in $L^1([0,s];\mathbb{R}^d)$, where

$$h(t) = \frac{g(x,t) - g(y,t)}{|x-y|}$$

As this convergence must be pointwise, then $|h_n(t)| \leq l_K$ for all $n \in \mathbb{N}$ implies $|h(t)| \leq l_K$. A similar argument shows $|g(x,t)| \leq m_K$. □

By our notation convention $\gamma^+(g,x_0)$ is the positive orbit through (g,x_0) and $L^+(g,x_0)$ is the positive limit set for the point (g,x_0) in the semidynamical system $(\mathcal{F} \times W,\pi)$. Denote by $\gamma^+(g)$ and $L^+(g)$ the positive orbit and the positive limit set respectively of the point g in the semidynamical system $(\mathcal{F},\pi^*)$. $H^+(f,x_0)$ and $H^+(f)$ are the positive hulls in $\mathcal{F} \times W$ and $\mathcal{F}$ respectively.

<u>Definition 4.2</u>. If $f^* \in L^+(f)$, then the ordinary differential equation

$$\dot{x} = f^*(x,t) \tag{4.2}$$

is called a *limiting equation* of Equation (4.1).

The conditions $\mathcal{F}_1'$ and $\mathcal{F}_2'$ which define $\mathcal{F}$ and the subsequent conclusion that $\mathcal{F}$ is compact (metric) ensure that Equation (4.2) is indeed an ordinary differential equation. See Section 9 for additional remarks on this.

<u>Definition 4.3</u>. Given a solution $\phi(f,x_0;\cdot)$ of Equation (4.1), its *positive trajectory* is given by

$$\Gamma_f^+(x_0) = \bigcup_{t \in \mathbb{R}^+} \phi(f,x_0;t),$$

and its *positive limit set* is given by

$$\Omega_f^+(x_0) = \{x^* \in \mathbb{R}^d: \phi(f,x_0;t_n) \to x^* \quad \text{for some} \quad \{t_n\} \subset \mathbb{R}^+ \text{ with } t_n \to \infty\}.$$

We say that the solution is *compactly contained* in W (or that just $\phi(f,x_0;\cdot)$ is *compact*) provided $\Gamma_f^+(x_0)$ lies in a compact subset of W.

Lemma 4.4.

(i) $H^+(f,x_0) = H^+(f) \times \overline{\Gamma_f^+(x_0)}$,

(ii) $L^+(f,x_0) = L^+(f) \times \Omega_f^+(x_0)$.

Proof: Suppose $(g,y) \in H^+(f,x_0) \subset \mathcal{F} \times W$. There must exist a sequence $\{t_n\} \subset \mathbb{R}^+$ so that $(f_{t_n}, \phi(f,x_0;t_n)) \to (g,y)$. It is obvious that $g \in H^+(f)$ and $y \in \overline{\Gamma_f^+(x_0)}$. Conversely, let $(g,y) \in H^+(f) \times \overline{\Gamma_f^+(x_0)}$. There exists a sequence $\{t_n\} \subset \mathbb{R}^+$ such that $\phi(f,x_0;t_n) \to y$. By choosing a subsequence if necessary, we may assume $f_{t_n} \to t$ by compactness of $H^+(f)$. Therefore $(g,y) \in H^+(f,x_0)$. The same holds for the factorization of $L^+(f,x_0)$ except for the requirement, $t_n \to \infty$. □

Define the (continuous) projections in $H^+(f) \times W$

$$P: H^+(f) \times W \to H^+(f) \quad \text{by} \quad P(g,x) = g,$$
$$Q: H^+(f) \times W \to W \quad \text{by} \quad Q(g,x) = x.$$

The proofs of the following corollaries are straightforward and therefore, are omitted.

Corollary 4.5. $P\gamma^+(f,x_0) = \gamma^+(f)$ and $PL^+(f,x_0) \subset L^+(f)$. We have equality in the second relation provided $\phi(f,x_0;\cdot)$ is compactly contained in W.

Corollary 4.6. $QL^+(f,x_0) = \Omega_f^+(x_0)$. In other words, $x^* \in \Omega_f^+(x_0)$ if and only if there exists a sequence $\{t_n\} \subset \mathbb{R}^+$ with $t_n \to \infty$ so that $\phi(f,x_0;t_n) \to x^*$.

Lemma 4.7. Suppose $\phi(f,x_0;\cdot)$ is compactly contained in W. Then $L^+(f,x_0)$ is a nonempty, compact, connected, and weakly invariant subset of $\mathcal{F} \times W$. Moreover, every orbit in $L^+(f,x_0)$ is principal.

Proof: The hypothesis of the lemma implies $\overline{\Gamma_f^+(x_0)}$ is a compact subset of W. In view of Lemma 4.4, $H^+(f,x_0)$ is the product of compact sets and hence is compact. The conclusion follows from Theorem 3.5 of Chapter II. □

Remark 4.8. Every $f^* \in L^+(f)$ is defined on $W \times \mathbb{R}$. Indeed, $f_{t_n} \to f^*$ for some sequence $\{t_n\} \subset \mathbb{R}^+$ with $t_n \to \infty$ implies that for any $t \in \mathbb{R}$ we must have $t_n+t \in \mathbb{R}^+$ for all sufficiently large $n \in \mathbb{N}$. Thus $f_{t_n}(x,t)$ is defined for all sufficiently large $n \in \mathbb{N}$, and so is $f^*(x,t)$.

As we are ultimately interested in the limiting behavior of $\phi(f,x_0;t)$, some kind of invariance of $\Omega_f^+(x_0)$ is called for. Weak invariance is the best we can hope for with respect to $L^+(f,x_0)$. In fact, Remark 4.14 will demonstrate that $L^+(f,x_0)$ need not be invariant (see Definition 2.1 of Chapter II). The next proposition suggests a sort of invariance for $\Omega_f^+(x_0)$.

Proposition 4.9. Suppose $\phi(f,x_0;\cdot)$ is compactly contained in W. Then $(f^*,x^*) \in L^+(f,x_0)$ implies $\phi(f^*,x^*;t) \in \Omega_f^+(x_0)$ for all $t \in \mathbb{R}$. If, additionally, the functions m_K and l_K are constant from each compact set $K \subset W$, then $\phi(f^*,x^*;\cdot)$ is a solution of the limiting equation $\dot{x} = f^*(x,t)$ for all $t \in \mathbb{R}$.

<u>Proof</u>: Let $(f^*,x^*) \in L^+(f,x_0)$. There exists a sequence $\{t_n\} \subset \mathbb{R}^+$ with $t_n \to \infty$ so that $f_{t_n} \to f^* \in L^+(f)$ and $\phi(f,x_0;t_n) \to x^* \in \Omega_f^+(x_0)$. We know that $f^*(x,t)$ is defined for all $x \in W$ and $t \in \mathbb{R}$ according to Remark 4.8. Fix any $t \in \mathbb{R}$. Then $\phi(f_{t_n},\phi(f,x_0;t_n);t) = \phi(f,x_0;t+t_n)$ is well defined for sufficiently large $n \in \mathbb{N}$. By continuity of ϕ we obtain $\phi(f_{t_n},\phi(f,x_0;t_n);t) \to \phi(f^*,x^*;t)$. In view of the definition of $\Omega_f^+(x_0)$ we must have

$$\phi(f^*,x^*;t) = \lim_{n\to\infty} \phi(f,x_0;t+t_n) \in \Omega_f^+(x_0).$$

Now in the event the functions m_K and l_K are constant, we see by Proposition 4.1 that $|f^*(x,t)| \le m_K$ for all $(x,t) \in W \times \mathbb{R}$ and f^* is locally Lipschitz in x with Lipschitz constant l_K. Choose a compact set K so that $\Omega_f^+(x_0) \subset K \subset W$. It follows from Remark 2.3 that $\phi(f^*,x^*;\cdot)$ must be a solution of $\dot{x} = f^*(x,t)$ for all $t \in \mathbb{R}$. □

The last proposition suggests some form of invariance for $\Omega_f^+(x_0)$ which we shall shortly name quasi-invariance. Eventually it will be shown that quasi-invariant sets play a major role in determining the asymptotic behavior of solutions $\phi(f,x_0;\cdot)$.

<u>Definition 4.10</u>. A subset $M \subset \mathbb{R}^d$ is called *quasi-invariant* (with respect to Equation (4.1)) if for every $x^* \in M$ there is some $f^* \in H^+(f)$ so that $\phi(f^*,x^*;t) \in M$ for all $t \in \mathbb{R}$.

In view of Proposition 4.9 we have that $\phi(f^*,x^*;t)$ is defined for all $t \in \mathbb{R}$ whenever $f^* \in L^+(f)$. The preceeding results can be summarized as follows.

<u>Theorem 4.11</u>. If $\phi(f,x_0;\cdot)$ is compactly contained in W, then $\Omega_f^+(x_0)$ is quasi-invariant.

Proof: Let $x^* \in \Omega_f^+(x_0)$. There exists a sequence $\{t_n\} \subset \mathbb{R}^+$ with $t_n \to \infty$ so that $\phi(f,x_0;t) \to x^*$. By the compactness of $H^+(f)$ we may assume (by choosing a subsequence if necessary) that $f_{t_n} \to f^* \in L^+(f)$. As this implies $(f^*,x^*) \in L^+(f,x_0)$, then Proposition 4.9 says $\phi(f^*,x^*;t) \in \Omega_f^+(x_0)$ for all $t \in \mathbb{R}$. □

The following corollary will be useful.

Corollary 4.12. If $x^* \in \Omega_f^+(x_0)$, there exists $f^* \in L^+(f)$ and a sequence $\{t_n\} \subset \mathbb{R}^+$ with $t_n \to \infty$ so that $\phi(f_{t_n},x^*;t)$ converges to $\phi(f^*,x^*;t)$ uniformly on compact subsets of $\mathbb{R}$.

It does not follow that Theorem 4.11 is true for all $f^* \in H^+(f)$ as the next example indicates.

Example 4.13. Consider the periodic system

$$(4.3) \qquad \begin{aligned} \dot{x} &= -y \\ \dot{y} &= x + m(t)g(x,y). \end{aligned}$$

where $m(t)$ and $g(x,y)$ are defined as follows. Fix $c > 0$ and set

$$g(x,y) = \begin{cases} 0, & x^2 + (y-1)^2 \geq 2c \\ 1, & x^2 + (y-1)^2 \leq c \end{cases}$$

Extend g as a $C^1(\mathbb{R}^2;\mathbb{R})$ function to all of $\mathbb{R}^2$. Choose c sufficiently small so that there exists a number α $(0 < \alpha < \frac{1}{4}\pi)$ with the property that

$$\{t \in [0,\pi] \Rightarrow \sin t \geq 1 - c\} \Longleftrightarrow |t-\tfrac{1}{2}\pi| \leq \alpha.$$

Then for $t \in [0,\pi]$ one has

$$\cos^2 t + (\sin t - 1)^2 \leq 2c \Longleftrightarrow |t-\tfrac{1}{2}\pi| \leq \alpha.$$

That is,

$$g(\cos t, \sin t) = 0 \Longleftrightarrow |t-\tfrac{1}{2}\pi| > \alpha.$$

Define

$$m(t) = \begin{cases} 0, & \tfrac{1}{2}\pi-\alpha \le t \le \tfrac{1}{2}\pi+\alpha \\ 1, & \tfrac{1}{2}\pi+2\alpha \le t \le 5\pi/2-2\alpha, \end{cases}$$

and extend m to be a continuous 2π-periodic function.

Now Equation (4.3) is periodic in t and defines an element of $\mathscr{F}$. Take $W = \mathbb{R}^2$. It is easy to verify that $(x(t),y(t)) = (\cos t, \sin t)$ is a bounded solution of Equation (4.3). The point $(0,1)$ belongs to the positive limit set $\Omega_f^+(0,1) = \{(x,y)\colon x^2+y^2 = 1\}$ of this solution. For any $t_0 \in \mathbb{R}^+$ let $T_0 = t_0 \pmod{2\pi}$. If $|T_0-\tfrac{1}{2}\pi| \ge 2\alpha$, then the solution through $(0,1)$ at time t_0 does not remain in $\Omega_f^+(0,1)$. Indeed, according to Equation (4.3) if $(x(t_0),y(t_0)) = (0,1)$, then

$$\begin{aligned} \dot{x}(t_0) &= -y(t_0) = -1 \\ \dot{y}(t_0) &= x(t_0) + m(t_0)g(x(t_0),y(t_0)) \\ &= 0 + 1\cdot g(0,1) = 1. \end{aligned}$$

But in order for the solution to remain in $\Omega_f^+(0,1)$ beyond $t = t_0$, it would be necessary for the direction $(\dot{x}(t_0),\dot{y}(t_0))$ to be $(-1,0)$ and not $(-1,1)$ as calculated above.

<u>Remark 4.14</u>. We may also conclude from Example 4.13 that $L^+(f,x_0)$ need not be invariant. For otherwise in view of the factorization $L^+(f,x_0) = L^+(f) \times \Omega_f^+(x_0)$ it would be necessary that for any $f^* \in L^+(f)$ and $x^* \in \Omega_f^+(x_0)$,

$$\pi(f^*,x^*,t) = (f_t^*,\phi(f^*,x^*;t)) \in L^+(f) \times \Omega_f^+(x_0)$$

for all $t \in \mathbb{R}$. That is, for any $f^* \in L^+(f)$ and $x^* \in \Omega_f^+(x_0)$, $\phi(f^*,x^*;t) \in \Omega_f^+(x_0)$ for every $t \in \mathbb{R}$. But

Example 4.13 provides a counterexample to this assertion.

In preparation for the Invariance Principle we develop some properties of quasi-invariant sets. The proof of the first lemma is immediate and so is omitted.

<u>Lemma 4.15</u>. The union of quasi-invariant sets is quasi-invariant.

<u>Lemma 4.16</u>. The closure of a quasi-invariant set is quasi-invariant.

<u>Proof</u>: Let M be quasi-invariant, and suppose $\{x_n\}$ is a sequence in M with $x_n \to x_0$. For each $n \in \mathbb{N}$ there exists a sequence $\{t_n^k\}_{k=1}^{\infty} \subset \mathbb{R}^+$ and some $f_n^* \in H^+(f)$ so that $\lim_{k\to\infty} \phi(f_{t_n^k}, x_n;t) = \phi(f_n^*, x_n;t)$. The convergence is uniform on compact subsets of $\mathbb{R}$. Choose a diagonalizing sequence $\{t_n\}$ by $t_n = t_n^n$ such that

$$|\phi(f_{t_n}, x_n;t) - \phi(f^*, x_n;t)| < \frac{1}{n}, \quad t \in [0,n].$$

We may assume (since $H^+(f)$ is compact) that $f_{t_n} \to f^* \in H^+(f)$. We will prove that $\phi(f^*, x_0;t) \in \overline{M}$ for all $t \in \mathbb{R}$. As $x_0 \in \overline{M}$, then $\overline{M}$ will be quasi-invariant.

Now

$$|\phi(f^*, x_0;t) - \phi(f_n^*, x_n;t)| \le |\phi(f^*, x_0;t) - \phi(f_{t_n}, x_0;t)|$$

$$+ |\phi(f_{t_n}, x_0;t) - \phi(f_{t_n}, x_n;t)| + |\phi(f_{t_n}, x_n;t) - \phi(f_n^*, x_n;t)|.$$

The continuity of ϕ implies that the first two terms on the right hand side tend to zero as $n \to \infty$, uniformly on compact subsets of $\mathbb{R}$. The third term also tends to zero as $n \to \infty$ uniformly on compact subsets of $\mathbb{R}$ by the special choice of the sequence $\{t_n\}$. As quasi-invariance implies

$\phi(f_n^*,x_n;t) \in M$ for all $t \in \mathbb{R}$ and $n \in \mathbb{N}$, the distance from $\phi(f^*,x_0;t)$ to M can be made arbitrarily small. It follows that $\phi(f^*,x_0;t) \in \overline{M}$ for every $t \in \mathbb{R}$. □

An immediate consequence of quasi-invariance is an extension of the generalized invariance principle of LaSalle (Theorem 3.11 of Chapter II).

Theorem 4.17. (Generalized Invariance Principle for Ordinary Differential Equations) Suppose there exists a closed set $E \subset W$ and a set $H \subset W$ with the following property: for each $x_0 \in H$, $\phi(f,x_0;t) \to E$ as $t \to \infty$. If $\phi(f,x_0;\cdot)$ is compactly contained in W, then $\phi(f,x_0;t) \to M$ as $t \to \infty$ where M is the largest quasi-invariant subset of E.

The set E is usually obtained from an auxiliary type function which is also called a Lyapunov function, but whose form differs from that given by Definition 8.1 of Chapter II. Rather than determine the (difficult to find) identity of the limit set $\Omega_f^+(x_0)$, it is usually easier to find M. As $\Omega_f^+(x_0)$ is also quasi-invariant, Theorem 4.17 makes it easier to find out where the solutions go as $t \to \infty$.

In contrast to quasi-invariance we have the usual concept of positive invariance for solutions of Equation (4.1).

Definition 4.18. A set $M \subset W$ is called *positively invariant* with respect to Equation (4.1) if $x_0 \in M$ implies $\Gamma_f^+(x_0) \subset M$.

Corollary 4.19. If the set $M \subset W$ is positively invariant with respect to Equation (4.1), then M is quasi-invariant with respect to Equation (4.1).

Example 4.13 shows that even a positive trajectory $\Gamma_f^+(0,1)$ need not be positively invariant. But positive trajectories will always be quasi-invariant.

Lemma 4.20. For any $x_0 \in W$, the positive trajectory $\Gamma_f^+(x_0)$ is quasi-invariant with respect to Equation (4.1).

Proof: Let $y_0 \in \Gamma_f^+(x_0)$. Then $y_0 = \phi(f,x_0;t_0)$ for some $t_0 \in \mathbb{R}^+$, hence $\phi(f_{t_0},y_0;t) \in \Gamma_f^+(x_0)$ for all $t \in \mathbb{R}^+$. Indeed, by Lemma 2.10, $\phi(f_{t_0},y_0;t) = \phi(f_{t_0},\phi(f,x_0;t_0);t) = \phi(f,x_0;t_0+t) \in \Gamma_f^+(x_0)$ for all $t \in \mathbb{R}^+$. This shows $\Gamma_f^+(x_0)$ is quasi-invariant with respect to Equation (4.1). □

Definition 4.21. Suppose $G \subset W$ with $\overline{G} \subset W$. A function $V: W \times \mathbb{R}^+ \to \mathbb{R}$ is called a *Lyapunov function* for Equation (4.1) on G if

(i) V is continuous,

(ii) given any compact set $K \subset W$ there is some $b_K \in \mathbb{R}$ so that $V(x,t) \geq b_K$ for all $(x,t) \in K \times \mathbb{R}^+$, and

(iii) $\Gamma_f^+(x) \subset G$ implies $V'(x,t) \leq 0$ for every $t \in \mathbb{R}^+$, where

$$V'(x,t) = \limsup_{h \downarrow 0} \frac{V[\phi(f_t,x;h),t+h] - V(x,t)}{h}$$

We recall that $\phi(f_t,x;h)$ is the solution of Equation (4.1) through (x,t) evaluated at time $t + h$. Thus $V'(x,t)$ is the derivative of V along solutions of Equation (4.1).

If V is a C^1 function on $\overline{G} \times \mathbb{R}^+$, then

$$V'(x,t) = \frac{\partial V(x,t)}{\partial t} + \sum_{i=1}^{d} \frac{\partial V(x,t)}{\partial x_i} f_i(x,t).$$

We have represented f and x here by $f = (f_1,f_2,\ldots,f_d)$ and $x = (x_1,x_2,\ldots,x_d)$ respectively.

Corollary 4.22. If V is a Lyapunov function for Equation (4.1) on G, then $\Gamma_f^+(x_0) \subset G$ implies $V(\phi(f,x_0;t),t)$ is nonincreasing in $t \in \mathbb{R}^+$.

Proof: Set $v(t) = V(\phi(f,x_0;t),t)$. As V is differentiable in t a.e. then for $0 \leq t_1 < t_2$

$$v(t_1) - v(t_2) \leq \int_{t_1}^{t_2} V'(\phi(f,x_0;s),s)ds \leq 0. \qquad \square$$

Suppose V is a Lyapunov function for Equation (4.1) on G and let $V_0\colon \overline{G} \to \mathbb{R}$ be continuous and satisfy $V'(x,t) \leq V_0(x) \leq 0$ for all $(x,t) \in G \times \mathbb{R}^+$. Define

$$E = \{x \in \overline{G}\colon V_0(x) = 0\}$$

and let

M = largest subset of E which is quasi-invariant (with respect to Equation (4.1)).

We are ready for an extension of LaSalle's invariance principle (Theorem 8.4 of Chapter II) for nonautonomous ordinary differential equations.

Theorem 4.23. (Invariance Principle) If V is a Lyapunov function for Equation (4.1) on G and V_0, E, and M are defined as above, then all solutions of Equation (4.1) which remain in G and are bounded on $\mathbb{R}^+$ must approach M as $t \to \infty$.

Proof: Let $x_0 \in W$ and suppose $\Gamma_f^+(x_0) \subset G$ is bounded. We will first demonstrate that V_0 is zero on $\Omega_f^+(x_0)$. In view of Corollary 4.22 the function $v(t) \stackrel{\text{def}}{=} V(\phi(f,x_0;t),t)$ is nonincreasing in $t \in \mathbb{R}^+$. As $v(t)$ is also bounded from below from property (ii) of Definition 4.21, it follows that $\lim_{t\to\infty} v(t) = c$ for some $c \in \mathbb{R}$. For any $\tau,t \in \mathbb{R}^+$ with $\tau < t$,

$$v(t) - v(\tau) \leq \int_\tau^t V'(\phi(f,x_0;s),s)ds \leq \int_\tau^t V_0(\phi(f,x_0;s))ds.$$

Let $\varepsilon > 0$ and pick $\tau, t \in \mathbb{R}^+$ sufficiently large so that

$$-\tfrac{1}{2}\varepsilon < v(t) - v(\tau) < \int_\tau^t V_0(\phi(f,x_0;s))ds \leq 0.$$

Suppose $x^* \in \Omega_f^+(x_0)$ and choose $\tau \in \mathbb{R}^+$ so that

$$|V_0(\phi(f,x_0;\tau)) - V_0(x^*)| < \tfrac{1}{2}\varepsilon.$$

This is possible in view of the continuity of ϕ and V_0. Thus $-\varepsilon < V_0(x^*) \leq 0$. As ε is arbitrary, we conclude $V_0(x^*) = 0$, so V_0 is zero on $\Omega_f^+(x_0)$. This means $\Omega_f^+(x_0) \subset E$. An application of Theorem 4.17 shows that $\phi(f,x_0;t) \to M$ as $t \to \infty$. □

As in the case of semidynamical systems (Lemma 8.2, Chapter II), Lyapunov functions are constant on positive limit sets in the following sense.

<u>Corollary 4.24</u>. If V is a Lyapunov function for Equation (4.1) on G and $\Gamma_f^+(x_0)$ is a bounded positive trajectory in G, then $x^* \in \Omega_f^+(x_0)$ implies there exists $f^* \in L^+(f)$ for which $V(\phi(f^*,x^*;t),t)$ is constant in $t \in \mathbb{R}^+$.

<u>Proof</u>: For $x^* \in \Omega_f^+(x_0)$ choose $f^* \in L^+(f)$ by quasi-invariance. We may assume there exists a sequence $\{t_n\} \subset \mathbb{R}^+$ with $t_n \to \infty$ so that $f_{t_n} \to f^*$ and $\phi(f,x_0;t_n) \to x^*$. As in the proof of Theorem 4.23, $\lim_{t\to\infty} v(t) = \lim_{t\to\infty} V(\phi(f,x_0;t),t) = c$ for some $c \in \mathbb{R}^+$. Let $t \in \mathbb{R}$. For all sufficiently large $n \in \mathbb{N}$, $\phi(f_{t_n},\phi(f,x_0;t_n);t) = \phi(f,x_0;t_n+t)$ is defined, and so $v(t+t_n) = V(\phi(f_{t_n},\phi(f,x_0;t_n);t),t)$. Letting $n \to \infty$ we get $c = V(\phi(f^*,x^*;t),t)$. □

As we saw in Chapter II (Sections 3 and 8), the Invariance Principle and its ally, the Lyapunov function, serve to locate the positive limit set $\Omega_f^+(x_0)$ of a solution $\phi(f,x_0;\cdot)$ of Equation (4.1). Furthermore, the Lyapunov function can be used to establish stability properties of the limit set. But unless the set is a critical point though, we cannot hope to find out much about the nature of the solutions of Equation (4.1) in $\Omega_f^+(x_0)$ from the Invariance Principle. The theorem which follows represents about the best we can do in this direction now. Later in Section 5 we will look at this same question without recourse to Laypunov functions, but with the aid of the limiting equations (4.2). The following stability concepts for closed sets in W are required. See Definitions 6.2 and 6.3 of Chapter III for comparison.

Definition 4.25. A closed set $M \subset W$ is called *stable* with respect to Equation (4.1) if given any $\varepsilon > 0$ and $t_0 \in \mathbb{R}^+$ there exists $\delta = \delta(\varepsilon,t_0) > 0$ so that $d(x_0,M) < \delta$ implies

$$d(\phi(f_{t_0},x_0;t),M) < \varepsilon \quad \text{for every} \quad t \in \mathbb{R}^+.$$

If δ can be chosen independently of $t_0 \in \mathbb{R}^+$, then M is called *uniformly stable* with respect to Equation (4.1).

We recall once again here that $\phi(f_{t_0},x_0;t)$ is the solution of Equation (4.1) through (x_0,t_0) evaluated at time $t_0 + t$. The following proposition is included for completeness of exposition. The result is classical and the proof is readily available (e.g., Hale [4], pp. 304-305). Recall Definition 8.6 of Chapter II for a positive definite function.

Proposition 4.26. Suppose M is a compact subset of W with $M \subset G \subset W$. If

(i) V is a Lyapunov function for Equation (4.1) on G, and

(ii) $U_0: G \to \mathbb{R}^+$ is a positive definite function with respect to M so that

$$U_0(x) \leq V(x,t) \quad \text{for every} \quad (x,t) \in G \times \mathbb{R}^+,$$

then the set M is stable with respect to Equation (4.1). If, in addition

(iii) $U_1: G \to \mathbb{R}^+$ is a positive definite function with respect to M so that

$$U_0(x) \leq V(x,t) \leq U_1(x) \text{ for every } (x,t) \in G \times \mathbb{R}^+,$$

then the set M is uniformly stable with respect to Equation (4.1).

The definition of an attractor for a semidynamical system (Definition 6.4, Chapter II) carries over nicely to ordinary differential equations.

Definition 4.27. The closed set $M \subset W$ is called an *attractor* with respect to Equation (4.1) if there exists a neighborhood U of M so that $(x_0,t_0) \in U \times \mathbb{R}^+$ implies $\phi(f_{t_0},x_0;t) \to M$ as $t \to \infty$. That is, for every $\varepsilon > 0$, $x_0 \in U$ and $t_0 \in \mathbb{R}^+$, there exists $T = T(\varepsilon,x_0,t_0) > 0$ so that

$$d(\phi(f_{t_0},x_0;t),M) < \varepsilon \quad \text{for all} \quad t \geq T.$$

The set U is called a *region of attraction* of M with respect to Equation (4.1). If for every $\varepsilon > 0$ and compact subset $K \subset U$ there exists $T = T(\varepsilon,K)$ so that $(x_0,t_0) \in K \times \mathbb{R}^+$ implies $d(\phi(f_{t_0},x_0;t),M) < \varepsilon$ for all $t \geq T$, then M is called a *uniform attractor* with respect to Equation

(4.1). In this case the set U is called a *region of uniform attraction* of M. If $\mathbb{R}^d$ is a (uniform) region of attraction of M, then M is called a *global (uniform) attractor*.

Definition 4.28. The closed set $M \subset W$ is called *asymptotically stable* with respect to Equation (4.1) if it is both uniformly stable and an attractor with respect to Equation (4.1). The closed set $M \subset W$ is called *uniformly asymptotically stable* with respect to Equation (4.1) if it is both uniformly stable and a uniform attractor with respect to Equation (4.1). M is called *globally uniformly asymptotically stable* if it is uniformly stable and a global uniform attractor.

Theorem 4.29. Suppose V is a Lyapunov function for Equation (4.1) on a bounded open set $G \subset W$. Moreover, suppose

(i) G is positively invariant with respect to Equation (4.1),

(ii) V is a C^1 function and $\frac{\partial V}{\partial x}$ is bounded on $\overline{G} \times \mathbb{R}^+$,

(iii) $V(x,t) = c(t)$ is a nondecreasing time dependent function on ∂M, where $M \subset G$ is the largest subset of E which is quasi-invariant with respect to Equation (4.1).

Then M is asymptotically stable with respect to Equation (4.1), and G lies in its region of attraction.

Proof: First we show that M is an attractor. As M must be closed from Lemma 4.16 and M lies in the bounded set G, then M is compact. Choose $\delta > 0$ so that $B_\delta(M) \subset G$. Then according to Theorem 4.23, M is an attractor with G contained in its region of attraction.

Now suppose M is not uniformly stable with respect to Equation (4.1). Then there exist $\varepsilon > 0$ and sequences $\{x_n\} \subset G$, $\{t_n\}$, $\{t_n'\} \subset \mathbb{R}^+$ with $d(x_n,M) \to 0$ as $n \to \infty$ but $d(y_n,M) = \varepsilon$ for all $n \in \mathbb{N}$ where $y_n = \phi(f_{t_n},x_n;t_n')$ is the solution of Equation (4.1) through (x_n,t_n) evaluated at time $t_n + t_n'$. We can assume that the sequence $\{t_n\}$ is increasing and $x_n \to x_0 \in M$, $y_n \to y_0$ with $d(y_0,M) = \varepsilon$. Now V and hence c must be bounded on $\overline{G} \times \mathbb{R}^+$. Thus $\lim_{n\to\infty} c(t_n)$ exists. But if $\lim_{n\to\infty} V(x_n,t_n)$ exists, then as $x_0 \in \partial M$, we have that

$$\lim_{n\to\infty} V(x_n,t_n) = \lim_{n\to\infty} c(t_n) = c_0$$

for some $c_0 \in \mathbb{R}$. Set $\tau_n = t_n+t_n'$ for every $n \in \mathbb{N}$. By choosing a subsequence if necessary, we may assume $f_{\tau_n} \to g \in H^+(f)$. We claim that $\Gamma_g^+(y_0) \subset M$. This would imply $y_0 \in M$, a contradiction. Thus it must be that M is uniformly stable.

In order to establish that $\Gamma_g^+(y_0) \subset M$, we need only show that $\Gamma_g^+(y_0) \subset E$ and $\Gamma_g^+(y_0)$ is quasi-invariant. Suppose $\Gamma_g^+(y_0) \not\subset E$. There exists $t_0 \in \mathbb{R}^+$ so that $V_0(\phi(g,y_0;t_0)) = \lambda < 0$. Choose $n_0 \in \mathbb{N}$ so $V_0(\phi(f_{\tau_n},y_n;t_0)) < \frac{1}{2}\lambda$ for all $n \geq n_0$. Thus $\phi(f_{\tau_n},y_n;t_0) \notin M$ for all $n \geq n_0$. There is some w_n, $0 < w_n \leq \infty$ such that $\phi(f_{\tau_n},y_n;t_0+t) \notin M$ for $t \in [0,w_n)$, and $\phi(f_{\tau_n},y_n;t_0+t) \to \partial M$ as $t \uparrow w_n$. This is because the solution of Equation (4.1) through (x_n,t_n) goes through $\phi(f_{\tau_n},y_n;t_0+t)$ at time $\tau_n + t_0 + t$, and so by Theorem 4.23, $\phi(f_{\tau_n},y_n;t_0+t) \to M$ as $t \uparrow w_n$. Write

$$V(\phi(f_{\tau_n},y_n;t_0+t),\tau_n+t_0+t)$$

$$= V(\phi(f_{\tau_n},y_n;t_0),\tau_n+t_0) + \int_0^t V'(\phi(f_{\tau_n},y_n;t_0+s),\tau_n+t_0+s)ds$$

$$\leq V(\phi(f_{\tau_n},y_n;t_0),\tau_n+t_0) + \int_0^t V_0(\phi(f_{\tau_n},y_n;t_0+s))ds$$

for all $n \geq n_0$. By continuity of $V_0(\phi(f_{\tau_n},y_n;\cdot))$ at t_0, we can find $\delta > 0$ so that $V_0(\phi(f_{\tau_n},y_n;t_0+s)) < \frac{1}{4}\lambda$ for every $s \in [0,\delta)$. Since $V_0(x) \leq 0$ for all $x \in \overline{G}$, we have

$$V(\phi(f_{\tau_n},y_n;t_0+t),\tau_n+t_0+t) \leq V(\phi(f_{\tau_n},y_n;t_0),\tau_n+t_0) + \tfrac{1}{4}\lambda\delta$$

for all $n \geq n_0$. As $\phi(f_{\tau_n},y_n;t_0+t) \to M$ as $t \uparrow w_n$, then the left hand side of the last inequality yields

$$\lim_{t\uparrow w_n} V(\phi(f_{\tau_n},y_n;t_0+t),\tau_n+t_0+t) = c(\tau_n+t_0+w_n), \quad n \geq n_0.$$

The hypothesis that $c(t)$ is inondecreasing insures $c(t_n) \leq c(\tau_n+t_0+w_n)$ for all $n \geq n_0$. Since $y_n = \phi(f_{t_n},x_n;t_n')$ and $V(x,t)$ is nonincreasing along solutions we get

$$V(\phi(f_{\tau_n},y_n;t_0),\tau_n+t_0)$$

$$= V(\phi(f_{t_n},x_n;t_n'+t_0),t_n'+t_0+t_n) \leq V(x_n,t_n)$$

for all $n \geq n_0$. Combining these results we have

$$c(t_n) \leq V(x_n,t_n) + \tfrac{1}{4}\lambda\delta \quad \text{for all} \quad n \geq n_0.$$

Now let $n \to \infty$. We must have the contradiction

$$c_0 \leq c_0 + \tfrac{1}{4}\lambda\delta$$

since $\lambda < 0$. Thus $\Gamma_g^+(y_0) \subset E$. As $\Gamma_g^+(y_0)$ is quasi-invariant from Lemma 4.20, then $\Gamma_g^+(y_0)$ must lie in M. In particular, $y_0 \in M$, a contradiction again. Thus M must be uniformly stable. □

Corollary 4.30. M is positively invariant with respect to Equation (4.1) under the hypotheses of Theorem 4.29.

Proof: As M is uniformly stable with respect to Equation (4.1), then M must be positively invariant with respect to Equation (4.1). □

Corollary 4.31. Suppose V is a Lyapunov function for Equation (4.1) on $G \subset W$. If

(i) V does not depend on $t \in \mathbb{R}^+$,

(ii) G is a bounded component of $\{x \in \mathbb{R}^d: V(x) < \alpha\}$,

(iii) $M = \{x_0\}$ (a single point), the largest subset of $E = \{x \in \overline{G}: V'(x) = 0\}$ which is quasi-invariant with respect to Equation (4.1),

then M is uniformly asymptotically stable.

Example 4.32. Consider the Liénard equation $\ddot{x} + h(x,\dot{x},t)\dot{x} + f(x) = 0$ or the equivalent system

$$\dot{x} = y \qquad (4.4)$$
$$\dot{y} = -h(x,y,t)y - f(x).$$

Assume

(i) f is C^1, $\frac{\partial h}{\partial x}$, $\frac{\partial h}{\partial y}$ exist, are continuous on $\mathbb{R}^2 \times \mathbb{R}^+$ and are uniformly bounded in sets of the form $G \times \mathbb{R}^+$ where $G \subset \mathbb{R}^2$ is bounded,

(ii) $h(x,y,t) \geq k(x,y)$ with $k(x,y) > 0$ if $y \neq 0$, $k(x,y)$ is continuous on $\mathbb{R}^2$,

(iii) for every bounded set $B \subset \mathbb{R}^2$ and $\varepsilon > 0$ there exists $\mu_B(\varepsilon) > 0$ so that $\int_t^{t+\delta} h(x,y,s)ds < \varepsilon$ whenever $|\delta| < \mu_B(\varepsilon)$ and $(x,y) \in B$,

(iv) $xf(x) > 0$ if $x \neq 0$,

(v) $S(x) \stackrel{\text{def}}{=} \int_0^x f(s)ds \to \infty$ as $|x| \to \infty$.

Then the origin $(0,0)$ is globally uniformly asymptotically stable for Equation (4.4). It is clear that conditions $\mathcal{F}_1$ and $\mathcal{F}_2$ are met. Define $V(x,y,t) = \frac{1}{2}y^2 + S(x)$. Then

$$V'(x,y,t) = -h(x,y,t)y^2 \leq -k(x,y)y^2.$$

Set $G_\alpha = \{(x,y) \in \mathbb{R}^2 : V(x,y,t) < \alpha\}$ for $\alpha > 0$. Then G_α is an open bounded set which is positively invariant. Thus all solutions of Equation (4.4) are bounded on $\mathbb{R}^+$. Now $E = G \cap \{(x,y): y = 0\}$. We claim that $\{(0,0)\}$ is the largest positively invariant subset of E. Indeed if $x \neq 0$, $y = 0$, then $\dot{y} = -f(x) \neq 0$, so every solution leaves the x-axis. In view of Corollary 4.31, $M \subset \{(0,0)\}$. As all solutions of Equation (4.4) are bounded, their positive limit sets (which are quasi-invariant) are non-empty and must lie in M. Thus $\emptyset \neq M = \{(0,0)\}$, and so $\{(0,0)\}$ is uniformly asymptotically stable. Moreover, as $V(x,y) \to \infty$ when $\sqrt{x^2+y^2} \to \infty$, $\mathbb{R}^2$ is the region of attraction for $\{(0,0)\}$.

5. Limiting Equations and Stability

We begin with some examples of ordinary differential equations and their limiting equations. As before, we assume

$$\dot{x} = f(x,t) \tag{5.1}$$

satisfies the conditions $\mathcal{F}_1$ and $\mathcal{F}_2$ set forth in Section 2 and that the space $\mathcal{F}$ is constructed accordingly.

Example 5.1. The limiting equations of

$$\dot{x} = \sin\sqrt{t}$$

consist of all the equations with constant right side

$$\dot{x} = c, \quad c \in [-1,1].$$

Example 5.2. The limiting equations of

$$\dot{x} = f(x) + g(x)\sin(t)^2$$

where f,g are C^1 functions on $\mathbb{R}$ consist of the single equation

$$\dot{x} = f(x).$$

Whereas in Section 4 we employed a Lyapunov function to establish a criterion for asymptotic stability of solutions of Equation (5.1), here we explore the relationship between the uniform stability and attraction of solutions Equation (5.1) and that of its limiting equations. A reason for the latter approach is that the limiting equations can be much simpler than the original equation. The previous examples illustrated that.

In practice the sets M for which the stability and attraction concepts are defined in Section 4 are usually the positive trajectory of a solution of Equation (5.1). In particular, M would be positively invariant. By a standard technique we can reduce the concepts of stability and attraction of arbitrary solutions of Equation (5.1) to the stability of the set $M = \{0\}$. This is done by a change of variables which maps a given solution to zero. That is, if $\psi(t)$ is a solution of $\dot{x} = g(x,t)$, then we consider the differential equation

$$\dot{x} = G(x,t) \tag{5.2}$$

where

$$(5.3) \qquad G(x,t) = g(x+\psi(t),t) - g(\psi(t),t).$$

This requires $G(0,t) = 0$ for all $t \in \mathbb{R}^+$ and therefore the function defined by $x(t) = 0$ for all $t \in \mathbb{R}^+$ is a solution of Equation (5.2). We need to know when $G \in \mathscr{F}$. This is handled by the next lemma whose proof is left as an exercise. We will assume as before that f fulfills the Carathéodory conditions and satisfies conditions $\mathscr{F}_1$ and $\mathscr{F}_2$.

Lemma 5.3. If the solution $\psi = \phi(g,x_0;\cdot)$ is compactly contained in W, then G defined by Equation (5.3) is a member of $\mathscr{F}$ and admits the same Lipschitz constant as g.

Definition 5.4. Suppose the set W contains the origin 0 and that $g(0,t) = 0$ for all $t \in \mathbb{R}^+$. The solution of $\dot{x} = g(x,t)$ defined by $\phi(g,0;t) = 0$ for all $t \in \mathbb{R}^+$ is called the *null solution.* The null solution is said to be *stable, uniformly stable, attracting, uniformly attracting, asymptotically stable,* or *uniformly asymptotically stable* provided the set $M = \{0\}$ possesses the corresponding property as given by Definitions 4.25, 4.27, and 4.28.

The argument which preceeds Lemma 5.3 provides a natural definition for stability and attraction of an arbitrary solution $\phi(g,x_0;\cdot)$ of $\dot{x} = g(x,t)$. If the solution is compactly contained in W, then it follows that W contains the origin 0. Hence $G(0,t) = 0$ for all $t \in \mathbb{R}^+$.

Definition 5.5. Suppose the solution $\phi = \phi(g,x_0;\cdot)$ of $\dot{x} = g(x,t)$ is compactly contained in W. We call ϕ *stable, uniformly stable, attracting, uniformly attracting, asymptotically stable,* or *uniformly asymptotically stable* provided the null solution of $\dot{x} = G(x,t)$ possesses the corresponding property.

Now suppose Equation (5.1) admits the null solution. The next lemma shows that every limiting equation of Equation (5.1) also admits the null solution. We wish to see how stability and attraction of the null solution of Equation (5.1) carries over to the limiting equations.

Lemma 5.6. Suppose the set W contains the origin 0. If $f(0,t) = 0$ for all $t \in \mathbb{R}^+$, then for each $f^* \in L^+(f)$ we have $f^*(0,t) = 0$ for all $t \in \mathbb{R}^+$.

Proof: Let $f^* \in L^+(f)$ and suppose $f^* = \lim_{n\to\infty} f_{t_n}$ for some sequence $\{t_n\} \subset \mathbb{R}^+$ with $t_n \to \infty$. We must have for each $t \in \mathbb{R}^+$

$$\int_0^t f^*(0,s)ds = \lim_{n\to\infty} \int_0^t f(0,s+t_n)ds = 0$$

Differentiate with respect to t; we get $f^*(0,t) = 0$. □

The converse of the lemma is false as the next example illustrates.

Example 5.7. The equation $\dot{x} = -2x + e^{-t}$ has as the solution through $(x_0,0)$ the function $\phi(t) = x_0e^{-t} + e^{-t} - e^{-2t}$. On the other hand the limiting equation $\dot{x} = -2x$ has as the solution through $(x_0,0)$ the function $\psi(t) = x_0e^{-2t}$. The latter admits the null solution; the former does not.

Theorem 5.8. Suppose the set W contains the origin 0 and that $f(0,t) = 0$ for all $t \in \mathbb{R}^+$. If the null solution of $\dot{x} = f(x,t)$ is uniformly stable or uniformly attracting, then the null solution of every limiting equation is also uniformly stable or uniformly attracting respectively.

Proof: Let $\varepsilon > 0$ and choose $\delta = \delta(\varepsilon)$ corresponding to the uniform stability of the null solution of $\dot{x} = f(x,t)$.

Suppose $f^* \in L^+(f)$. Then $f^* = \lim_{n\to\infty} f_{t_n}$ for some sequence $\{t_n\} \subset \mathbb{R}^+$ with $t_n \to \infty$. Let $t_0 \in \mathbb{R}^+$. We have $f^*_{t_0} = \lim_{n\to\infty} f_{t_0+t_n}$. Therefore $\lim_{n\to\infty} \phi(f_{t_0+t_n}, x_0; t) = \phi(f^*_{t_0}, x_0; t)$ uniformly for t in compact subsets of $\mathbb{R}^+$. By selecting a subsequence of $\{t_n\}$ if necessary, we may assume for each $n \in \mathbb{N}$

$$|\phi(f^*_{t_0}, x_0; t) - \phi(f_{t_0+t_n}, x_0; t)| < \frac{1}{n} \text{ for every } t \in [0,n].$$

Let $|x_0| < \delta$. The hypothesis insures that $|\phi(f_{t_0+t_n}, x_0; t)| < \varepsilon$ for every $t \in \mathbb{R}^+$ and $n \in \mathbb{N}$. Hence for $|x_0| < \delta$

$$|\phi(f^*_{t_0}, x_0; t)| < \frac{1}{n} + \varepsilon \quad \text{for every } n \in \mathbb{N} \text{ and } t \in [0,n].$$

Letting $n \to \infty$ we get $|\phi(f^*_{t_0}, x_0; t)| < \varepsilon$ for all $t \in \mathbb{R}^+$. This establishes the uniform stability of the null solutions of $\dot{x} = f(x,t)$.

Now suppose U is a region of uniform attraction of the null solution of $\dot{x} = f(x,t)$. Let $f^* \in L^+(f)$. There exists a sequence $\{t_n\} \subset \mathbb{R}^+$ with $t_n \to \infty$ so that $f^* = \lim_{n\to\infty} f_{t_n}$. We will show that U is also a region of uniform attraction of $\dot{x} = f^*(x,t)$ with the same estimate T as for $\dot{x} = f(x,t)$. So let $\varepsilon > 0$ and $K \subset U$ be compact. Choose $T = T(\varepsilon, K)$ according to the uniform attraction of the null solution of $\dot{x} = f(x,t)$. Suppose $(x_0, t_0) \in K \times \mathbb{R}^+$. As before we may assume $|\phi(f^*_{t_0}, x_0; t) - \phi(f_{t_0+t_n}, x_0; t)| < \frac{1}{n}$ for every $n \in \mathbb{N}$ and $t \in [0,n]$. As $\phi(f_{t_0+t_n}, x_0; \cdot)$ may be thought of as the solution of $\dot{x} = f(x,t)$ through (x_0, t_0+t_n), our hypothesis insures $|\phi(f_{t_0+t_n}, x_0; t)| < \varepsilon$ for every $n \in \mathbb{N}$ and $t \geq T$. Consequently

$$|\phi(f^*_{t_0},x_0;t)| < \frac{1}{n} + \varepsilon \quad \text{for every} \quad n \in \mathbb{N} \quad \text{and} \quad t \in [T,n].$$

Letting $n \to \infty$ we obtain the desired result. □

The last theorem has a partial converse.

Theorem 5.9. Suppose the set W contains the origin 0 and that $f(0,t) = 0$ for all $t \in \mathbb{R}^+$. If the null solution of $\dot{x} = f(x,t)$ is uniformly stable and U is a region of attraction of the null solution of every limiting equation $\dot{x} = f^*(x,t)$, $f^* \in L^+(f)$, then U is also a region of uniform attraction of the null solution of $\dot{x} = f(x,t)$.

Proof: Let U be a region of attraction of the null section of every limiting equation of $\dot{x} = f(x,t)$. If U is not a region of uniform attraction of the null solution of $\dot{x} = f(x,t)$, there must exist $\varepsilon > 0$, a compact set $K \subset U$, a sequence $\{x_n\} \subset K$ (which we may assume converges to some $x_0 \in K$), sequences $\{t_n\}$, $\{T_n\} \subset \mathbb{R}^+$ with both $t_n \to \infty$, $T_n \to \infty$, so that $|\phi(f_{t_n},x_n;T_n)| \geq \varepsilon$. Choose $\delta = \delta(\frac{1}{2}\varepsilon)$ corresponding to the uniform stability of the null solution of $\dot{x} = f(x,t)$. Then we must have

$$|\phi(f_{t_n},x_n;t)| \geq \delta \quad \text{for every} \quad t \in [0,T_n].$$

Otherwise, if $|\phi(f_{t_n},x_n;\tau)| < \delta$ for some $\tau \in [0,T_n]$, then letting $y_n \stackrel{\text{def}}{=} |\phi(f_{t_n},x_n;\tau)|$ we get from uniform stability of the null solution of $\dot{x} = f(x,t)$

$$|\phi(f_{t_n},x_n;\tau+t)| = |\phi(f_{t_n+\tau},y_n;t)| < \tfrac{1}{2}\varepsilon \quad \text{for every} \quad t \in \mathbb{R}^+.$$

If we select $t = T_n - \tau$, then $|\phi(f_{t_n},x_n;T_n)| < \frac{1}{2}\varepsilon$, a contradiction.

Because $H^+(f)$ is compact we may assume (by choosing a subsequence if necessary) that $\lim_{n\to\infty} f_{t_n} = f^* \in L^+(f)$. As $\lim_{n\to\infty} \phi(f_{t_n}, x_n; t) = \phi(f^*, x_0; t)$, uniformly in t belonging to compact subsets of $\mathbb{R}^+$, and $|\phi(f_{t_n}, x_n; t)| \geq \delta$ for every $t \in [0, T_n]$, we conclude that $|\phi(f^*, x_0; t)| \geq \delta$ for all $t \in \mathbb{R}^+$. But this contradicts the assumption that U is a region of attraction of the null solution of $\dot{x} = f^*(x,t)$. □

Remark 5.10. Observe that the estimates $\delta(\varepsilon)$ and $T(\varepsilon, K)$ for stability and uniform attraction of the null solution of $\dot{x} = f(x,t)$ carry over to every limiting equation according to Theorem 5.8. Also note that we require the same region of attraction U for every limiting equation in the hypothesis of Theorem 5.9 but that U need not be a region of uniform attraction for every limiting equation.

We summarize the last two results as follows.

Corollary 5.11. Suppose the set W contains the origin 0 and that $\dot{x} = f(x,t)$ admits a uniformly stable null solution. Then the null solution is uniformly asymptotically stable if and only if there exists a neighborhood U of the origin 0 such that U is a region of attraction of 0 for every limiting equation $\dot{x} = f^*(x,t)$, $f^* \in L^+(f)$.

Remark 5.12. Theorems 5.8, 5.9 and Corollary 5.11 begin with the assumption that $f(0,t) = 0$ for every $t \in \mathbb{R}^+$, or equivalently, the function ϕ given by $\phi(t) = 0$ for every $t \in \mathbb{R}^+$ is a solution of $\dot{x} = f(x,t)$ (the null solution). But each of the theorems can be proved without this assumption. We need only refer to the uniform stability and (uniform) attraction of the set $M = \{0\}$ according to Definition

4.27. Moreover, we do not even use the uniqueness of solutions of the initial value problems associated with $\dot{x} = f(x,t)$. Theorem 5.14 to follow does, though, require the existence of a unique null solution to $\dot{x} = f(x,t)$.

In order to illustrate the corollary we consider once again the Liénard equation of Example 4.32. We will develop a necessary and sufficient condition for the null solution of the linear differential equation $\ddot{x} + h(t)\dot{x} + x = 0$ to be uniformly asymptotically stable.

<u>Example 5.13</u>. Consider the above equation in system form

$$\begin{aligned} \dot{x} &= y \\ \dot{y} &= -h(t)y - x. \end{aligned} \tag{5.4}$$

Assume (i) h is measurable and nonnegative in $t \in \mathbb{R}^+$,

(ii) $H(t) \overset{\text{def}}{=} \int_0^t h(s)ds$ is uniformly continuous in $t \in \mathbb{R}^+$.

It is clear that conditions $\mathcal{F}_1$ and $\mathcal{F}_2$ are met. All the limiting equations of Equation (5.4) have the same form, namely

$$\begin{aligned} \dot{x} &= y \\ \dot{y} &= -g(t)y - x, \end{aligned} \tag{5.5}$$

where $\int_0^t g(s)ds = \lim_{n\to\infty} \int_0^t h(s+t_n)ds$ for some sequence $\{t_n\} \subset \mathbb{R}^+$ with $t_n \to \infty$. Use for a Lyapunov function $V(x,y,t) = \frac{1}{2}x^2 + \frac{1}{2}y^2$. Then $V'(x,y,t) = -h(t)y^2 \leq 0$, so $V(x,y,t)$ must be nonincreasing along solutions. It follows that if $(x(\cdot),y(\cdot))$ is a solution of Equation (5.4), then $x^2(t) + y^2(t) \leq x^2(t_0) + y^2(t_0)$ for every $t \geq t_0 \geq 0$. This implies the null solution of Equation (5.4) is uniformly stable. We can now state the promised condition:

A necessary and sufficient condition for the null solution of Equation (5.4) to be uniformly asymptotically stable is that the system

$$\text{(5.6)} \qquad \begin{aligned} \dot{x} &= y \\ \dot{y} &= -x \end{aligned}$$

is not a limiting equation of Equation (5.4).

Proof: The necessity follows directly from Corollary 5.11 as each non null solution of Equation (5.6) is bounded away from zero. Conversely, suppose the null solution of Equation (5.4) is not uniformly asymptotically stable. According to the second part of the proof of Theorem 5.9, some solution $\phi = (x(\cdot),y(\cdot))$ of the limiting Equation (5.5) is bounded away from the origin $(0,0)$ for all $t \in \mathbb{R}^+$. As the Lyapunov function for Equation (5.4) is the same as for Equation (5.5), then ϕ must be bounded. It follows that $\phi(t)$ must approach its positive limit set, which we denote here just by Ω. As Ω is quasi-invariant (with respect to Equation (5.5)), there is some limiting equation of Equation (5.5) and a solution ψ thereof for which $\psi(t) \in \Omega$ for all $t \in \mathbb{R}$. This limiting equation must be a limiting equation of Equation (5.4) as well; hence it has the form $\dot{x} = y$, $\dot{y} = -k(t)y - x$. According to Corollary 4.24, V is constant along $\psi(t)$. In view of the form of V, then Ω must be the set $\{(x,t): x^2+y^2 = \alpha\}$ for some constant $\alpha > 0$. Unless $k(t) = 0$ a.e. we would have $V'(x,y,t) = -k(t)y^2 < 0$ on some t-set of positive measure. This would force the solution ψ to leave Ω, a contradiction. Hence the null solution of Equation (5.4) must be uniformly asymptotically stable. □

A drawback to Theorem 5.9 is that one must first establish uniform stability of the null solution of $\dot{x} = f(x,t)$ before inferring the null solution is a uniform attractor as well. The next theorem eliminates the requirement of uniform stability of the null solution of $\dot{x} = f(x,t)$, though at the expense of requiring U to be a region of *uniform* attraction for the null solution of each limiting equation.

Theorem 5.14. Suppose the set W contains the origin 0 and that $f(0,t) = 0$ for every $t \in \mathbb{R}^+$. Then the null solution of $\dot{x} = f(x,t)$ is uniformly asymptotically stable if and only if there exists a neighborhood U of the origin 0 which is a region of uniform attraction with respect to the null solution of every limiting equation $\dot{x} = f^*(x,t)$, $f^* \in L^+(f)$.

Proof: If the null solution of $\dot{x} = f(x,t)$ is uniformly asymptotically stable, then such a neighborhood U exists according to Theorem 5.9 and Remark 5.10. Conversely, suppose such a neighborhood U exists. In view of Theorem 5.9 it will be sufficient to demonstrate that the null solution of $\dot{x} = f(x,t)$ is uniformly stable. So assume the null solution of $\dot{x} = f(x,t)$ is not uniformly stable. There must exist $\varepsilon > 0$ and sequences $\{x_n\} \subset W$, $\{t_n\}$, $\{t_n'\} \subset \mathbb{R}^+$ with $x_n \to 0$ so that $|\phi(f_{t_n},x_n;t_n')| = \varepsilon$. It follows that $t_n+t_n' \to \infty$. Otherwise $\{t_n+t_n'\}$ is bounded and hence so are $\{t_n\}$, $\{t_n'\}$. We may assume (by choosing a subsequence if necessary) that $t_n \to t_0$ and $t_n' \to t_0'$. Consequently

$$\varepsilon = |\phi(f_{t_n},x_n;t_n')| \to |\phi(f_{t_0},0;t_0')|.$$

But the null solution is the only solution through $(0,t_0)$.

Thus we must have that $\{t_n+t_n'\}$ is unbounded. Without loss of generality we may assume $t_n \to \infty$. We distinguish between two cases:

A. $\{t_n'\}$ is bounded. By choosing subsequences if necessary we may assume $f_{t_n} \to f^* \in L^+(f)$ and $t_n' \to t_0'$. Consequently $\lim_{n\to\infty} \phi(f_{t_n}, x_n; t_n') = \phi(f^*, 0; t_0')$. But $|\phi(f^*,0;t_0')| = \varepsilon$ as before and hence violates the uniqueness of solutions of $\dot{x} = f^*(x,t)$ through $(0,0)$.

B. $\{t_n'\}$ is unbounded. Without loss of generality we may assume $|\phi(f_{t_n}, x_n; t)| < \varepsilon$ for every $t \in [0, t_n')$. Let $\varepsilon > 0$ be such that $K \overset{\text{def}}{=} \overline{B_\varepsilon(0)} \subset U$. Choose $T = T(\tfrac{1}{2}\varepsilon, K)$ according to the uniform attraction of the null solution of every limiting equation. Set $s_n = t_n + t_n' - T$. We may suppose (by choosing a subsequence if necessary) that $f_{s_n} \to f^* \in L^+(f)$. Set $y_n = \phi(f_{t_n}, x_n; t_n' - T)$. Then $|y_n| < \varepsilon$, hence we may assume $\{y_n\}$ converges to $y_0 \in K$. Also $\phi(f_{s_n}, y_n; t) \to \phi(f^*, y_0; t)$ uniformly in $t \in [0,T]$. The uniform attraction condition implies $\phi(f^*, y_0; T) < \tfrac{1}{2}\varepsilon$. But

$$|\phi(f_{s_n}, y_n; T)| = |\phi(f_{t_n}, x_n; t_n')| = \varepsilon \quad \text{for every } n \in \mathbb{N},$$

hence letting $n \to \infty$ we get $|\phi(f^*, y_0; T)| = \varepsilon$, a contradiction. □

As an application of the preceeding result, consider the differential equation

$$(5.7) \qquad \dot{x} = f(x,t)$$

with its perturbation

$$(5.8) \qquad \dot{x} = f(x,t) + h(x,t),$$

where we assume $h_t \to 0$ in $\mathcal{F}$ as $t \to \infty$. It is clear that

Equations (5.7) and (5.8) share the same limiting equations.

Theorem 5.15. Suppose W contains the origin 0 and that $f(0,t) = 0$, $h(0,t) = 0$ for all $t \in \mathbb{R}^+$. Assume the null solution of Equation (5.7) is uniformly asymptotically stable and the null solution of Equation (5.8) is uniformly stable. If $\lim_{t\to\infty} h_t = 0$ in $\mathcal{F}$, then the null solution of Equation (5.8) is also uniformly asymptotically stable.

Proof: The uniform attraction of the null solution of Equation (5.7) is characterized by the existence of a fixed region of uniform attraction for every limiting equation of Equation (5.7), or equivalently, Equation (5.8). Consequently, the null solution of Equation (5.8) is uniformly asymptotically stable. □

A special case of Theorem 5.15 arises when the limiting equations are autonomous or periodic.

Definition 5.16. The function $f \in \mathcal{F}$ is called *asymptotically autonomous* if $L^+(f) = \{f^*\}$; i.e., a point f is called *asymptotically periodic* if $L^+(f)$ consists of a periodic orbit; i.e., $f^* \in L^+(f)$ implies f^* is periodic in $t \in \mathbb{R}$.

Remark 5.17. f is asymptotically autonomous if and only if $f^*_t = f^*$ for every $t \in \mathbb{R}$. This is because $L^+(f) = \{f^*\}$ must be invariant. Consequently, f is asymptotically autonomous if and only if f^* is independent of t (autonomous).

Theorem 5.13 motivates a new definition which in turn establishes a computable criterion for f to be asymptotically autonomous.

Definition 5.18. A function $h \in \mathcal{F}$ is called *diminishing* if for every compact set $K \subset W$, there is a function

$\mu_K: \mathbb{R}^+ \to \mathbb{R}^+$ so that $\lim_{t\to\infty} \mu_K(t) = 0$ and $|\int_t^{t+\sigma} h(x,s)ds| \le \mu_K(t)$ for every $(x,\sigma) \in K \times [0,1]$ and $t \in \mathbb{R}^+$.

A function may be diminishing even though its Euclidean norm is unbounded.

Example 5.19. Consider the function

$$h(x,t) = xe^t \cos e^{2t}.$$

Compute

$$\int_t^{t+\sigma} e^s \cos e^{2s} ds = \int_t^{t+\sigma} [e^s \cos e^{2s} - \tfrac{1}{2}e^{-s} \sin e^{2s}]ds$$
$$+ \int_t^{t+\sigma} \tfrac{1}{2}e^{-s} \sin e^{2s} ds.$$

$$|\int_t^{t+\sigma} e^s \cos e^{2s} ds| \le |[\tfrac{1}{2}e^{-s} \sin e^{2s}]_t^{t+\sigma}| + \int_t^{t+\sigma} \tfrac{1}{2}e^{-s} ds$$

$$\le e^{-t} + e^{-t} \quad \text{for every} \quad \sigma \in [0,1].$$

Set $\mu_K(t) = 2Ke^{-t}$. For $|x| \le K$ we see that Definition 5.18 is satisfied.

Proposition 5.20. The function f is asymptotically autonomous if and only if there exist functions $g, h \in \mathscr{F}$ so that

(i) $f = g + h$,

(ii) g is autonomous, and

(iii) h is diminishing.

Proof: Sufficiency. It is straightforward to verify that condition (iii) implies $h_t \to 0$ as $t \to \infty$ in $\mathscr{F}$. Thus $L^+(f) = \{g\}$, so f is asymptotically autonomous.

Necessity. Suppose $L^+(f) = \{g\}$ where g is autonomous. Set $h = f - g$. For any sequence $t_n \to \infty$ it follows that $h_{t_n} = f_{t_n} - g \to 0$. Thus $h_t \to 0$ as $t \to \infty$. So for any $s \in \mathbb{R}^+$, $\int_0^s h_t(x,u)du \to 0$ as $t \to \infty$. In particular this convergence is uniform on $K \times [0,1]$ for compact $K \subset W$. Define

$$\mu_K(t) = \sup_{\substack{x \in K \\ \sigma \in [0,1]}} \left|\int_0^\sigma h_t(x,u)du\right|$$

Then $\mu_K(t) \to 0$ as $t \to \infty$. Moreover

$$\int_t^{t+\sigma} h(x,s)ds = \int_0^\sigma h_t(x,u)du \le \mu_K(t)$$

for every $(x,\sigma) \in K \times [0,1]$ and $t \in \mathbb{R}^+$. □

Even if there is only one limiting equation whose region of uniform attraction is global (i.e., all of $\mathbb{R}^d$), the uniform asymptotic stability of the original equation need not be global.

Example 5.21. Consider the differential equation

$$\dot{x} = -x + x^2 e^{-t}. \tag{5.9}$$

as $x^2 e^{-t}$ is diminishing, then the limiting equation

$$\dot{x} = -x \tag{5.10}$$

is autonomous. Now the null solution of Equation (5.10) is globally uniformly asymptotically stable. Consequently by Theorem 5.12 the null solution of Equation (5.9) is uniformly asymptotically stable. This stability is not global as $x(t) = 2e^t$ is a solution of Equation (5.9).

We conclude our discussion of asymptotically autonomous and asymptotically periodic equations with a characterization of their positive limit sets.

Theorem 5.22. Consider the autonomous equation

(5.11) $$\dot{x} = f(x)$$

and its perturbation

(5.12) $$\dot{x} = f(x) + h(x,t)$$

where both $f, h \in \mathscr{F}$ and h is diminishing. If the solution $\phi(f+h,x_0;\cdot)$ is compactly contained in W, then the positive limit set $\Omega^+_{f+h}(x_0)$ is invariant with respect to solutions of Equation (5.11).

Proof: The proof is obvious in view of the quasi-invariance of $\Omega^+_{f+h}(x_0)$ with respect to the single limiting equation $\dot{x} = f(x)$. □

Theorem 5.23. Consider the periodic equation

(5.13) $$\dot{x} = f(x,t)$$

where f is periodic in t with period $\tau > 0$. Let h be diminishing and consider the perturbed system

(5.14) $$\dot{x} = f(x,t) + h(x,t).$$

Suppose the solution $\phi(f+h,x_0;\cdot)$ is compactly contained in W. For each $y_0 \in \Omega^+_{f+h}(x_0)$ there exist $t_0 \in [0,\tau]$, a sequence $\{k_n\} \subset \mathbb{N}$ with $k_n \to \infty$, and a solution $y(\cdot)$ of Equation (5.13) through (y_0,t_0) so that $y(t) \in \Omega^+_{f+h}(x_0)$ for every $t \in \mathbb{R}$ and

$$\lim_{n\to\infty} \phi(f+h,x_0;k_n\tau+t) = y(t)$$

uniformly in t belonging to bounded intervals of $\mathbb{R}$.

We conclude this section on limiting equations with an application of the recurrence property of semidynamical systems to solutions of the limiting equations of $\dot{x} = f(x,t)$.

Definition 5.24. The solution $\phi(f,x_0;\cdot)$ of $\dot{x} = f(x,t)$ is called *uniformly recurrent* if for every $\varepsilon > 0$ there exists $L = L(\varepsilon) > 0$ such that for any $s \in \mathbb{R}^+$ and any interval $J \subset \mathbb{R}^+$ of length L there is some $\tau \in J$ such that $|\phi(f,x_0;s) - \phi(f,x_0;\tau)| < \varepsilon$. The solution $\phi(f,x_0;\cdot)$ is just called *recurrent* if L depends on s as well as ε.

Theorem 5.25. Suppose $\phi(f,x_0;\cdot)$ is compactly contained in W. Then there exists a nonempty subset $M \subset L^+(f)$ so that if $f^* \in M$, the limiting equation admits a uniformly recurrent solution.

Proof: The positive limit set $L^+(f,x_0)$ is nonempty, compact and weakly invariant by Lemma 4.7. Then according to Theorem 4.3 of Chapter II $L^+(f,x_0)$ contains a nonempty positively minimal set V. We may write $V = H^+(f^*,x^*)$ for some $(f^*,x^*) \in V$ from Theorem 4.2 of Chapter II. In view of Theorem 3.10 of Chapter III we have that the positive motion $\pi_{(f^*,x^*)}$ is uniformly recurrent. Thus for each $\varepsilon > 0$ there exists $L = L(\varepsilon) > 0$ so that for any $s \in \mathbb{R}^+$, every interval in $\mathbb{R}^+$ of length L contains a point τ with

$$\hat{\rho}(\pi(f^*,x^*,s),\pi(f^*,x^*,\tau)) < \varepsilon.$$

According to the definition of the skew product semidynamical system of Section 2 and Remark 2.18, this inequality implies

$$|\phi(f^*,x^*;s) - \phi(f^*,x^*,\tau)| < \varepsilon.$$

Thus the solution $\phi(f^*,x^*;\cdot)$ is uniformly recurrent. Finally choose for M the set PV, where P is the projection operator defined in Section 4. □

Corollary 5.26. If $\phi(f,x_0;\cdot)$ is compactly contained in W and $L^+(f,x_0)$ is positively minimal, then every limiting equation $\dot{x} = f^*(x,t)$ has a uniformly recurrent solution.

Two observations are in order. The first shows that the hypothesis of Theorem 5.25 is not sufficient to guarantee even recurrent solutions of $\dot{x} = f(x,t)$ itself.

Example 5.27. Let $f(x,t) = 1/(1+t^2)$. Then $f \in \mathcal{F}$, but $\phi(f,x;t) = x + \tan^{-1}t$ is clearly not recurrent.

The next observation shows that not every limiting equation admits a recurrent solution.

Example 5.28. Let $f(x,t) = f(t)$ be continuous from $\mathbb{R}^+$ to $\mathbb{R}$ which is zero everywhere except for isosceles triangular pulses of height one and base one, the left vertex at $t = 2^j$, $j = 0,1,2,\ldots$. The pulses are alternately positive and negative. Then $f \in \mathcal{F}$. For $t_j = 3(2^j)$ we have $f_{t_j} \to 0$ in $\mathcal{F}$. All solutions to $\dot{x} = 0$ are uniformly recurrent, and for this sequence of translates, $\pi(f,0,t_j)$ converges to some point $(0,v) \in V \subset L^+(f,0)$ where V is the positively minimal set described in the proof of Theorem 5.25. However, if $t_j = 2^{2j} - 1$, f_{t_j} converges to $g \in L^+(f)$ which has exactly one positive pulse at $t = 1$. Clearly this g cannot give rise to a recurrent solution.

6. Differential Equations without Uniqueness

It is possible to generate a semidynamical system for the differential equation

$$\dot{x} = f(x,t) \tag{6.1}$$

when no uniqueness conditions are imposed on the solutions. Even though we could establish these results under local existence conditions, we shall assume the existence of a global solution to Equation (6.1) through every $(x_0,t_0) \in W \times \mathbb{R}^+$.

Let $f: W \times \mathbb{R}^+ \to \mathbb{R}^d$ be continuous in $x \in W$ and measurable in $t \in \mathbb{R}^+$. W is open in $\mathbb{R}^d$. Suppose f satisfies the following condition: for every compact subset $K \subset W$, there exists a locally integrable function m_K so that

(6.2) $\quad |f(x,t)| \le m_K(t)$ for every $(x,t) \in K \times \mathbb{R}^+$,

(6.3) $\quad$ for each $\varepsilon > 0$ there exists $\delta = \delta_K(\varepsilon) > 0$ such that if E is a measurable set in $\mathbb{R}^+$, contained in an interval $[s,s+1]$ and with measure less than δ, then $\int_E m_K(\tau)d\tau < \varepsilon$.

As in Section 2 (but without the Lipschitz condition), let $\mathcal{G}$ consist of all functions $g: W \times \mathbb{R}^+ \to \mathbb{R}^d$, continuous in $x \in W$ and measurable in $t \in \mathbb{R}^+$. For each compact set $K \subset W$ there exists a locally integrable function $m_{K,g}$ so that

(6.4) $\quad |g(x,t)| \le m_{K,g}(t)$ for every $(x,t) \in K \times \mathbb{R}^+$

where for each $\varepsilon > 0$ and measurable subset $E \subset [s,s+1]$ with measure less than $\delta_K(\varepsilon)$,

$$\int_E m_{K,g}(t)dt < \varepsilon.$$

As before $\mathcal{G}$ is closed under t-translations. Endow $\mathcal{G}$ with the topology given by the convergence property of Lemma 2.9. Note that this topology need not be metrizable, in fact the metric topology of Equation (2.4) is weaker than the convergence topology of Lemma 2.9. Henceforth every solution of $\dot{x} = g(x,t)$, $g \in \mathcal{G}$, will be assumed to be defined on all of $\mathbb{R}^+$.

Let $\mathscr{S}(g)$ denote the family of all solutions of $\dot{x} = g(x,t)$. For $\phi \in \mathscr{S}(g)$, let ϕ_t denote the t-translate of ϕ, $t \in \mathbb{R}^+$. Again for $t \in \mathbb{R}^+$ define $\mathscr{S}_t(g) = \{\phi_t : \phi \in \mathscr{S}(g)\}$. The following lemma is proved by an easy change of variables.

Lemma 6.1. $\mathscr{S}_t(g) = \mathscr{S}(g_t)$ for every $t \in \mathbb{R}^+$ and $g \in H^+(f)$.

The phase space of the semidynamical system will be

$$X = \{(\phi,g): g \in H^+(f),\ \phi \in \mathscr{S}(g)\}.$$

Note that $H^+(f)$ need not be compact as in Theorem 3.4. Endow X with the topology given by the following convergence structure.

Definition 6.2. $\{(\phi^n,g^n)\}$ converges to (ϕ,g) in X if and only if

(i) $g^n \to g$ in the topology of $\mathscr{G}$,

(ii) $\phi^n \to \phi$ uniformly on compact subsets of $\mathbb{R}$.

Theorem 6.3. The pair (X,π) is a semidynamical system where $\pi(\phi,g,t) = (\phi_t,g_t)$, $t \in \mathbb{R}^+$.

Proof: The proof is an immediate consequence of Lemma 6.1 and the definition of the topology of X. □

Remark 6.4. The topology on X is quite natural in view of Kamke [1]. Kamke's theorem says in essence (for ordinary differential equations without uniqueness) that if $g^n \to g$ (in some topology), and ϕ^n is a solution of $\dot{x} = g^n(x,t)$ with $\phi^n(0) \to x_0$, there is some solution $\phi \in \mathscr{S}(g)$ and a subsequence $\{\phi^{n_k}\}$ so that $\phi^{n_k} \to \phi$ uniformly on compact subsets of $\mathbb{R}^+$.

Remark 6.5. Unlike the space $\mathcal{F}$ defined in Section 2, the space $\mathcal{G}$ need not be compact. It can be proved that it is complete. Therefore the space X is complete. Then most of the theorems of Sections 4 and 5 now hold without uniqueness. It is only necessary to assume (where applicable) that $H^+(f)$ is compact.

7. Volterra Integral Equations

The skew product semidynamical system developed for the nonautonomous ordinary differential equation $\dot{x} = f(x,t)$ can be extended to (nonlinear) Volterra integral equations of the form

$$(7.1) \qquad x(t) = f(t) + \int_0^t k(t,s)g(x(s),s)ds, \quad t \in \mathbb{R}^+.$$

Our development here will be sketchy; the reader should consult Miller and Sell [4] for more details. Our objective is only to provide another example of a semidynamical system and to illustrate some of its properties.

We shall assume the functions f, g, and k belong to appropriately defined function spaces $\mathcal{L}$, $\mathcal{G}$, and $\mathcal{K}$ respectively. In particular, $f\colon \mathbb{R}^+ \to \mathbb{R}^d$, $g\colon \mathbb{R}^d \times \mathbb{R}^+ \to \mathbb{R}^d$, and $k\colon \mathbb{R}^+ \times \mathbb{R}^+ \to \mathbb{R}^{d^2}$; that is, k is $d \times d$ matrix valued and defined to be zero whenever $s > t$. The spaces $\mathcal{L}$, $\mathcal{G}$, and $\mathcal{K}$ shall be chosen so that for each triple (f,g,k) there is a unique solution $\phi(\cdot) = \phi(f,g,k;\cdot)$ of Equation (7.1) which depends continuously on f, g, h, and t. With ϕ so determined, we define the function $T_\tau f = T_\tau(f,g,k)$ by

$$(7.2) \qquad T_\tau f(\theta) = f(\tau+\theta) + \int_0^\tau k(\tau+\theta,s)g(\phi(s),s)ds$$

for every $\theta \in \mathbb{R}^+$ and τ in the maximal interval of definition of ϕ. As in the case for ordinary differential equations, we will assume this interval is $\mathbb{R}^+$; that is, the solution ϕ is global. Set

$$(7.3) \qquad g_\tau(x,s) = g(x,\tau+s) \quad \text{for } x \in \mathbb{R}^d,\ \tau,s \in \mathbb{R}^+,$$

$$(7.4) \qquad k_\tau(t,s) = k(\tau+t,\tau+s) \quad \text{for } \tau \in \mathbb{R}^+,\ 0 \le s \le t < \infty.$$

The spaces $\mathscr{L}$, $\mathscr{G}$, and $\mathscr{K}$ must be chosen so that they are closed under the τ-translations defined by Equations (7.2), (7.3) and (7.4). Finally define $\pi: \mathscr{L} \times \mathscr{G} \times \mathscr{K} \times \mathbb{R}^+ \to \mathscr{L} \times \mathscr{G} \times \mathscr{K}$ by

$$\pi(f,g,k,\tau) = (T_\tau f, g_\tau, k_\tau).$$

The unusual definition of $T_\tau f$ insures the semigroup property holds for this candidate $(\mathscr{L} \times \mathscr{G} \times \mathscr{K}, \pi)$ for the semidynamical system. Let $\mathscr{L}$ be the set $C(\mathbb{R}^+;\mathbb{R}^d)$ with the topology of uniform convergence on compact sets. $\mathscr{L}$ is metrizable under this topology. Metric topologies on $\mathscr{G}$ and $\mathscr{K}$ are chosen so that

(a) the mapping $(g,\tau) \to g_\tau$ of $\mathscr{G} \times \mathbb{R}^+$ into $\mathscr{G}$ is continuous,

(b) the mapping $(k,\tau) \to k_\tau$ of $\mathscr{K} \times \mathbb{R}^+$ into $\mathscr{K}$ is continuous,

(c) the mapping $\tau \to k(\tau+\cdot,\cdot)$ of $\mathbb{R}^+$ into $\mathscr{K}$ is continuous,

(d) the mapping $(x,g,k) \to y$, $y(t) = \int_0^t k(t,s)g(x(s),s)ds$, is a continuous mapping of $\mathscr{L} \times \mathscr{G} \times \mathscr{K}$ into $\mathscr{L}$, and

(e) for every $(f,g,k) \in \mathscr{L} \times \mathscr{G} \times \mathscr{K}$, Equation (7.1) admits a uniquely defined solution $\phi(t) = \phi(f,g,k;t)$

for all $t \in \mathbb{R}^+$ and which depends continuously on f, g, k, and t. The continuity is uniform with respect to t in compact subsets of $\mathbb{R}^+$.

Definition 7.1. The pair of function spaces $(\mathscr{G},\mathscr{K})$ is called *compatible* if conditions (a) through (e) above hold.

Theorem 7.2. If $(\mathscr{G},\mathscr{K})$ is a compatible pair, then $(\mathscr{C}\times\mathscr{G}\times\mathscr{K},\pi)$ is a semidynamical system.

Proof: Firstly, $\pi(f,g,k;0) = (f,g,k)$ is obvious. Secondly, an appropriate change of variables establishes the semigroup property. Thirdly, as ϕ depends continuously upon f, g, k, and τ, then so does $T_\tau f$, g_τ, and k_τ. □

Corollary 7.3. If $(\mathscr{G},\mathscr{K})$ is a compatible pair, then the mappings $(f,\tau) \to f_\tau$ and $(f,\tau) \to T_\tau f$ of $\mathscr{C}\times\mathbb{R}^+$ into $\mathscr{C}$ are semidynamical systems in the phase space $\mathscr{C}$. Additionally, the mappings $(g,\tau) \to g_\tau$ of $\mathscr{G}\times\mathbb{R}^+$ into $\mathscr{G}$ and $(k,\tau) \to k_\tau$ of $\mathscr{K}\times\mathbb{R}^+$ into $\mathscr{K}$ define semidynamical systems in the phase spaces $\mathscr{G}$ and $\mathscr{K}$, respectively.

We briefly give an example of a compatible pair.

Example 7.4. $(\mathscr{G}_p,\mathscr{K}_p)$, $1 < p < \infty$. $\mathscr{G}_p$ consists of all measurable functions $g\colon \mathbb{R}^d \times \mathbb{R}^+ \times \mathbb{R}^d$ so that for every compact set $K \subset \mathbb{R}^d$ there exist functions m_K and l_K in $L^p_{loc}(\mathbb{R}^+;\mathbb{R}^+)$ so that

$$|g(x,t)| \le m_K(t) \quad \text{for every} \quad (x,t) \in K \times \mathbb{R}^+,$$

$$|g(x,t)-g(y,t)| \le l_K(t)|x-y| \quad \text{for every} \quad (x,t),\ (x,t) \in K \times \mathbb{R}^+.$$

The topology on $\mathscr{G}_p$ is defined by saying a net $\{g_\alpha\}$ converges to g_0 if for each compact interval $I \subset \mathbb{R}^+$ and each compact set $\mathscr{A} \subset C(I; \mathbb{R}^d)$

$$\sup_{\psi \in \mathscr{A}} \int_I |g_\alpha(\psi(s),s) - g_0(\psi(s),s)|^p ds \to 0.$$

$\mathscr{K}_p$ consists of all matrix valued measurable functions $k(t,s)$ defined for $0 \le s \le t < \infty$ so that

(i) for each $t \in \mathbb{R}^+$, $k(t,\cdot) \in L^q_{loc}(\mathbb{R}^+; \mathbb{R}^{d^2})$ where $p^{-1} + q^{-1} = 1$, and

(ii) $\lim_{h \to 0} \int_I |k(t+h,s) - k(t,s)|^q ds = 0.$

The topology on $\mathscr{K}_p$ is defined by saying a net $\{k_\alpha\}$ converges to k_0 if for each compact interval $I \subset \mathbb{R}^+$,

$$\lim_\alpha \int_I |k_\alpha(t,s) - k_0(t,s)|^q ds = 0,$$

uniformly for t in compact subsets of $\mathbb{R}^+$.

Remark 7.5. Set $\theta = 0$ in Equation (7.2) to obtain

$$(7.5) \qquad T_\tau f(0) = f(\tau) + \int_0^\tau k(\tau,s)g(\phi(s),s)ds = \phi(\tau),$$

the solution to Equation (7.1). On the other hand, if k depends only upon s, and f is constant, then for every $\tau \in \mathbb{R}^+$, $T_\tau f$ is the constant function,

$$T_\tau f(\theta) = \phi(\tau).$$

Indeed we can write

$$T_\tau f(\theta) = f(0) + \int_0^\tau k(s)g(\phi(s),s)ds,$$

which is independent of $\theta \in \mathbb{R}^+$. This situation occurs when the initial value problem

$$\dot{x} = g(x,t), \quad x(0) = x_0$$

is expressed as an integral equation

$$x(t) = x_0 + \int_0^t g(x(s),s)ds.$$

Thus, the semidynamical system $(\mathscr{L} \times \mathscr{G} \times \mathscr{K},\pi)$ is an extension of the skew product semidynamical system for ordinary differential equations defined in Section 2.

It is interesting to examine the relationship between the positive motions in $\mathscr{L} \times \mathscr{G} \times \mathscr{K}$ and the solutions of Equation (7.1). We begin by characterizing critical motions.

Theorem 7.6. Suppose (f,g,k) is a critical point for $(\mathscr{L} \times \mathscr{G} \times \mathscr{K},\pi)$. Then g and k must have the form

$$g(x,s) = g(x), \tag{7.6}$$

$$k(t,s) = k(t-s), \tag{7.7}$$

so Equation (7.1) becomes

$$x(t) = f(t) + \int_0^t k(t-s)g(x(s))ds. \tag{7.8}$$

Moreover, the corresponding solution ϕ of Equation (7.8) is a constant function

$$\phi(t) = x_0 \overset{\text{def}}{=} f(0) \quad \text{for every} \quad t \in \mathbb{R}^+,$$

and so f satisfies

$$f(t) = x_0 - \int_0^t k(t-s)g(x_0)ds \quad \text{for every} \quad t \in \mathbb{R}^+. \tag{7.9}$$

Conversely, if (f,g,k) satisfy Equation (7.9) for some x_0, then

$$\phi(f,g,k;t) = x_0 \quad \text{for every} \quad t \in \mathbb{R}^+$$

is a solution of Equation (7.8), and (f,g,k) is a critical point for $(\mathscr{L} \times \mathscr{G} \times \mathscr{K},\pi)$.

Proof: If (f,g,k) is a critical point, then Equations

(7.6), (7.7), and hence (7.8) are obvious. In view of Equation (7.5) of Remark 7.5, $T_\tau f = f$ for every $\tau \in \mathbb{R}^+$ implies $\phi(\tau) = \phi(f,g,k;\tau) = T_\tau f(0) = f(0)$ for every $\tau \in \mathbb{R}^+$. Thus the solution of Equation (7.8) is a constant function which we denote by x_0. Consequently we obtain Equation (7.9).

Conversely, if f satisfies Equation (7.9) for some x_0, then $\phi(f,g,k;\cdot) \equiv x_0$ is surely a solution of Equation (7.8). Moreover,

$$\begin{aligned} T_\tau f(\theta) &= x_0 - \int_0^{\tau+\theta} k(\tau+\theta-s)g(x_0)ds + \int_0^{\tau} k(\tau+\theta-s)g(x_0)ds \\ &= x_0 - \int_\tau^{\tau+\theta} k(\tau+\theta-s)g(x_0)ds \\ &= x_0 - \int_0^{\theta} k(\theta-s)g(x_0)ds \\ &= f(\theta) \quad \text{for every } \tau,\theta \in \mathbb{R}^+. \end{aligned}$$

Thus $T_\tau f = f$ for every $\tau \in \mathbb{R}^+$. It is clear from the form of g and k that $g_\tau = g$, $k_\tau = k$ for every $\tau \in \mathbb{R}^+$. Hence, (f,g,k) is a critical point. □

Equation (7.8) is called the *nonlinear renewal equation.* It occurs in numerous applications from reactor dynamics to queuing models. The following lemma expresses some alternate formulations of the operator T_τ.

Lemma 7.7. If $\phi(\cdot) = \phi(f,g,k;\cdot)$ is a solution of Equation (7.1), then

$$\text{(i)} \quad T_\tau f(\theta) = f_\tau(\theta) + \int_{-\tau}^{0} k_\tau(\theta,s)g_\tau(\phi_\tau(s),s)ds$$

$$\text{(ii)} \quad T_\tau f(\theta) = \phi_\tau(\theta) - \int_0^{\theta} k_\tau(\theta,s)g_\tau(\phi_\tau(s),s)ds,$$

where f_τ and ϕ_τ are τ-translations namely, $f_\tau(t) = f(\tau+t)$

and $\phi_\tau(t) = \phi(\tau+t)$.

Proof: A change of variables from s to $\tau+s$ in Equation (7.2) yields (i). As ϕ is a solution of Equation (7.1) we can write

$$\begin{aligned}\phi(\tau+\theta) &= f(\tau+\theta) + \int_0^{\tau+\theta} k(\tau+\theta,s)g(\phi(s),s)ds \\ &= f_\tau(\theta) + \int_0^{\tau} k(\tau+\theta,s)g(\phi(s),s)ds \\ &\quad + \int_\tau^{\tau+\theta} k(\tau+\theta,s)g(\phi(s),s)ds.\end{aligned}$$

Replace s with $\tau+s$ in the last two integrals. We get

$$\begin{aligned}\phi_\tau(\theta) = f_\tau(\theta) &+ \int_{-\tau}^{0} k_\tau(\theta,s)g_\tau(\phi_\tau(s),s)ds \\ &+ \int_0^{\theta} k_\tau(\theta,s)g_\tau(\phi_\tau(s),s)ds.\end{aligned}$$

Combine this with (i) to obtain (ii). □

When is the positive motion through $(f,g,k) \in \mathscr{C} \times \mathscr{G} \times \mathscr{K}$ compact? In the case of ordinary differential equations the function space $\mathscr{F}$ was itself compact, and so we only required the solution of $\dot{x} = f(x,t)$ to be compactly contained in W for the positive motion through $(f,x_0) \in \mathscr{F} \times W$ to be compact. The situation for Volterra integral equations though, is not as simple. In order to determine what conditions might be sufficient, we first look at some of the necessary conditions for compactness of $\pi_{(f,g,k)}$. Certainly we must have compactness of the positive motions $\tau \to T_\tau f$ and $\tau \to f_\tau$ in $\mathscr{C}$, and compactness of the positive motions $\tau \to g_\tau$ in $\mathscr{G}$ and $\tau \to k_\tau$ in $\mathscr{K}$. By compactness of the positive motion $\tau \to T_\tau f$ we mean the family $\{T_\tau f: \tau \in \mathbb{R}^+\} \subset \mathscr{C}$ is precompact in the topology of $\mathscr{C}$. As the evaluation at $\theta = 0$, $T_\tau f \to T_\tau f(0) = \phi(\tau)$, is continuous at each $\tau \in \mathbb{R}^+$, it preserves compactness, hence $\phi(\tau)$ must be bounded for $\tau \in \mathbb{R}^+$. It can also be shown that ϕ is uniformly continuous in $\tau \in \mathbb{R}^+$. These conditions are also sufficient for compactness of $\pi_{(f,g,k)}$.

<u>Theorem 7.8</u>. Suppose $(f,g,k) \in \mathscr{L} \times \mathscr{G} \times \mathscr{K}$, where $(\mathscr{G},\mathscr{K})$ is a compatible pair. Assume the positive motions through $f \in \mathscr{L}$, $g \in \mathscr{G}$, and $k \in \mathscr{K}$, namely $\tau \to f_\tau$, $\tau \to g_\tau$, and $\tau \to k_\tau$ respectively, are compact. If the solution $\phi(\cdot) = \phi(f,g,k;\cdot)$ is uniformly continuous on $\mathbb{R}^+$ and lies in a compact subset of $\mathbb{R}^d$ for each $t \in \mathbb{R}^+$, then the positive motion $\pi_{(f,g,k)}$ is compact.

<u>Proof</u>: Consider the sequence $\{\pi(f,g,k,\tau_n)\}$ for some sequence $\{\tau_n\} \subset \mathbb{R}^+$. We will show that $\{\pi(f,g,k,\tau_n)\}$ contains a convergent subsequence. The compactness assumptions on the positive motions through f, g, and k imply that there exist limiting functions $f^* \in \mathscr{L}$, $g^* \in \mathscr{G}$, $k^* \in \mathscr{K}$ so that (by choosing a subsequence if necessary)

$$\lim_{n\to\infty} f_{\tau_n} = f^*, \quad \lim_{n\to\infty} g_{\tau_n} = g^*, \quad \lim_{n\to\infty} k_{\tau_n} = k^*.$$

The uniform continuity and boundedness of ϕ imply there exists a limiting function $\phi^* \in \mathscr{L}$ so that (again, choosing a subsequence of $\{\tau_n\}$ if necessary) $\lim_{n\to\infty} \phi_{\tau_n}(t) = \phi^*(t)$, uniformly in t belonging to compact subsets of $\mathbb{R}^+$. Thus,

$$\lim_{n\to\infty} \int_0^\theta k_{\tau_n}(\theta,s) g_{\tau_n}(\phi_{\tau_n}(s),s)ds = \int_0^\theta k^*(\theta,s)g^*(\phi^*(s),s)ds$$

uniformly in θ belonging to compact subsets of $\mathbb{R}^+$. From (ii) of Lemma 7.7 we obtain $\lim_{n\to\infty} T_{\tau_n} f(\theta) = f_0^*(\theta)$, where

$$(7.10) \qquad f_0^*(\theta) = \phi^*(\theta) - \int_0^\theta k^*(\theta,s)g^*(\phi^*(s),s)ds,$$

and the convergence is uniform in θ belonging to compact subsets of $\mathbb{R}^+$. It follows that $\{\pi(f,g,k,\tau_n)\}$ converges to $(f_0^*,g^*,k^*) \in \mathscr{L} \times \mathscr{G} \times \mathscr{K}$. □

A point (f_0^*,g^*,k^*) in the positive limit set $L^+(f,g,k)$ can be characterized as the limit of $\{\pi(f,g,k,\tau_n)\}$ for some sequence $\{\tau_n\} \subset \mathbb{R}^+$ with $\tau_n \to \infty$. If $\pi_{(f,g,k)}$ is compact, we obtain the limiting Volterra integral equation for Equation (7.1),

$$(7.11) \qquad y(t) = f_0^*(t) + \int_0^t k^*(t,s)g^*(y(s),s)ds.$$

A comparison of Equation (7.10) with Equation (7.11) shows that $y(t) = \phi^*(t)$ for every $t \in \mathbb{R}^+$, where $y = \phi(f_0^*,g^*,k^*;\cdot)$ is the solution of Equation (7.11). Also, $\lim_{n\to\infty} \phi_{\tau_n}(t) = y(t)$, uniformly in t belonging to compact subsets of $\mathbb{R}^+$. Note that $\phi^*(t)$ is defined for all $t \in \mathbb{R}$, so ϕ^* is an extension of y. This compares with the quasi-invariance property for ordinary differential equations. Indeed, the next corollary provides a limiting integral equation for $\phi^*(t)$ for every $t \in \mathbb{R}^+$. The proof is a direct consequence of Lemma 7.7.

Corollary 7.9. Assume the hypotheses of Theorem 7.8 are satisfied. If there exists a sequence $\{\tau_n\} \subset \mathbb{R}^+$ with $\tau_n \to \infty$ so that $\lim_{n\to\infty} \pi(f,g,k,\tau_n) = (f_0^*,g^*,k^*) \in L^+(f,g,k)$, then

$$(7.12) \qquad \phi^*(t) = f^*(t) + \lim_{n\to\infty} \int_{-\tau_n}^t k^*(t,s)g^*(\phi^*(s),s)ds$$

for every $t \in \mathbb{R}^+$.

It would be desirable for the right side of Equation (7.12) to be independent of the sequence $\{\tau_n\}$. This occurs in the following situation. The proof is left as an exercise.

Corollary 7.10. Assume the hypotheses of Theorem 7.8 are satisfied. In addition, suppose

(i) $k(t,s) = \alpha(t-s)$, where $\int_0^\infty |\alpha(r)|dr < \infty$,

(ii) g is uniformly continuous and bounded on sets of the form $K \times \mathbb{R}^+$ for every compact set $K \subset \mathbb{R}^d$.

Then Equation (7.12) reduces to

$$(7.13) \quad \phi^*(t) = f^*(t) + \int_{-\infty}^{t} \alpha(t-s)g^*(\phi^*(s),s)ds \text{ for all } t \in \mathbb{R}^+.$$

We conclude this section with a characterization of the limiting solutions of the Volterra integral equation

$$(7.14) \qquad x(t) = f(t) + \int_0^t k(t-s)g(x(s),s)ds.$$

Proposition 7.11. Suppose $f \in \mathscr{C}$ is bounded and uniformly continuous. Let $k \in L^q(\mathbb{R}^+; \mathbb{R}^{d^2})$ for $1 < q < \infty$, and suppose $g \in \mathscr{G}_p$ for $1 < p < \infty$ with $p^{-1} + q^{-1} = 1$. Assume

(i) $\int_0^\infty |g(\psi(s),s)|^p ds < \infty$ for all bounded $\psi \in \mathscr{C}$,

(ii) the solution $\phi(\cdot) = \phi(f,g,k;\cdot)$ is bounded on $\mathbb{R}^+$.

Then the positive motion $\pi_{(f,g,k)}$ is compact and the limiting equations corresponding to Equation (7.14) have the form

$$y(t) = f^*(t),$$

where f^* is a limit point of the compact motion $\tau \to f_\tau$ in $\mathscr{C}$. Consequently

$$\lim_{t\to\infty} |\phi(t) - f(t)| = 0.$$

Remark 7.12. In the event solutions of Equation (7.1) are not unique, we can take the approach of Section 6. Indeed, the mapping $T_\tau f$ also depends upon the particular solution ϕ, so that one could write

$$T_\tau f = T_\tau(\phi,f,g,k).$$

The corresponding semidynamical system is $(\mathscr{C}\times\mathscr{C}\times\mathscr{G}\times\mathscr{K},\pi)$, where

$$\pi(\phi,f,g,k,\tau) = (\phi_\tau,T_\tau f,g_\tau,k_\tau).$$

8. Exercises

8.1. Show that $\mathscr{F}$ is closed under t-translations.

8.2. Prove that ρ is a metric on $\mathscr{F}$.

8.3. Prove that if $\{g_n\} \subset \mathscr{F}$ converges to $g_0 \in \mathscr{F}$, then $\int_0^t g_n(x,s)ds \to \int_0^t g_0(x,s)ds$ uniformly for (x,t) in compact subsets of $W \times \mathbb{R}^+$.

8.4. Show that (f,x_0) is a critical point of $(\mathscr{F}\times W,\pi)$ if and only if $f_t = f$ and $\phi(f,x_0;t) = x_0$ for every $t \in \mathbb{R}^+$.

8.5. Prove that $L^+(g)$ is an invariant set in $(\mathscr{F},\pi^*)$.

8.6. Prove that the union of quasi-invariant sets is quasi-invariant.

8.7. Show that if $\lim_{s\to\infty} \phi(f,x_0;s) = y_0$, then $\phi(f^*,y_0;t) = y_0$ for every $t \in \mathbb{R}$.

8.8. Show that the limiting equations of $\dot{x} = \sin\sqrt{t}$ consist of all the equations with constant right side, $\dot{x} = c$, $c \in [-1,1]$.

8.9. Prove Lemma 5.3.

8.10. Suppose the set W contains the origin 0 and that $\dot{x} = f(x,t)$ admits a uniformly stable null solution. If the null solution is uniformly attracting, prove there is a neighborhood U of 0 such that for every $\delta > 0$ there are numbers $a > 0$ and $b \in \mathbb{R}^+$ such that $\int_{t_0}^{t} |f(x,s)|ds \geq a(t-t_0) + b$ for every $t \geq t_0 \geq 0$ and every $x \in U$ such that $|x| \geq \delta$.

8.11. Show that each of the following implies $h \in \mathcal{F}$ is diminishing.

(i) $\int_0^\infty h(x,s)ds < \infty$ for every $x \in W$.

(ii) $\lim_{t\to\infty} h(x,t) = 0$ uniformly in x belonging to compact subsets of W.

(iii) for every $\varepsilon > 0$ and compact set $K \subset W$ there exists $T = T(\varepsilon,K) > 0$ so that $\sup_{x\in K} \int_T^\infty |h(x,s)|ds < \varepsilon$.

8.12. Suppose $f,h \in \mathcal{F}$ with f autonomous and h diminishing. If $\Omega^+_{f+h}(x_0)$ is a periodic orbit for $\dot{x} = f(x)$, show that there exists $T > 0$ so that $|\phi(f+h,x_0;t) - \phi(f+h,x_0;t+T)| \to 0$ as $t \to \infty$.

8.13. Suppose $f \in \mathcal{F}$ can be written as $f = g + h$, where both g and h belong to $\mathcal{F}$, g is periodic in t and h is diminishing. Prove that f is asymptotically periodic.

8.14. Show that the positive motion $t \to f_t$ in $(\mathcal{F},\pi^*)$ is uniformly recurrent (Theorem 3.10, Chapter III) if and only if for each $\varepsilon > 0$ and $t \in \mathbb{R}^+$ there exists $L = L(\varepsilon) > 0$ so that any interval in $\mathbb{R}^+$ of length L contains a point τ so that

$\left|\int_0^s f(x,t+r)dr - \int_0^s f(x,\tau+r)dr\right| < \varepsilon$ for every $(x,s) \in W \times \mathbb{R}^+$.

8.15. Prove Lemma 6.1.

8.16. Prove Theorem 6.3.

8.17. Prove that (f,g,k) is a critical point of $(\mathscr{L} \times \mathscr{G} \times \mathscr{K}, \pi)$ if and only if f is absolutely continuous and $\dot{f}(t) = -k(t)g(x_0)$ a.e. for some $x_0 \in \mathbb{R}^d$.

8.18. Suppose (f,g,k) is an ω-periodic point of $(\mathscr{L} \times \mathscr{G} \times \mathscr{K}, \pi)$. If $\phi(\cdot) = \phi(f,g,k;\cdot)$ is the corresponding solution of Equation (7.1), show that

(i) $f(\theta) = f(\omega+\theta) + \int_0^\omega k(\omega+\theta,s)g(\phi(s),s)ds$

(ii) ϕ is ω-periodic.

8.19. Show that if the positive motion $\pi_{(f,g,k)}$ in $\mathscr{L} \times \mathscr{G} \times \mathscr{K}$ is compact, then $\phi(\tau) = \phi(f,g,k;\tau)$ is uniformly continuous in $\tau \in \mathbb{R}^+$.

8.20. Consider the equation (7.13) where $f^*(t) = f_0$ is independent of t, g is also independent of t and satisfies a global Lipschitz condition $|g(x) - g(y)| \leq L|x-y|$ for some $L > 0$ and all $x,y \in \mathbb{R}^d$. If $\int_0^\infty |\alpha(r)|dr = A$ where $AL < 1$, prove that $x(t) = f_0 + \int_{-\infty}^t \alpha(t-s)g(x(s))ds$ has a unique solution $x(t) = x_0$ for all $t \in \mathbb{R}$ where x_0 satisfies $x_0 = f_0 + Ag(x_0)$.

9. Notes and Comments

The development of the *skew product* semidynamical system for solutions of nonautonomous ordinary differential equations was begun by Miller [3] and Sell [2]. Their approach assumed $f \in C(W \times \mathbb{R}; \mathbb{R}^d)$ is regular; that is f fulfills the Carathéodory conditions and for every $(x_0, t_0) \in W \times \mathbb{R}^+$, there exists exactly one noncontinuable solution of $\dot{x} = f(x,t)$ which satisfies $x(t_0) = x_0$. Endow $C(W \times \mathbb{R}; \mathbb{R}^d)$ with the compact open topology and set $\mathscr{A}(f)$ to be the closure of $\{f_\tau : \tau \in \mathbb{R}\}$ in $C(W \times \mathbb{R}; \mathbb{R}^d)$. Then $(\mathscr{A}(f) \times W, \pi)$, is a local dynamical system where π is given by Equation (2.1). Unfortunately these motions may have empty limit sets in $C(W \times \mathbb{R}; \mathbb{R}^d)$. That is, the limiting equatons may not be regular. Thus it is desirable, amongst other reasons, for $\mathscr{A}(f)$ to be compact. Sell [2] proves $\mathscr{A}(f)$ is compact if and only if f is bounded and uniformly continuous on sets of the form $K \times \mathbb{R}$ for every compact set $K \subset W$. A major accomplishment now was a generalization of LaSalle's Invariance Principle to nonautonomous equations. Most of the other results in Sections 4 and 5 can also be obtained with this additional restriction on f. Miller [1] and others restricted f to be almost periodic in t, uniformly in $x \in W$. Wakeman [1] removed this restriction by weakening the topology on $\mathscr{A}(f)$. In fact, his topology is the compact open topology on the functions $F(x,t) = \int_0^t f(x,s)ds$, where f satisfies conditions $\mathscr{F}_1$ and $\mathscr{F}_2$ of Section 2 for functions m_K independent of t. Then $(\mathscr{A}(f) \times W, \pi)$ is a local dynamical system. If the Lipschitz constant l_K is also independent of t, then $\mathscr{A}(f)$ is compact. The next step was taken by

Artstein [2]. It is his approach which we use in Section 2, where m_K and l_K are allowed to depend upon t. The topology is the same as Wakeman's, but the resulting function space $\mathcal{F}$ is larger. For an excellent survey and highly motivated account of these ideas, see Artstein's Appendix A in LaSalle [8].

<u>Sections 1-3</u>. The construction of the semidynamical system $(\mathcal{F} \times W, \pi)$ and the resulting proof that $\mathcal{F}$ is a compact metric space is due to Artstein [2]. See Artstein [3,4] in the event the conditions on $\mathcal{F}$ are relaxed. One ends up with the class of Kurzweil equations.

<u>Section 4</u>. The results in this section do not depend on the specific topology on $\mathcal{A}(f)$. What's important though, is that the family $\{f_\tau : \tau \in \mathbb{R}^+\}$ is relatively sequentially compact in $\mathcal{A}(f)$ so that the limiting equations are regular in the sense of Sell. Autonomous limiting equations were first studied (in a less general setting) by Levin and Nohel [1], Levin [1], Markus [1], Strauss and Yorke [2], and Yoshizawa [1]. The case of periodic limiting equations was first treated by LaSalle [3] and Deysach and Sell [1]. Indeed, LaSalle [3] establishes a stronger invariance principle than Theorem 4.17 for periodic equations. He shows that if f is periodic in t, and $\phi(f,x_0;\cdot)$ is compactly contained in W, then $\phi(f,y_0;t) \in \Omega_f^+(x_0)$ for every $t \in \mathbb{R}$; that is, $\Omega_f^+(x_0)$ is actually invariant with respect to Equation (4.1). The present formulation of limiting equations goes back to Miller [1]. Lemma 4.4, Corollaries 4.5 and 4.6 can be deduced from Miller [1,2] or Sell [2]. Likewise, Proposition 4.9, Definition 4.10 of quasi-invariance, Example 4.13 and the

Generalized Invariance Principle, Theorem 4.17, are all due to Miller [1]. Lemma 4.16 is from Wakeman [1]. The definition of the Lyapunov function is standard; c.f. LaSalle [10] or Yoshizawa [3]. Other invariance results are established by Peng [1], Onuchic [1,2], and Rouché [1]. The definitions of stability and attraction are those of LaSalle [10] or Yoshizawa [3]. The reader should be advised that there are numerous definitions of these terms. Theorem 4.29, which is due to Wakeman [1], is a generalization of a result of LaSalle [5] for autonomous equations. The attraction here is not uniform though.

Section 5. Theorems 5.8 and 5.9 are originally due to Sell [2]. The statements here, as well as the proofs are from Artstein [5]. Example 5.13 is also due to Artstein [5]. Another such example is in Artstein and Infante [1]. Theorem 5.14 which is also due to Artstein [5], is a generalization of a result of Markus [1] and is a solution to the Inverse Limit Problem of Sell [4]. Also Bondi, Moauzo, and Visentin [1] proved Theorem 5.14 but with stricter hypotheses than Artstein. Artstein [5] has also used these ideas to characterize uniform asymptotic stability of the null solution of Equation (5.1) in terms of a "growth" condition, namely $\int_{t_0}^{t} |f(x,s)|ds \geq a(t-t_0) + b$. This generalizes results for linear equations by Morgan and Narenda [1]. The concept of diminishing functions has been used by many people including Chow [1], Onuchic [1,2], Strauss and Yorke [1,2,3] and Yorke [1]. Example 5.19 is based on an example of Strauss and Yorke [1]. Proposition 5.20 is from Sell [2]. Finally Theorem 5.25 is due to Bender [1], as are Examples 5.26 and

5.27. See Seifert [2] for a more comprehensive view of recurrent solutions.

Section 6. This work is based on Sell [5]. It has an excellent bibliography. The function space used here is Artstein's, not Sell's. The local Lipschitz condition has been dropped, though.

Section 7. The development here is almost entirely due to Miller and Sell [1,4]. Theorem 7.6 can also be found in Sell [6]. The question of nonuniqueness in Remark 7.12 is treated by Sell [5]. Existence, uniqueness, and continuous dependence of solutions may be found in Miller and Sell [2,3]. Also see Artstein [1]. For a more general formulation of the Volterra integral equation, see Miller [4]. Miller and Sell [5] surveys the role of dynamical systems in the study of nonautonomous ordinary differential equations and Volterra integral equations. The reference list is extensive.

CHAPTER V
SEMIDYNAMICAL SYSTEMS IN BANACH SPACE

1. Introduction

Material fundamental to the existence and qualitative behavior of partial differential equations and differential delay equations (to name just two areas) are developed in this chapter. The general formulation is an evolution equation in a Banach space. The work of Crandall and Liggett on the nonlinear version of the Hille - Yosida - Phillips theorem for linear semigroups has spawned an elegant analysis of nonlinear evolution equations in Banach spaces. One special feature of the semigroup generation theorem (linear or nonlinear) is that we obtain a representation of the solutions to $\frac{du}{dt} + Au = 0$ in terms of the operator A. The classical approach was to establish the existence, uniqueness, and continuous dependence of the solutions of the particular partial differential equation, for example, and then demonstrate that the solutions generate a semigroup. This was essentially the approach we also took in Chapter IV.

We motivate the semigroup generation theorem by examining the nonlinear diffusion equation

$$\frac{\partial u}{\partial t}(x,t) = \Delta\phi(u(x,t)), \quad (x,t) \in \Lambda \times (0,\infty) \tag{1.1}$$

where Δ is the operator $\sum_{i=1}^{d} \frac{\partial^2}{\partial x_i^2}$, $\Lambda \subset \mathbb{R}^d$ is a bounded open set with smooth boundary $\partial\Lambda$, and $\phi\colon \mathbb{R} \to \mathbb{R}$ is continuous, strictly increasing with $\phi(0) = 0$, $\phi(\mathbb{R}) = \mathbb{R}$, and ϕ^{-1} Lipschitz. In addition we impose the boundary condition

$$u(x,t) = 0 \quad \text{for} \quad (x,t) \in \partial\Lambda \times (0,\infty) \tag{1.2}$$

and the initial condition

$$u(x,0) = u_0(x) \quad \text{for} \quad x \in \Lambda. \tag{1.3}$$

This is reasonable when $u(\cdot,t)$ represents a density. Choose for the Banach space $X = L^1(\Lambda;\mathbb{R})$. We interpret the unknown function $u(x,t)$ as the map $t \to u(\cdot,t)$ with values in X. For simplicity, we write the function $u(\cdot,t)$ as $u(t)$. Thus we may think of $\frac{\partial u}{\partial t}(x,t)$ as $\frac{du}{dt}(t)$, the derivative of the X-valued function $u(t)$. The nature of this derivative is discussed in Section 2. Next define an operator $A\colon \mathscr{D}(A) \subset X \to X$ by $Au = -\Delta\phi(u)$. The domain $\mathscr{D}(A)$ of A must be chosen so that $-\Delta\phi(u)$ makes sense and lies in X for $u \in \mathscr{D}(A)$. Also we build into $\mathscr{D}(A)$ the boundary condition $u = 0$ on $\partial\Lambda$. Summarizing, we may write

$$\frac{du}{dt} + Au = 0, \quad u(0) = u_0. \tag{1.4}$$

Now consider the discrete problem

$$\begin{aligned} \frac{u_\lambda(t) - u_\lambda(t-\lambda)}{\lambda} + Au_\lambda(t) &= 0, \quad t > 0 \\ u_\lambda(t) &= u_0, \quad t \le 0 \end{aligned} \tag{1.5}$$

where $\lambda > 0$. We solve Equation (1.5) formally by writing

$u_\lambda(t) + \lambda A u_\lambda(t) = u_\lambda(t-\lambda)$ or, $u_\lambda(t) = (I + \lambda A)^{-1} u_\lambda(t-\lambda)$. Set $J_\lambda = (I + \lambda A)^{-1}$ and iterate to get

$$u_\lambda(t) = J_\lambda^{[t/\lambda]+1} u_0, \quad t > 0, \tag{1.6}$$

where $[\tau]$ is the largest integer in $(-\infty, \tau]$. Note that A may even be multivalued, yet J_λ is still a function. To ensure that u_λ is defined, it is necessary that the iterates $\{J_\lambda^k\}$ have a sufficiently large domain. Thus the range of $I + \lambda A$ needs to be large. It is also reasonable for the iterates $\{J_\lambda^k\}$ be equicontinuous. This will be so if J_λ is a contraction. Indeed, these properties are all we need in order to establish that $\lim_{\lambda \downarrow 0} u_\lambda(t)$ exists and defines a semidynamical system. The corresponding operator A will be called accretive.

The purpose in presenting the nonlinear semigroup theory is to provide a unifying generalization for stability problems in certain partial differential equations and functional differential equations. The former is addressed in this chapter; the latter is addressed in the next chapter. In applications the operator A is typically nonlinear, not everywhere defined, discontinuous or not smooth. Moreover, the space X may not be a Hilbert space nor even reflexive. It is fortunate though, that some interesting problems can be realized in the form of Equation (1.4) for accretive operators A.

The semigroup generated by the above scheme need not be differentiable in t with respect to the strong (norm) topology on X. It may be differentiable with respect to some weaker topology though. In Section 3 we find a generalized domain $\hat{\mathscr{D}}(A) \supset \mathscr{D}(A)$ so that the semigroup generated by A is at least Lipschitz continuous in t there. This generalized domain is used in Section 4 to establish criteria for compact positive motions in X. In practice, compact positive motions may be obtained in a number of ways. The structure of some particular systems (differential delay equations, Chapter VI) ensures that the solutions are smoother than their initial functions. As positive orbits are families of functions, this guarantees equicontinuity of solutions. Another way is to embed the phase space into a larger Banach space in which the positive orbits are now precompact. This approach is used in Section 4. Section 5 is devoted to a discussion of the abstract Cauchy problem and the nature of solutions. In Section 6 we apply the semidynamical theory developed in the first three chapters to characterize the positive limit sets of contracting semigroups. The details of the nonlinear semigroup generation theorem are left for Section 8.

2. Nonlinear Semigroups and Their Generators

Definition 2.1. A *(strongly continuous) semigroup* of operators on a Banach space X is a family $T = \{T(t): t \in \mathbb{R}^+\}$ of operators on X, each with domain X and which satisfies

(i) $T(0) = I$ (the identity operator)

(ii) $T(s+t) = T(s)T(t)$ for all $s,t \in \mathbb{R}^+$

(iii) $T(\cdot)x: \mathbb{R}^+ \to X$ is continuous for all $x \in X$.

If in addition there exists $\omega \in \mathbb{R}$ so that

(iv) $\|T(t)x - T(t)y\| \leq e^{\omega t}\|x-y\|$ for all $t \in \mathbb{R}^+$, $x,y \in X$, we call T *quasi-contracting* and write $T \in Q_\omega$. If $\omega = 0$, T is called *contracting*.

Henceforth let X denote a real Banach space. The formal development in Section 1 called for the existence of operators $J_\lambda = (I + \lambda A)^{-1}$. This is possible when A has the following property.

Definition 2.2. An operator $A: \mathcal{D}(A) \subset X \to X$ is called *accretive* if

$$\|x-y + \lambda(Ax-Ay)\| \geq \|x-y\| \text{ for all } \lambda > 0 \text{ and } x,y \in \mathcal{D}(A).$$

Remark 2.3. In the case $X = \mathbb{R}$ it is easy to see that A is accretive if and only if A is a nondecreasing function. Indeed, if A is accretive, then for each $\lambda > 0$,

$$(x-y)^2 + \lambda(x-y)(Ax-Ay) + \lambda^2(Ax-Ay)^2 \geq (x-y)^2$$

or

$$(x-y)(Ax-Ay) \geq -\lambda(Ax-Ay)^2$$

for every $x,y \in \mathbb{R}$. Let $\lambda \downarrow 0$ to obtain

$$(x-y)(Ax-Ay) \geq 0.$$

This demonstrates that A is nondecreasing. Conversely, we can reverse our steps to prove A is accretive whenever it is nondecreasing. It should also be clear from the graph of such a function that the addition of λI to A ensures $A + \lambda I$ is one-to-one for every $\lambda > 0$. Hence $(I + \lambda A)^{-1}$ exists.

In order to better understand the nature of the accretive operators A which generate semigroups $T \in Q_\omega$, consider the case of an autonomous ordinary differential equation in $\mathbb{R}^d$,

(2.1) $$\dot{x} = f(x), \quad x(0) = x_0.$$

It is true (see the opening paragraph of Chapter IV or Hirsch and Smale [1], p. 175) that if f is Lipschitz, then Equation (2.1) possesses a unique solution $x(t) = T(t)x_0$, where $T = \{T(t): t \in \mathbb{R}^+\}$ forms a semigroup (indeed, a group). Though it is unnecessarily stringent, suppose for instance that f were globally ω-Lipschitz. Then the operator $-f + \omega I$ is accretive. Indeed, for $\lambda > 0$,

$$\begin{aligned}\|x-y + \lambda(-f(x)+\omega x+f(y)-\omega y)\| &= \|(1+\lambda\omega)(x-y) + \lambda(-f(x)+f(y))\| \\ &\geq \|(1+\lambda\omega)(x-y)\| - \|\lambda(f(x)-f(y))\| \\ &\geq (1+\lambda\omega)\|x-y\| - \lambda\omega\|x-y\| \\ &= \|x-y\|.\end{aligned}$$

In the scalar case the accretiveness of $-f + \omega I$ is readily seen. Because f can grow no faster than the rate ω, the addition of ωI to $-f$ ensures that $-f + \omega I$ is nondecreasing, hence accretive by the last remark.

Another property of f can be deduced. Even if the range of f is not all of $\mathbb{R}^d$, the range of $I + \lambda(-f)$ is all of $\mathbb{R}^d$ for all sufficiently small $\lambda > 0$. Indeed, for $\lambda\omega < 1$ and every $y \in \mathbb{R}^d$, there exists a unique $x \in \mathbb{R}^d$ with $x - \lambda f(x) = y$. To see this set

$$F(x) = y + \lambda f(x).$$

An easy computation shows F is $\lambda\omega$-Lipschitz. So for $\lambda\omega < 1$, the contraction mapping principle (Appendix A) shows that F has a unique fixed point. Consequently, $(I - \lambda f)(x) = y$ has a unique solution x. In particular, the operator $(I + \lambda(-f))^{-1}$ is well defined.

With these properties in mind we turn to the statement of the (nonlinear) semigroup generation theorem. Only an outline of its proof is given here; the details are postponed until Section 8. Some of the lemmas established there will be required in the following sections, though.

Theorem 2.4. Let $A: \mathscr{D}(A) \subset X \to X$ with $A + \omega I$ accretive for some $\omega \in \mathbb{R}$. If $\mathscr{R}(I + \lambda A) = X$ for all sufficiently small $\lambda > 0$, then for every $x \in \overline{\mathscr{D}(A)}$

$$T(t)x \overset{\text{def}}{=} \lim_{n\to\infty} (I + \tfrac{t}{n}A)^{-n}x \tag{2.2}$$

exists uniformly in t belonging to bounded subsets of $\mathbb{R}^+$. Moreover, $T = \{T(t): t \in \mathbb{R}^+\} \in Q_\omega$ is a strongly continuous semigroup of operators in X.

Outline of Proof: Define the operators $J_\lambda = (I + \lambda A)^{-1}$ for $\lambda\omega < 1$, and show that J_λ is $(1 - \lambda\omega)^{-1}$-Lipschitz on $\mathscr{D}(J_\lambda) = \mathscr{R}(I + \lambda A)$. Letting $x \in \mathscr{D}(A)$ and replacing λ by t/n ($t \in \mathbb{R}^+$ and $n \in \mathbb{N}$), we can prove that $\{J^n_{t/n}x\}_{n=1}^\infty$ is Cauchy. Thus $T(t)x = \lim_{n\to\infty} J^n_{t/n}x$ exists and extends to $x \in \overline{\mathscr{D}(A)}$ by continuity. Moreover,

$$\|J^n_{t/n}x - J^n_{t/n}y\| \le (1 - \tfrac{t}{n}\omega)^{-n}\|x-y\|$$

so,

$$\|T(t)x - T(t)y\| \le e^{\omega t}\|x-y\| .$$ □

We return to a discussion of accretive operators.

Lemma 2.5. If $A + \omega I$ is accretive for some $\omega \in \mathbb{R}$, then $(I + \lambda A)^{-1}$ exists for $\lambda\omega < 1$ and is $(1 - \lambda\omega)^{-1}$-Lipschitz on its domain.

Proof: As $A + \omega I$ is accretive, then

$$\|x-y + \lambda(Ax+\omega x-Ay-\omega y)\| \geq \|x-y\| .$$

So

$$\|(I+\lambda A)x - (I+\lambda A)y\| \geq (1-\lambda\omega)^{-1}\|x-y\| .$$

This establishes that the operator $I + \lambda A$ is one-to-one, so $(I + \lambda A)^{-1}$ exists. This yields the desired result. □

Definition 2.6. Suppose $A: \mathscr{D}(A) \subset X \to X$ is an accretive operator. The operator $J_\lambda \overset{\text{def}}{=} (I + \lambda A)^{-1}$ is well defined for all $\lambda > 0$ and is called the *resolvent* of A. Note that $\mathscr{D}(J_\lambda) = \mathscr{R}(I + \lambda A)$. If $\mathscr{D}(J_\lambda) = X$ for all sufficiently small $\lambda > 0$, then A is called *m-accretive*.

We return to Equation (1.1) and show that the operator A given by $Au = -\Delta\phi(u)$ is an m-accretive operator in the space $X = L^1(\Lambda; \mathbb{R})$. The choice of an L^1 space is appropriate in view of the mass interpretation of $\|u\|_1$.

Example 2.7. For simplicity we wil take the domain Λ to be the open interval $(0,1)$ and X to be the space $L^1([0,1];\mathbb{R})$. We will assume $\phi: \mathbb{R} \to \mathbb{R}$ is continuous, strictly increasing with $\phi(0) = 0$, $\phi(\mathbb{R}) = \mathbb{R}$, and ϕ^{-1} is Lipschitz. Let the symbol ' denote $\frac{d}{dx}$. Set

$$\mathscr{D}(A) = \{u \in C([0,1]; \mathbb{R}): u(0) = u(1) = 0,$$
$$\phi(u), \phi(u)' \in AC([0,1]; \mathbb{R})\}.$$

First we show that $Au = -\Delta\phi(u)$ is accretive. Let $u,v \in \mathscr{D}(A)$ and define

$$\Lambda_+ = \{x \in [0,1]: u(x) > v(x)\},$$

$$\Lambda_- = \{x \in [0,1]: u(x) < v(x)\}.$$

For $\lambda > 0$,

$$\int_0^1 |u-v + \lambda(Au-Av)|dx$$
$$\geq \int_{\Lambda_+} |u-v + \lambda(Au-Av)|dx + \int_{\Lambda_-} |u-v + \lambda(Au-Av)|dx$$
$$\geq \int_{\Lambda_+} [u-v + \lambda(Au-Av)]dx + \int_{\Lambda_-} [v-u + \lambda(Av-Au)]dx$$
$$= \int_0^1 |u-v|dx + \lambda\int_{\Lambda_+} (Au-Av)dx + \lambda\int_{\Lambda_-} (Av-Au)dx.$$

Let (a,b) be any component of Λ_+ and set $h = \phi(v) - \phi(u)$. Then $h(a) = h(b) = 0$, while $h(x) \leq 0$ for $x \in [a,b]$. Consequently, $h'(a) \leq 0$ and $h'(b) \geq 0$, so

$$\int_a^b (Au-Av)dx = \int_a^b h''(x)dx = h'(b) - h'(a) \geq 0.$$

In this fashion we find

$$\int_{\Lambda_+} (Au-Av)dx \geq 0, \quad \int_{\Lambda_-} (Av-Au)dx \geq 0.$$

We conclude that $\|u-v + \lambda(Au-Av)\|_1 \geq \|u-v\|_1$.

Next we establish that $\mathscr{R}(I + A) = X$. (The reader will see that there is no loss in generality in taking $\lambda = 1$.) Let $h \in X$. We must find $u \in \mathscr{D}(A)$ so that $u + Au = h$. Set $v = \phi(u)$, $\beta = \phi^{-1}$. We solve instead the differential equation $\beta(v)-v'' = h$ subject to $v(0) = v(1) = 0$. There are two cases.

A. β is bounded. Let $|\beta(\xi)| \leq M$ for all $\xi \in \mathbb{R}$. Define the operator $T: L^1([0,1];\mathbb{R}) \to L^1([0,1];\mathbb{R})$ by

$$Tv(x) = \int_0^1 g(x,y)[\beta(v(y))-h(y)]dy$$

where $g(x,y) = y(x-1)$ for $0 \leq y \leq x \leq 1$, and $g(x,y) = g(y,x)$ for $x,y \in [0,1]$. We can write $Tv(x)$ as

$$\int_0^x y(x-1)[\beta(v(y))-h(y)]dy + \int_x^1 x(y-1)[\beta(v(y))-h(y)]dy.$$

An easy computation shows $w \overset{def}{=} Tv$ satisfies $w'' = \beta(v)-h$, $w(0) = w(1) = 0$. Also,

$$|Tv(x)| \le \int_0^1 |g(x,y)||\beta(v(y))-h(y)|dy \le \|\beta(v)\|_1 + \|h\|_1$$

and

$$(Tv(x))' = \int_0^x y[\beta(v(y))-h(y)]dy + \int_x^1 (y-1)[\beta(v(y))-h(y)]dy$$

$$|(Tv(x))'| \le \int_0^1 |\beta(v(y))-h(y)|dy \le \|\beta(v)\|_1 + \|h\|_1.$$

Then $\|Tv\|_\infty \le M_1$, $\|(Tv)'\|_\infty \le M_1$, where $M_1 = M + \|h\|_1$. Moreover, T is continuous as we may write

$$|Tv_n(x) - Tv(x)| \le \int_0^1 |\beta(v_n(y))-\beta(v(y))|dy \le B\|v_n-v\|_1,$$

where B is the Lipschitz constant for $\beta = \phi^{-1}$. Hence $\|Tv_n-Tv\|_1 \le B\|v_n-v\|_1$. Thus T is a continuous mapping of X into

$$W = \{w \in X\colon w \in AC([0,1];\mathbb{R}) \text{ with } \|w\|_\infty \le M_1, \|w'\|_\infty \le M_1\}.$$

W is convex and compact according to Ascoli's theorem. Therefore T must have a fixed point v_0 by the Schauder fixed point theorem (see Appendix A). It follows that $u_0 = \beta(v_0)$ solves $u + Au = h$.

B. β is unbounded. Since A is accretive, if $\beta(v) \in \mathscr{D}(A)$ and $\beta(v) - v'' = h$, we can write

$$\|\beta(v)\|_1 = \|\beta(v)-0\|_1 \le \|\beta(v)-0+(v''-0'')\|_1 = \|h\|_1.$$

Moreover, $v(0) = v(1) = 0$ implies $v'(\xi) = 0$ for some $\xi \in (0,1)$. Thus,

$$|v'(x)| \le \int_\xi^x |v''(y)|dy \le \int_0^1 |\beta(v(y))-h(y)|dy \le 2\|h\|_1,$$

so

$$|v(x)| \le \int_0^x |v'(y)|dy \le \int_0^1 2\|h\|_1 dy = 2\|h\|_1, \quad x \in [0,1].$$

Set

$$\hat{\beta}(\xi) = \begin{cases} 2\|h\|_1, & \text{if } \beta(\xi) > 2\|h\|_1 \\ \beta(\xi), & \text{if } |\beta(\xi)| \le 2\|h\|_1 \\ -2\|h\|_1, & \text{if } \beta(\xi) < -2\|h\|_1. \end{cases}$$

Then $\hat{\beta}$ is bounded so there is a solution v_0 of $\hat{\beta}(v) - v'' = h$ with $v(0) = v(1) = 0$. Also we have $\|v\|_\infty \le 2\|h\|_1$. Consequently, $\hat{\beta}(v) = \beta(v)$, so $u_0 = \beta(v_0)$ solves $u + Au = h$.

Theorem 2.4 demonstrates how an accretive operator generates a semigroup $T \in Q_\omega$. Conversely, given a semigroup T we can produce an operator A_T.

Definition 2.8. Given a semigroup $T \in Q_\omega$ we define its *(infinitesimal) generator* to be the operator $A_T: \mathscr{D}(A) \subset X \to X$ by

$$A_T x = \lim_{t \downarrow 0} \frac{x - T(t)x}{t}$$

whenever this limit exists. In this event, we say $x \in \mathscr{D}(A_T)$.

If A generates T it need not be true that $A_T = A$. Indeed, there are cases when $\mathscr{D}(A_T) = \emptyset$. (See Notes and Comments, Section 9.) In the event A is linear though, we can characterize those operators A which generate quasi-contracting semigroups. This is accomplished via the Hille-Yoshida-Phillips generation theorem. (For a proof, see

Barbu [1], p. 26.)

Theorem 2.9. Suppose $A: \mathcal{D}(A) \subset X \to X$ is a densely defined linear operator. Then A is the generator of a uniquely determined semigroup $T \in Q_\omega$ for some $\omega \in \mathbb{R}$ if and only if

(i) A is closed,

(ii) J_λ is well defined with $\mathcal{D}(J_\lambda) = X$ and

(iii) $\|J_\lambda^n\| \le (1 - \lambda\omega)^{-n}$

for every positive λ satisfying $\lambda\omega < 1$.

Corollary 2.10. If T is a semigroup of linear operators, then for each $x \in \mathcal{D}(A_T)$, the mapping $t \to T(t)x$ is differentiable with $\frac{d}{dt}T(t)x = -A_T T(t)x = -T(t)A_T x$.

In general one does not obtain differentiability of nonlinear semigroups $T \in Q_\omega$. This is especially unfortunate in view of the fact that $T(\cdot)u_0$, $u_0 \in \overline{\mathcal{D}(A)}$, given by Equation (2.2) appears to be the obvious candidate for a solution of the evolution equation (also called the Cauchy problem),

$$\frac{du}{dt} + Au = 0, \quad u(0) = u_0. \tag{2.3}$$

Example 2.11. Take for X the space $C([0,1]:\mathbb{R})$ and let $\mathcal{K} = \{f \in X: 0 \le f(x) \le x \text{ for } 0 \le x \le 1\}$. It is easy to check that $\mathcal{K}$ is a closed convex subset of X and that $T(t): \mathcal{K} \to \mathcal{K}$ defined by

$$(T(t)f)(x) = \min(t+f(x),x)$$

is a contraction semigroup on $\mathcal{K}$. We compute the derivative of $T(t)f$ at $t = 0$:

$$\frac{f(x)-(T(t)f)(x)}{t} = \begin{cases} -1, & t + f(x) < x \\ \dfrac{f(x)-x}{t}, & t + f(x) \ge x. \end{cases}$$

Thus it is apparent that $T(t)f$ is differentiable at $t = 0$ if and only if $f(x) = x$ for all $x \in [0,1]$.

Definition 2.12. A function $u: \mathbb{R}^+ \to X$ is called a *strong solution* of Equation (2.3) provided

(i) u is continuous,

(ii) there exists $v \in L^1_{loc}(\mathbb{R}^+;X)$ so that

$$u(t) - u(s) = \int_s^t v(\tau)d\tau, \quad 0 \le s \le t < \infty,$$

(iii) u satisfies Equation (2.3) a.e. in $t \in \mathbb{R}^+$.

Fortunately, if Equation (2.3) has a strong solution then (under an additional mild restriction on A), it is the semigroup generated by A. The proof is deferred to Section 5 where strong solutions are characterized by Theorem 5.2.

Theorem 2.13. Suppose $A: \mathcal{D}(A) \subset X \to X$ is a closed operator with $A + \omega I$ accretive for some $\omega \in \mathbb{R}$ and $\mathcal{R}(I + \lambda A) = X$ for all sufficiently small $\lambda > 0$. If u is a strong solution of Equation (2.3) for $u_0 \in \mathcal{D}(A)$, then $u(t) = T(t)u_0$ for every $t \in \mathbb{R}^+$, where $T \in Q_\omega$ is generated by A.

In the event A is linear we have the following existence and uniqueness result for strong solutions of Equation (2.3). The proof is an easy consequence of Theorem 2.9 and Corollary 2.10.

Corollary 2.14. Suppose $A: \mathcal{D}(A) \subset X \to X$ is a closed, densely defined linear operator in a real Banach space X. Furthermore suppose

(i) $\mathcal{D}(J_\lambda) = X$, ($J_\lambda$ is the resolvent of A)

(ii) $\|J_\lambda^n\| \le (1 - \lambda\omega)^{-n}$ for some $\omega \in \mathbb{R}$ and all positive λ satisfying $\lambda\omega < 1$.

Then Equation (2.3) has the unique strong solution $u(t) = T(t)u_0$, where T is the semigroup generated by A.

There are other types of solutions of Equation (2.3) which do not require strong differentiability. For example, the derivative $\frac{du}{dt}$ may be taken in the weak topology of X. Another approach entirely is to approximate T by a (strongly) differentiable semigroup. This result will be required in Chapter VI for differential delay equations. We state it here and only outline the proof; the details can be found in Section 8. But first we require a technical lemma which will be used frequently.

<u>Lemma 2.15</u>. Let $\lambda \in \mathbb{R}^+$ so that $\lambda\omega < 1$. Then

(i) $\|J_\lambda x - x\| \le \lambda(1-\lambda\omega)^{-1}\|Ax\|$, $x \in \mathcal{D}(J_\lambda) \cap \mathcal{D}(A)$,

(ii) $\|J_\lambda^n x - x\| \le n(1-\lambda\omega)^{-n+1}\|J_\lambda x - x\|$, $x \in \mathcal{D}(J_\lambda^n)$, $n \in \mathbb{N}$,

(iii) $J_\lambda x = J_\mu(\frac{\mu}{\lambda}x + \frac{\lambda-\mu}{\lambda} J_\lambda x)$, $\lambda,\mu > 0$, $x \in \mathcal{D}(J_\lambda)$,

(iv) $(1-\lambda\omega)\|AJ_\lambda x\| \le (1-\mu\omega)\|AJ_\mu x\|$, $0 < \mu \le \lambda$, $x \in \mathcal{D}(J_\lambda) \cap \mathcal{D}(A)$.

<u>Proof</u>: Use Lemma 2.5 to obtain (i); namely,

$$\|J_\lambda x - x\| = \|J_\lambda x - J_\lambda(I + \lambda A)x\| \le \lambda(1 - \lambda\omega)^{-1}\|Ax\|.$$

To get (ii) consider

$$\|J_\lambda^n x - x\| = \|\sum_{i=0}^{n-1} (J_\lambda^{n-i}x - J_\lambda^{n-(i+1)}x)\|$$

$$\le \sum_{i=0}^{n-1} (1-\lambda\omega)^{-n+(i+1)}\|J_\lambda x - x\| \le n(1-\lambda\omega)^{-n+1}\|J_\lambda x - x\|.$$

For (iii) let $y = J_\lambda x$. Then

$$\frac{\mu}{\lambda}x + \frac{\lambda-\mu}{\lambda} J_\lambda x = \frac{\mu}{\lambda}(I + \lambda A)y + \frac{\lambda-\mu}{\lambda}y = (I + \mu A)y.$$

So

$$J_\mu(\frac{\mu}{\lambda}x + \frac{\lambda-\mu}{\lambda} J_\lambda x) = y = J_\lambda x.$$

Now for (iv). Since $AJ_\lambda x = J_\lambda Ax = \lambda^{-1}(I-J_\lambda)x$,

$$\|AJ_\lambda x\| = \lambda^{-1}\|x-J_\lambda x\| \leq \lambda^{-1}(\|x-J_\mu x\| + \|J_\mu x-J_\lambda x\|$$

$$\leq \frac{\mu}{\lambda} \|AJ_\mu x\| + \frac{1}{\lambda} \|J_\mu x - J_\mu(\frac{\mu}{\lambda}x + \frac{\lambda-\mu}{\lambda} J_\lambda x)\|$$

$$\leq \frac{\mu}{\lambda} \|AJ_\mu x\| + (\lambda-\mu)\lambda^{-1}(1-\mu\omega)^{-1} \|AJ_\lambda x\|.$$

Rearrange to obtain the desired inequality. □

Theorem 2.16. Suppose A is densely defined in X and $A + \omega I$ is m-accretive for some $\omega \in \mathbb{R}$. Let $T \in Q_\omega$ be the semigroup generated by A and define the operators $A_\lambda = \lambda^{-1}(I - J_\lambda)$ for all $\lambda > 0$ satisfying $\lambda\omega < 1$. Then there exist strongly continuous semigroups $T_\lambda = T_\lambda(t): t \in \mathbb{R}^+\}$ which satisfy for every $x \in X$

(i) $T_\lambda(\cdot)x: \mathbb{R}^+ \to X$ is differentiable

(ii) $\frac{d}{dt} T_\lambda(t)x + A_\lambda T_\lambda(t)x = 0$

(iii) $\lim_{\lambda\downarrow 0} T_\lambda(t)x = T(t)x.$

Proof: $\mathcal{D}(A_\lambda) = X$ by the definition of A_λ. Note that $A_\lambda = AJ_\lambda = J_\lambda A$. Also if $x,y \in X$, then $\|A_\lambda x - A_\lambda y\| = \lambda^{-1}\|(x-J_\lambda x) - (y-J_\lambda y)\| \leq (2-\lambda\omega)\lambda^{-1}(1-\lambda\omega)^{-1}\|x-y\|$. Thus A_λ is Lipschitz continuous on X and for $\lambda\omega < 1$, $A_\lambda + \omega' I$ is accretive where $\omega' = \omega(1-\lambda\omega)^{-1}$. Therefore A_λ generates a semigroup $T_\lambda \in Q_{\omega'}$.

Next we show that T_λ is weakly differentiable at $t = 0$. Let $J_{\lambda,\mu}$ denote $(I+ \mu A_\lambda)^{-1}$. If $x \in X$, then

$$\frac{T_\lambda(t)x-x}{t} = \lim_{n\to\infty} \frac{J^n_{\lambda,t/n}x-x}{t}$$

$$= \lim_{n\to\infty} \frac{1}{n} \sum_{i=0}^{n-1} \frac{J_{\lambda,t/n}(J^i_{\lambda,t/n}x)-J^i_{\lambda,t/n}x}{t/n}$$

$$= -\lim_{n\to\infty} \frac{1}{n} \sum_{i=0}^{n-1} A_\lambda J^i_{\lambda,t/n}x.$$

Let $0 \leq i \leq n-1$. Then $\|J^i_{\lambda,t/n}x-x\| \leq t(1-\frac{t}{n}\omega')^i\|Ax\|$ from (ii) of Lemma 2.15. Let U be any weak convex neighborhood of zero in X. For sufficiently small t we have $A_\lambda J^i_{\lambda,t/n}x - A_\lambda x \in U$ as A_λ is continuous from X to X_w (X_w is the set X with the weak topology). Taking the strong limit as $n \to \infty$ we have

$$\frac{T_\lambda(t)x-x}{t} + A_\lambda x \in U$$

for t sufficiently small. Now take the weak limit as $t \downarrow 0$. Thus

$$w - \lim_{t\downarrow 0} \frac{T_\lambda(t)x-x}{t} + A_\lambda x = 0.$$

This implies that $T_\lambda(\cdot)x$ is weakly differentiable at $t = 0$ (from the right). This extends to weak differentiability at all $t \in \mathbb{R}^+$ via property (ii) of Definition 2.1.

We now prove that $T_\lambda(\cdot)x$ is strongly differentiable. For any $t \in \mathbb{R}^+$ and $h \in \mathbb{R}$ so that $t + h \in \mathbb{R}^+$ observe that

$$\|T_\lambda(t+h)x - T_\lambda(t)x + hA_\lambda T_\lambda(t)x\|$$

$$= \|-\int_t^{t+h} A_\lambda T_\lambda(s)x\,ds + \int_t^{t+h} A_\lambda T_\lambda(t)x\,ds\|$$

$$\leq \int_t^{t+h} \|A_\lambda T_\lambda(s)x - A_\lambda T_\lambda(t)x\|ds \to 0 \quad \text{as} \quad h \to 0.$$

The integral used here is obviously a weak limit in X.

The proof of the convergence of T_λ to T is rather involved. It is accomplished by a sequence of estimates. See Section 8 for the details. □

3. The Generalized Domain for Accretive Operators

As pointed out in the last section the semigroup $T \in Q_\omega$ generated by A need not be differentiable (in t). We shall here find a domain which is invariant under each $T(t)$ and on which $T(t)x$ is Lipschitz continuous in t. This domain will be needed in succeeding sections on asymptotic behavior of $T(t)x$.

We begin with a discussion of a generalized inner product for Banach spaces. As usual, X denotes a real Banach space, X^* its dual, and (x,x^*) the value of $x^* \in X^*$ at $x \in X$.

Definition 3.1. For every $x \in X$ let

(3.1) $F(x) = \{x^* \in X^*: (x,x^*) = \|x\|^2 = \|x^*\|^2\}$, the *dual map*.

(3.2) $\langle x,y\rangle_s = \max\{(x,y^*): y^* \in F(y)\}$, the *semi-scalar product*.

Both $F(x)$ and $\langle x,y\rangle_s$ are well defined. The Hahn-Banach theorem guarantees that $F(x) \neq \emptyset$. Moreover $F(y)$ is w*-compact so there exists $y^* \in F(y)$ with $\langle x,y\rangle_s = (x,y^*)$.

Lemma 3.2. Suppose $x,y,z \in X$, $\alpha \in \mathbb{R}$. Then

(i) $\langle x+\alpha y,y\rangle_s = \langle x,y\rangle_s + \alpha\|y\|^2$

(ii) $\langle x+z,y\rangle_s \leq \langle x,y\rangle_s + \|z\|\,\|y\|$

(iii) $\langle\cdot,\cdot\rangle_s: X \times X \to \mathbb{R}$ is upper semicontinuous.

Proof: (i) and (ii) are left as an exercise. We prove (iii). So let $x_n \to x$, $y_n \to y$. We show that $\limsup_{n\to\infty} \langle x_n, y_n\rangle_s \le \langle x,y\rangle_s$. Choose $y_n^* \in F(y_n)$ so that $\langle x_n, y_n\rangle_s = (x_n, y_n^*)$. We can assume that $\lim_{n\to\infty} \langle x_n, y_n\rangle_s$ exists. Suppose (by choosing a subsequence if necessary) that $w^*\text{-}\lim y_n^* = y^*$. Then $\|y^*\| \le \liminf_{n\to\infty} \|y_n^*\| = \liminf_{n\to\infty} \|y_n\| = \|y\|$. As $|(y_n - y, y_n^*)| \le \|y_n - y\| \, \|y_n^*\| = \|y_n - y\| \, \|y_n\| \to 0$, then $\|y\|^2 = \lim_{n\to\infty}(y_n, y_n^*) = \lim_{n\to\infty}(y, y_n^*) = (y, y^*)$. Thus $y^* \in F(y)$ so $\lim_{n\to\infty} \langle x_n, y_n\rangle_s = \lim_{n\to\infty} (x_n, y_n^*) = (x, y^*) \le \langle x, y\rangle_s$. □

Lemma 3.3. Let $x,y \in X$. Then $\|x\| \le \|x+\alpha y\|$ for all $\alpha > 0$ if and only if there exists $x^* \in F(x)$ such that $(y,x^*) \ge 0$.

Proof: Assume $x \neq 0$; otherwise the statement is trivial. Suppose $(y,x^*) \ge 0$ for some $x^* \in F(x)$. Then $\|x\|^2 = (x,x^*) \le (x+\alpha y, x^*) \le \|x+\alpha y\| \, \|x^*\| = \|x+\alpha y\| \, \|x\|$ for all $\alpha > 0$. Thus $\|x\| \le \|x+\alpha y\|$.

Conversely, let $\|x\| \le \|x+\alpha y\|$ for all $\alpha > 0$. Let $v_\alpha^* \in F(x+\alpha y)$ and set $u_\alpha^* = v_\alpha^*/\|v_\alpha\|$. Then $\|x\| \le \|x+\alpha y\| = (x+\alpha y, u_\alpha^*) = (x, u_\alpha^*) + \alpha(y, u_\alpha^*) \le \|x\| + \alpha(y, u_\alpha^*)$. Hence $\liminf_{\alpha\downarrow 0} (x, u_\alpha^*) \ge \|x\|$, and $(y, u_\alpha^*) \ge 0$. Since the closed unit ball of X^* is w^*-compact, then the net $\{u_\alpha^*\}$ has a w^*-limit point u^*, $\|u^*\| \le 1$. Thus $(x,u^*) \ge \|x\|$, and $(y,u^*) \ge 0$. This implies $\|u^*\| = 1$ and $(x,u^*) = \|x\|$. Then $x^* \overset{\text{def}}{=} \|x\| u^* \in F(x)$ and satisfies $(y,x^*) \ge 0$. □

Lemma 3.4. $A + \omega I$ is accretive if and only if $\langle Ax - Ay, x-y\rangle_s \ge -\omega\|x-y\|^2$ for every $x,y \in \mathscr{D}(A)$.

Proof: A is accretive if and only if $\|x-y\| \le \|x-y + \alpha(Ax-Ay)\|$ for all $\alpha > 0$, $x,y \in \mathscr{D}(A)$. Using Lemma 3.3 and Definition 3.1, this is equivalent to $\langle Ax-Ay, x-y\rangle_s \ge 0$.

Replace A by $A + \omega I$ and use Lemma 3.2 to complete the proof. □

Remark 3.5. The last lemma allows us a nice geometrical interpretation of an accretive operator. In the event $\omega = 0$, A is accretive if and only if $\langle Ax-Ay,x-y\rangle_s \geq 0$ for every $x,y \in \mathscr{D}(A)$. Thus, if X were a Hilbert space, then the angle between the vectors $x-y$ and $Ax-Ay$ is between $-\frac{1}{2}\pi$ and $+\frac{1}{2}\pi$. If $X = \mathbb{R}$, the condition merely states that A is monotonic.

Lemma 3.6. Suppose A is densely defined with $A + \omega I$ m-accretive for some $\omega \in \mathbb{R}$. If T is the semigroup generated by A,

$$(3.3)\qquad \sup_{\zeta^*\in F(x-x_0)} \limsup_{t\downarrow 0} \left(\frac{T(t)x-x}{t},\zeta^*\right) \leq \langle Ax_0,x_0-x\rangle_s + \omega\|x_0-x\|^2$$

for every $x \in X$ and $x_0 \in \mathscr{D}(A)$.

Proof: As $A + \omega I$ is accretive, there exists $\eta^* \in F(x_0-J_\lambda^k x)$ so that $(Ax_0-AJ_\lambda^k x,\eta^*) \geq \omega\|x_0-J_\lambda^k x\|^2$. Then

$$\begin{aligned}
(AJ_\lambda^k x,\eta^*) &= \lambda^{-1}(J_\lambda^{k-1}x - J_\lambda^k x,\eta^*) \\
&= \lambda^{-1}([x_0-J_\lambda^k x] - [x_0-J_\lambda^{k-1}x],\eta^*) \\
&= \lambda^{-1}[\|x_0-J_\lambda^k x\|^2 - (x_0-J_\lambda^{k-1}x,\eta^*)] \\
&\geq \lambda^{-1}[\|x_0-J_\lambda^k x\|^2 - \|x_0-J_\lambda^{k-1}x\|\,\|x_0-J_\lambda^k x\|] \\
&\geq (2\lambda)^{-1}[\|x_0-J_\lambda^k x\|^2 - \|x_0-J_\lambda^{k-1}x\|^2].
\end{aligned}$$

Thus

$$<Ax_0+\omega[x_0-J_\lambda^k x],\ x_0-J_\lambda^k x>_s = (Ax_0,\eta^*) + \omega\|x_0-J_\lambda^k x\|^2$$

$$\geq (AJ_\lambda^k x,\eta^*) \geq (2\lambda)^{-1}[\|x_0-J_\lambda^k x\|^2 - \|x_0-J_\lambda^{k-1}x\|^2].$$

Sum this expression over k, $1 \leq k \leq [t/\lambda]$. Then

$$2\lambda \sum_{k=1}^{[t/\lambda]} <Ax_0+\omega[x_0-J_\lambda^k x],\ x_0-J_\lambda^k x>_s \geq \|x_0-J_\lambda^{[t/\lambda]}x\|^2 - \|x_0-x\|^2.$$

Let $f_\lambda(s) = <Ax_0+\omega[x_0+J_\lambda^k x],\ x_0-J_\lambda^k x>_s$ for $k\lambda \leq s < (k+1)\lambda$. Then

$$2\int_0^{[t/\lambda]\lambda} f_\lambda(s)ds \geq \|x_0-J_\lambda^{[t/\lambda]}x\|^2 - \|x_0-x\|^2.$$

Lemma 2.15(ii) shows that f_λ is bounded as $\lambda \downarrow 0$. As f_λ is also integrable we can apply the Lebesgue convergence theorem to obtain

$$2\int_0^t <Ax_0+\omega[x_0-T(s)x],\ x_0-T(s)x>_s ds \geq \|x_0-T(t)x\|^2 - \|x_0-x\|^2$$

$$\geq 2(T(t)x-x,\zeta^*) \quad \text{for each} \quad \zeta^* \in F(x-x_0).$$

The map $t \to <Ax_0 + \omega[x_0-T(t)x],\ x_0-T(s)x>_s$ is upper semicontinuous in view of Lemma 3.2(iii). For each $\varepsilon > 0$ there exists $\delta > 0$ so that

$$<Ax_0+\omega[x_0-T(s)x],\ x_0-T(s)x>_s \leq <Ax_0+\omega[x_0-x],\ x_0-x>_s + \varepsilon$$

for $0 \leq s < \delta$. So if $0 < t < \delta$, then

$$(\frac{T(t)x-x}{t},\zeta^*) \leq <Ax_0,x_0-x>_s + \omega\|x_0-x\|^2 + \varepsilon.$$

As $\varepsilon > 0$ was arbitrary then the inequality (3.3) holds. □

We have chosen to assume $A + \omega I$ is m-accretive and $\overline{\mathcal{D}(A)} = X$. This simplifies the analysis without any loss in generality. To just assume accretiveness of $A + \omega I$, we need only keep track of the various domains of J_λ^k, A_λ, and A.

The special domain of $T(t)$ referred to earlier is now defined.

Definition 3.7. Suppose A is densely defined with $A + \omega I$ m-accretive for some $\omega \in \mathbb{R}$. Set

$$|Ax| = \lim_{\lambda \downarrow 0} \|A_\lambda x\| \quad \text{for } x \in X,$$
$$\hat{\mathcal{D}}(A) = \{x \in X: |Ax| < \infty\}.$$

$\hat{\mathcal{D}}(A)$ is called the *generalized domain* of A.

$|Ax|$ is well defined in view of Lemma 2.15(iv). Henceforth throughout this section we will always assume A is densely defined with $A + \omega I$ m-accretive for some $\omega \in \mathbb{R}$. The proof of the next lemma is straightforward and is omitted.

Lemma 3.8.

(i) $|Ax| \le \|Ax\|$ for all $x \in \mathcal{D}(A)$,

(ii) $\mathcal{D}(A) \subset \hat{\mathcal{D}}(A)$,

(iii) $|Ax| = \sup_{0<\lambda<\omega^{-1}} (1-\lambda\omega)\,\|A_\lambda x\|$, for all $x \in X$.

Lemma 3.9. Suppose A generates $T \in Q_\omega$. Let

$$H(x) \overset{\text{def}}{=} \liminf_{s \downarrow 0} \frac{T(s)x-x}{s}, \quad x \in X. \tag{3.4}$$

Then $H(x) = |Ax|$ for every $x \in X$.

Proof: We first show $H(x) \le |Ax|$. For $t > 0$,

$$\|T(t)x-x\| = \lim_{n\to\infty} \|J^n_{t/n}x-x\| \le \limsup_{n\to\infty} \sum_{k=1}^{n} \|J^n_{t/n}x-J^{k-1}_{t/n}x\|$$
$$\le \limsup_{n\to\infty} \sum_{k=1}^{n} (1-\tfrac{t}{n}\omega)^{-k+1}\|J_{t/n}x-x\|,$$

where the last inequality follows from Lemma 2.5. Then by the definition of A_λ,

$$||T(t)x-x|| \leq t \limsup_{n\to\infty} \frac{1}{n} \sum_{k=1}^{n} (1-\frac{t}{n}\omega)^{-k+1} ||A_{t/n}x||$$

$$\leq t|Ax| \lim_{n\to\infty} \frac{1}{n} \sum_{k=1}^{n} (1-\frac{t}{n}\omega)^{-k}$$

from Lemma 3.8 (iii). Therefore

$$||T(t)x-x|| \leq t \frac{e^{\omega t}-1}{\omega t}|Ax|,$$

where we set $(e^{\omega t}-1)/\omega = t$ if $\omega = 0$.

Now we show $|Ax| \leq H(x)$. Let $\zeta^* \in F(x-x_0)$ for any $x \in X$, $x_0 \in \mathscr{D}(A)$. Using Lemma 3.6 we get

$$-H(x)||x-x_0|| \leq \langle Ax_0, x_0-x\rangle_s + \omega||x_0-x||^2.$$

In particular choose $x_0 = J_\lambda x$, and so

$$-H(x)||A_\lambda x|| \leq ||A_\lambda x||^2 + \lambda\omega||A_\lambda x||^2.$$

If $\lambda \downarrow 0$, then $-|Ax| \leq H(x)$. □

The next theorem is useful. It characterizes the smoothness of the maps $T(\cdot)x: \mathbb{R}^+ \to X$ in terms of the generalized domain $\hat{\mathscr{D}}(A)$.

<u>Theorem 3.10</u>. Suppose A is densely defined with $A + \omega I$ m-accretive for some $\omega \in \mathbb{R}$. Then $T(\cdot)x: \mathbb{R}^+ \to X$ is uniformly Lipschitz continuous on bounded subsets of $\mathbb{R}^+$ if and only if $x \in \hat{\mathscr{D}}(A)$. Moreover $\hat{\mathscr{D}}(A)$ is positively invariant under $T(t)$.

<u>Proof</u>: We establish for every $x \in X$ and $t,h \in \mathbb{R}^+$ that

$$(3.5) \qquad ||T(t+s)x - T(t)x|| \leq e^{\omega t}[\frac{e^{\omega s}-1}{\omega}]H(x)$$

where $H(x)$ is defined by Equation (3.4). As $T \in Q_\omega$, we need only prove that Inequality (3.5) holds when $t = 0$.

Furthermore, we need only establish this when $H(x) < \infty$. When $H(x) = \infty$, Equation (3.4) tells us that $T(\cdot)x$ cannot be Lipschitz continuous.

Choose any number $M > H(x)$, and let $\{t_k\} \subset \mathbb{R}^+$, $t_k \downarrow 0$ so that $\|T(t_k)x-x\| \le Mt_k$. Let $\{n_k\}$ be positive integers with $n_k t_k \to s$. Then

$$\begin{aligned}\|T(s)x-x\| &= \lim_{k\to\infty} \|T(n_k t_k)x-x\| \\ &\le \limsup_{k\to\infty} \sum_{j=1}^{n_k} \|T(jt_k)x - T((j-1)t_k)x\| \\ &\le \lim_{k\to\infty} M \sum_{j=1}^{n_k} e^{\omega(j-1)t_k} t_k = M\,\frac{e^{\omega s}-1}{\omega}.\end{aligned}$$

As $M > H(x)$ was arbitrary, the proof of Inequality (3.5) is concluded. (Note this is a refinement of the Inequality (8.4).)

To see that $\hat{\mathscr{D}}(A)$ is positively invariant, let $x \in \hat{\mathscr{D}}(A)$. Then

$$|AT(t)x| = H(T(t)x) \le e^{\omega t}H(x) = e^{\omega t}|Ax|,$$

which is finite as $|Ax|$ is finite. □

Corollary 3.11. Assume the hypotheses of Theorem 3.10. If $x \in \hat{\mathscr{D}}(A)$ then $|AT(t)x| \le e^{\omega t}|Ax|$ for every $t \in \mathbb{R}^+$.

4. Precompactness of Positive Orbits

In order to establish an invariance principle for semi-dynamical systems in Banach spaces, one requires the existence of positive limit sets. This obtains when positive orbits are precompact. In applications the boundedness of positive orbits is relatively easy to demonstrate. It is generally more difficult to show when they are precompact.

To motivate the general approach we take in this section, we indicate how to formulate the classical (undamped) wave equation as a differential equation in a Hilbert space and show the corresponding positive orbits are precompact. In particular, this tells us that even in the absence of any damping mechanism in the wave equation, there are limiting solutions. First we make the obvious, yet significant remark.

Remark 4.1. Let T be a quasi-contracting semigroup on a subset $\mathscr{C}$ of a real Banach space X so that $T(t)\mathscr{C} \subset \mathscr{C}$ for all $t \in \mathbb{R}^+$. Then the map $\pi: \mathscr{C} \times \mathbb{R}^+ \to \mathscr{C}$ given by $\pi(x,t) = T(t)x$ satisfies the axioms for a semidynamical system on $\mathscr{C}$. Henceforth we shall dispense with the π-notation and refer to the semidynamical system as the pair $(\mathscr{C},T)$. In the event T is generated by an operator A for which $A + \omega I$ is m-accretive, then $(\hat{\mathscr{D}}(A),T)$ is a semidynamical system.

Example 4.2. We consider the wave equation in a single space variable. (The results for higher dimensions are the same; the analysis, though, requires the use of distributional derivatives.) This may represent the motion of a vibrating string. If $u(x,t)$ denotes the vertical displacement at time t of the point of the string with abscissa x, then u satisfies the partial differential equation

$$\frac{\partial^2 u}{\partial t^2} = \frac{\partial^2 u}{\partial x^2}, \quad 0 \leq x \leq 1, \quad t > 0, \tag{4.1}$$

with boundary conditions

$$u(0,t) = u(1,t) = 0, \text{ for } t > 0,$$

and initial values

$$u(x,0) = u_0(x), \quad 0 \le x \le 1$$

$$\frac{\partial u}{\partial t}(x,0) = v_0(x), \; 0 \le x \le 1.$$

The boundary conditions arise from the assumption that the endpoints of the string are fixed. Moreover, the string is subject to both an initial displacement $u_0(x)$ and an initial velocity $v_0(x)$. The solution of Equation (4.1), subject to the given boundary conditions and initial values admits a Fourier sine series

$$u(x,t) = \sum_{n=1}^{\infty} c_n(t) \sin nx,$$

where $c_n(t)$ satisfies the ordinary differential equation $\ddot{c}_n + n^2 c_n = 0$. If both $u_0(x)$ and $v_0(x)$ admit Fourier sine series, then both $c_n(0)$ and $\dot{c}_n(0)$ can be determined so that in turn, $c_n(t)$ is uniquely determined (cf. John [1], p. 42). All this suggests the space $L^2 \stackrel{\text{def}}{=} L^2([0,1]; \mathbb{R})$ as a candidate for solutions $u(\cdot,t)$.

Rewrite Equation (4.1) as a first order system

$$\frac{\partial u}{\partial t} = v$$

$$\frac{\partial v}{\partial t} = \frac{\partial^2 u}{\partial x^2}.$$

Interpret the unknown function u as the map $t \mapsto u(\cdot,t)$ from $\mathbb{R}^+$ to L^2. When considered as an element of L^2 we will write u or $u(t)$ for $u(\cdot,t)$ and $\frac{du}{dt}$ or just $\dot{u}$ for $\frac{\partial u(\cdot,t)}{\partial t}$. Similarly, we write $\frac{du}{dx}$ for $\frac{\partial u(\cdot,t)}{\partial x}$. Finally, let Δ denote the (Laplacian) operator $\frac{d^2}{dx^2}$. Then the first order system can be written as

$$\text{(4.2)} \qquad \frac{du}{dt} = v, \qquad \frac{dv}{dt} = \Delta u$$

with boundary conditions $u = 0$ at $x = 0, 1$, and initial values

$$u(0) = u_0, \quad v(0) = v_0.$$

Motivated by the choice of L^2 as the solution space for Equation (4.1), we choose $X = H_0^1 \times L^2$ as the state space for the system (4.2). Here, $H_0^1 \overset{\text{def}}{=} H_0^1([0,1]; \mathbb{R})$ consists of all those functions $u \in L^2$ which are absolutely continuous with $u' \in L^2$ and satisfy the boundary conditions $u(0) = u(1) = 0$. With the inner product

$$\langle (u,v),(\tilde{u},\tilde{v})\rangle = \int_0^1 [u'\tilde{u}' + v\tilde{v}]\,dx,$$

X is a Hilbert space. Finally we can express the System (4.2) as a first order differential equation in X. Set

$$A = \begin{bmatrix} 0 & -I \\ -\Delta & 0 \end{bmatrix},$$

where I is the identity operator. For $w = (u,v) \in X$, write the System (4.2) as

$$\frac{dw}{dt} + Aw = 0.$$

Because the boundary conditions have been incorporated into the space H_0^1, we need only specify the initial values, say $w(0) = w_0 \overset{\text{def}}{=} (u_0,v_0)$. Now w_0 must belong to the domain $\mathcal{D}(A)$ which is characterized by

$$\begin{aligned} \mathcal{D}(A) &= \{w \in X\colon Aw \in X \text{ and } w \text{ satisfies boundary conditions}\} \\ &= \{(u,v) \in H_0^1 \times L^2\colon (-v,-u'') \in H_0^1 \times L^2\} \\ &= (H_0^1 \cap H^2) \times H_0^1. \end{aligned}$$

Here H^2 consists of all those functions $u \in L^2$ which have absolutely continuous derivative u' with $u'' \in L^2$.

Now A is a densely defined m-accretive linear operator. As such it generates a contraction semigroup T in X according to Theorem 2.4. Indeed, the Sobolev Embedding Theorem (see Appendix A) ensures that $\overset{\bullet}{H^2}$ is dense in H^1 which, in turn, is dense in L^2. Consequently, $\mathscr{D}(A)$ is dense in X. A calculation shows that for $w = (u,v) \in \mathscr{D}(A)$

$$\langle Aw,w\rangle = \langle(-v,-u''),(u,v)\rangle = \int_0^1 [-u'v' - u''v]dx = 0$$

using integration by parts. This shows A is accretive. That A is m-accretive follows readily from the technique used in Example 2.7.

It is clear that A is linear. It can be shown (with the aid of Corollary 2.14) that A is the infinitesimal generator of T and $w(t) = T(t)w_0$ is the unique strong solution of Equation (4.2) which satisfies $w(0) = w_0 \in \mathscr{D}(A)$. This fact will be established in Section 5 by another argument. In the mean time we are only interested in showing the positive orbit $\gamma^+(w_0)$ is precompact. To this end we will prove that A^{-1} is a compact operator in X. First observe that

$$A^{-1} = \begin{bmatrix} 0 & -\Delta^{-1} \\ -I & 0 \end{bmatrix}.$$

Next we show $\Delta\colon \mathscr{D}(\Delta) = H_0^1 \times H^2 \to L^2$ has a compact inverse. Indeed, it is readily verified that $\Delta^{-1}\colon L^2 \to L^2$ is given by

$$(\Delta^{-1}f)(x) = \int_0^1 K(\xi,x)f(x)dx$$

where

$$K(\xi,x) = \begin{cases} (1-x)\xi, & 0 \leq \xi \leq x \leq 1 \\ (1-\xi)x, & 0 \leq x \leq \xi \leq 1. \end{cases}$$

Since $K(\xi,x) = K(x,\xi)$, then Δ^{-1} is self adjoint. Next we show that Δ^{-1} is a compact operator. So let $\{f_n\}$ be a bounded sequence in L^2. We may assume (by choosing a subsequence if necessary) that $\{f_n\}$ converges weakly to some $f \in L^2$. We will demonstrate that $\{\Delta^{-1}f_n\}$ converges to $\Delta^{-1}f$ in L^2. Now Δ^{-1} is a bounded linear operator in L^2 of norm one. Indeed, for any $h \in L^2$

$$\|\Delta^{-1}(h)\|_2^2 = \int_0^1\left[\int_0^1 K(\xi,x)h(x)dx\right]^2 d\xi \leq \int_0^1 [h(x)]^2 dx = \|h\|_2^2,$$

where the norm here arises from the usual inner product in L^2, namely

$$\langle u,v\rangle_2 = \int_0^1 uv\, dx.$$

Thus Δ^{-1} is continuous. In view of the fact that Δ^{-1} is self adjoint, then

$$\langle\Delta^{-1}f_n,h\rangle_2 = \langle f_n,\Delta^{-1}h\rangle_2 \to \langle f,\Delta^{-1}h\rangle_2 = \langle\Delta^{-1}f,h\rangle_2$$

for every $h \in L^2$. This calculation shows $\{\Delta^{-1}f_n\}$ converges weakly to $\Delta^{-1}f$. Now we show $\|\Delta^{-1}f_n\|_2 \to \|\Delta^{-1}f\|_2^2$. These last two properties are sufficient to establish that $\{\Delta^{-1}f_n\}$ converges (strongly) to $\Delta^{-1}f$ in L^2 (see Appendix A). So without loss of generality assume $\|f_n\|_2 \leq 1$. Then

$$|\Delta^{-1}(f_n)(\xi)|^2 \leq \|f_n\|^2 \leq 1 \quad \text{for all } \xi \in [0,1].$$

Also as $\{f_n\}$ converges weakly to f and K is continuous, then

$$(\Delta^{-1}f_n)(\xi) = \int_0^1 K(\xi,x)f_n(x)dx \to \int_0^1 K(\xi,x)f(x)dx = (\Delta^{-1}f)(\xi).$$

Therefore $\{|(\Delta^{-1}f_n)(\xi)|\}$ converges pointwise to $|(\Delta^{-1}f)(\xi)|$ in $\xi \in [0,1]$. The Lebesgue dominated convergence theorem implies

$$\lim_{n\to\infty} \int_0^1 |(\Delta^{-1}f_n)(\xi)|^2 d\xi = \int_0^1 |(\Delta^{-1}f)(\xi)|^2 d\xi;$$

that is, $\|\Delta^{-1}f_n\|_2^2 \to \|\Delta^{-1}f\|_2^2$. This completes the proof of compactness of Δ^{-1}.

To conclude the demonstration that A^{-1} is compact, we let $\{(u_n,v_n)\}$ be a bounded sequence in $H_0^1 \times L^2$. From what we have just proved it follows that $\{\Delta^{-1}v_n\}$ converges (strongly) in L^2. As H^1 is compactly embedded in L^2, then we may assume $\{u_n\}$ also converges (strongly) in L^2. Thus $\{A^{-1}(u_n,v_n)\} = \{(-\Delta^{-1}v_n,-u_n)\}$ converges (strongly) in $L^2 \times L^2$.

Now we are able to show that the positive orbits $\gamma^+(w_0)$ for all $w_0 \in X$ are precompact. Indeed, the linearity of A and Corollary 2.10 imply that for every $w_0 \in \mathscr{D}(A)$,

$$\gamma^+(w_0) = \bigcup_{t\geq 0} T(t)w_0 = \bigcup_{t\geq 0} A^{-1}T(t)Aw_0$$

$$= A^{-1} \bigcup_{t\geq 0} T(t)Aw_0 = A^{-1}\gamma^+(Aw_0).$$

As $\|T(t)Aw_0\| \leq \|Aw_0\|$ for all $t \in \mathbb{R}^+$, then $\gamma^+(Aw_0)$ is bounded. The compactness of A^{-1} ensures $\gamma^+(w_0) = A^{-1}\gamma^+(Aw_0)$ is precompact in X. Because A is densely defined we may extend the result to every initial value $w_0 \in X$. This concludes the example.

The result of Example 4.2 suggests that if we have a compact operator $P: X \to X$ which commutes with $T(t)$, then the positive orbit $\gamma^+(Px)$ for $x \in \mathscr{D}(A)$ is precompact, whenever $\gamma^+(x)$ is bounded. Of course, we are now dealing with a semidynamical system on the range of P. This can be formulated more precisely as follows.

<u>Theorem 4.3</u>. Suppose T is a quasi-contraction semigroup on a subset $\mathscr{C}$ of a real Banach space X so that $T(t)\mathscr{C} \subset \mathscr{C}$ for every $t \in \mathbb{R}^+$. Let $P: \mathscr{C} \to X$ be a compact operator which satisfies

(i) $\mathscr{R}(P) \subset \mathscr{C}$,

(ii) $PT(t)x = T(t)Px$ for every $x \in \mathscr{C}$.

Then $\mathscr{R}(P)$ is positively invariant. If, in addition, $\gamma^+(x)$ is bounded for some $x \in \mathscr{C}$, then $\gamma^+(Px)$ is precompact. Finally, if T is a contraction, has a critical point $x_0 \in \mathscr{C}$ and (i) is replaced by

(i)' $\mathscr{R}(P) \subset \mathscr{C} \subset \overline{\mathscr{R}(P)}$,

then all positive orbits $\gamma^+(x)$, $x \in \mathscr{C}$, are precompact.

<u>Proof</u>: For any $Px \in \mathscr{R}(P)$, $x \in \mathscr{C}$, write

$$\gamma^+(Px) = \bigcup_{t \geq 0} T(t)Px = \bigcup_{t \geq 0} PT(t)x = P\gamma^+(x) \subset \mathscr{R}(P).$$

This shows $\mathscr{R}(P)$ is positively invariant. As P is a compact operator, we see that $\gamma^+(Px)$ is precompact whenever $\gamma^+(x)$ is bounded.

Now suppose $T(t)x_0 = x_0$ for all $t \in \mathbb{R}^+$. For every $x \in \mathscr{C}$ we get

$$\|T(t)x\| \leq \|T(t)x - T(t)x_0\| + \|T(t)x_0\| \leq \|x - x_0\| + \|x_0\|,$$

so $\gamma^+(x)$ is bounded. Define the set

$$\mathcal{S} = \{x \in \mathcal{C}: \gamma^+(x) \text{ is precompact}\}.$$

We will show that $\mathcal{S}$ is closed. So let $x \in \bar{\mathcal{S}}$ and $\{x_k\} \subset \mathcal{S}$ with $x_k \to x$. Suppose $\{t_m\}$ be any sequence in $\mathbb{R}^+$. As each $\gamma^+(x_k)$ is precompact, we can assume the sequence $\{T(t_m)x_k\}$ is Cauchy for each fixed k. Given any $\varepsilon > 0$, choose k_0 so that $\|x-x_{k_0}\| < \frac{1}{4}\varepsilon$. Also there exists m_0 so that $\|T(t_m)x_{k_0} - T(t_n)x_{k_0}\| < \frac{1}{2}\varepsilon$ for all $m,n \geq m_0$. Thus for $m,n \geq m_0$

$$\begin{aligned}
&\|T(t_m)x-T(t_n)x\| \\
&\quad \leq \|T(t_m)x-T(t_m)x_{k_0}\| + \|T(t_m)x_{k_0}-T(t_n)x_{k_0}\| \\
&\qquad\qquad + \|T(t_n)x_{k_0}-T(t_n)x\| \\
&\quad \leq \|x-x_{k_0}\| + \|T(t_m)x_{k_0}-T(t_n)x_{k_0}\| + \|x-x_{k_0}\| < \varepsilon.
\end{aligned}$$

This establishes that $\{T(t_m)x\}$ is Cauchy so $\gamma^+(x)$ is precompact. Consequently $x \in \mathcal{S}$ and $\mathcal{S}$ is therefore closed. Assumption (i)' ensures $\mathcal{S} \subset \mathcal{C}$ as we have already seen that $\mathcal{R}(P) \subset \mathcal{S}$. □

Observe that the formulation of Theorem 4.3 is independent of A. Criteria involving A are to be found in Exercises 7.9 and 7.11. The role of $\mathcal{C}$ can be played by $\hat{\mathcal{D}}(A)$, the generalized domain of A. Also note that P need not be linear. Compact operators P which commute with A are uncommon, Example 4.2 not withstanding. We have, though, the following result which is similar to Theorem 4.3.

<u>Theorem 4.4</u>. Let T be a quasi-contraction semigroup in the real Banach space X which leaves a subset $\mathcal{C} \subset X$ positively invariant. Suppose there exists a sequence $\{P_n\}$ of compact

operators, $P_n: \mathscr{C} \to X$, and a bounded operator, $P: \mathscr{C} \to \mathscr{C}$ for which

$$PT(t)x = T(t)Px \quad \text{for all} \quad (x,t) \in \mathscr{C} \times \mathbb{R}^+,$$

$$\lim_{n\to\infty} P_n T(t)x = PT(t)x \quad \text{for all} \quad x \in \mathscr{K}$$

uniformly in $t \in \mathbb{R}^+$ for some subset $\mathscr{K} \subset \mathscr{C}$. Then

(a) $\gamma^+(Px)$ is precompact whenever $x \in \mathscr{K}$ and $\gamma^+(x)$ is bounded,

(b) if $x_0 \in \mathscr{C}$ is a critical point of T, $\overline{P\mathscr{K}} = \mathscr{C}$, and T is contracting, then all positive orbits $\gamma^+(x)$, $x \in \mathscr{C}$, are precompact.

<u>Proof</u>: (a) Let $x \in \mathscr{K}$ with $\gamma^+(x)$ bounded, and suppose $\{t_m\} \subset \mathbb{R}^+$ with $t_m \to \infty$. For each $n \in \mathbb{N}$ there is a subsequence $\{t_m^n\} \subset \{t_m^{n-1}\}$ $(t_m^0 = t_m)$ so that $\{P_n T(t_m^n)x\}_{m=1}^\infty$ is Cauchy. Let $\tau_m = t_m^m$. Then for each n, the sequence $\{P_n T(\tau_m)x\}_{m=1}^\infty$ is Cauchy. Let $\varepsilon > 0$ and choose positive integers n, N so that $\|P_n T(t)x - PT(t)x\| < \varepsilon/3$ for all $t \in \mathbb{R}^+$, and

$$\|P_n T(\tau_i)x - P_n T(\tau_j)x\| < \varepsilon/3 \quad \text{for all} \quad i,j > N.$$

As

$$\|T(\tau_i)Px - T(\tau_j)Px\| \le \|PT(\tau_i)x - P_n T(\tau_i)x\|$$
$$+ \|P_n T(\tau_i)x - P_n T(\tau_j)x\| + \|P_n T(\tau_j)x - PT(\tau_j)x\|,$$

we see that $\{T(\tau_m)Px\}_{m=1}^\infty$ is Cauchy, hence $\gamma^+(Px)$ is precompact.

(b) The argument used in the proof of Theorem 4.3 shows that every $\gamma^+(x)$, $x \in \mathscr{C}$, is bounded. The same proof also shows the set $\mathscr{S} = \{x \in \mathscr{C}: \gamma^+(x) \text{ is precompact}\}$ is closed.

From part (a) we can conclude $P\mathcal{K} \subset \mathcal{S}$, so it follows that $\mathcal{S} = \mathcal{C}$. Thus every positive orbit in $\mathcal{C}$ is precompact. □

The advantage of Theorem 4.4 is that P need not be compact. It might be possible to choose P as the identity for a suitable selection of $\{P_n\}$ and $\mathcal{K}$. This, in fact, is the point of the next theorem. It is phrased in terms of the results on accretive operators in the last two sections.

Theorem 4.5. Suppose A is a densely defined m-accretive operator in X. If $0 \in \mathcal{R}(A)$ and J_λ is compact for some $\lambda > 0$, then for every $x \in X$ the positive orbit $\gamma^+(x)$ of the contraction semigroup generated by A is precompact.

Proof: Let $\mathcal{K} = \hat{\mathcal{D}}(A)$, the generalized domain of A (Definition 3.7). Define $P: X \to X$ by $Px = x$, the identity on X, and let $P_n = J_{1/n}$. If T is the contraction semigroup generated by A, then

$$\|J_{1/n}T(t)x - T(t)x\| \leq \frac{1}{n}\|AT(t)x\| \leq \frac{1}{n}|Ax|,$$

where the first inequality comes from Lemma 2.15(i) and the second from Corollary 3.11 whenever $x \in \mathcal{S}$. As J_λ is compact for some $\lambda > 0$, Lemma 2.15(iii) shows that $J_{1/n}$ is compact for all n. Thus $\{P_n\}$ and P chosen here satisfy the hypotheses of Theorem 4.4.

Since $0 \in \mathcal{R}(A)$, then there exists $x_0 \in \mathcal{D}(A)$ with $J_\lambda x_0 = x_0$ for every small $\lambda > 0$. This leads to $T(t)x_0 = x_0$ for all $t \in \mathbb{R}^+$. Finally, $P\mathcal{K} = \hat{\mathcal{D}}(A)$ is dense in X by Lemma 3.8(ii) and the definition of A. The conclusion follows from Theorem 4.4. □

We complete this section with another example which illustrates precompactness of positive orbits. Though the equation is linear, the boundary condition is not, hence the domain of the generator of the semigroup is nonlinear. We shall return to this example in Section 6 to investigate the structure of its limit sets.

Example 4.6. We consider the equations of motion of an elastic membrane with interior elastic support and boundary viscous support.

$$(4.3)\qquad \frac{\partial^2 u}{\partial t^2}(x,t) = \Delta u(x,t) - u(x,t), \quad (x,t) \in \Lambda \times \mathbb{R}^+,$$

where $\Lambda \subset \mathbb{R}^2$ is a bounded open set with smooth boundary $\partial\Lambda$. The damping mechanism is induced by the boundary condition

$$(4.4)\qquad \frac{\partial u}{\partial n}(x,t) = -\phi\left(\frac{\partial u}{\partial t}(x,t)\right), \quad (x,t) \in \partial\Lambda \times \mathbb{R}^+,$$

where $\frac{\partial}{\partial n}$ denotes the derivative with respect to the outward normal on $\partial\Lambda$ and where $\phi \in C^1(\mathbb{R};\mathbb{R})$, ϕ is strictly increasing with bounded derivative and $\phi(0) = 0$.

We rewrite Equation (4.3) as a first order system in an appropriate phase space:

$$(4.5)\qquad \frac{du}{dt} = v, \qquad \frac{dv}{dt} = \Delta u - u.$$

The analysis of Example 4.2 suggests we choose the phase space X to be $H^1 \times L^2$, where $H^1 \overset{\text{def}}{=} H^1(\Lambda;\mathbb{R})$ and $L^2 = L^2(\Lambda;\mathbb{R})$. Think of H^1 as consisting of all those functions on Λ whose partial derivatives belong to L^2. (See Yoshida [1] for details concerning distributional derivatives.) H^1 is a Hilbert space with inner product

$$\langle u,v\rangle_{H^1} = \int_\Lambda [uv + \nabla u\cdot\nabla v]\,dxdy$$

where ∇ is the gradient operator, $\nabla u = (\frac{\partial u}{\partial x}, \frac{\partial u}{\partial y})$. L^2 has the usual inner product

$$\langle u,v\rangle_2 = \int_\Lambda uv\, dxdy.$$

Then $X = H^1 \times L^2$ is a Hilbert space with inner product

$$\langle (u,v),(\tilde{u},\tilde{v})\rangle = \langle u,\tilde{u}\rangle_{H^1} + \langle v,\tilde{v}\rangle_2.$$

The System (4.5) can now be written in the form

$$\frac{dw}{dt} + Aw = 0,$$

where $w = (u,v) \in X$ and A is given by $A(u,v) = (-v,u-\Delta u)$. Its domain $\mathcal{D}(A)$ is the set of $(u,v) \in X$ with $\frac{\partial u}{\partial n} = -\phi(v)$ on $\partial\Lambda$ in $H^{\frac{1}{2}}(\Lambda)$. We will show that A satisfies the hypotheses of Theorem 4.5.

For $(u,v) \in \mathcal{D}(A)$ compute

$$\begin{aligned}\langle A(u,v),(u,v)\rangle &= \langle(-v,u-\Delta u),(u,v)\rangle = -\int_\Lambda [\nabla v\cdot\nabla u + v\Delta u]\,dxdy \\ &= -\int_{\partial\Lambda} v\frac{\partial u}{\partial n}\, d\sigma = \int_{\partial\Lambda} \phi(v)v\, d\sigma \geq 0,\end{aligned}$$

where the third equality follows from Green's Identity (cf. John [1]) and the last inequality comes from the assumption on ϕ. Since A operates linearly, we conclude that A is accretive. To show A is m-accretive we must prove for every $(f,g) \in H^1 \times L^2$ there is $(u,v) \in \mathcal{D}(A)$ with $(I+A)(u,v) = (f,g)$. This means we must solve the system $u-\Delta u = f+g$, $u+v = f$, subject to the boundary condition $\frac{\partial u}{\partial n} = -\phi(v)$. For the existence of a solution to this problem, see John [1], Chapter 4.

Finally, we show that $(I+A)^{-1}$ is a compact operator. It was proved in Example 4.2 that Δ^{-1} is a compact operator in $L^2([0,1];\mathbb{R})$. The extension to $\Lambda \subset \mathbb{R}^2$ presents no problem (cf. Oden [1], p. 395). Consequently, the representation

$$(4.6) \qquad (I+A)^{-1} = A^{-1} - (I+A)^{-1}A^{-1}$$

shows that $(I+A)^{-1}$ is compact, since by definition $(I+A)^{-1}$ is a bounded operator, and A^{-1} is compact. This follows as A^{-1} can be written as

$$A^{-1} = \begin{bmatrix} 0 & (I-\Delta)^{-1} \\ -I & 0 \end{bmatrix},$$

where also

$$(I-\Delta)^{-1} = (I-\Delta)^{-1}\Delta^{-1} - \Delta^{-1}.$$

So according to Theorem 4.5, all positive orbits $\gamma^+(u_0,v_0)$ with $(u_0,v_0) \in \overline{\mathscr{D}(A)}$ are precompact. Of course, we have not demonstrated that $T(t)(u_0,v_0)$ is the unique strong solution to System (4.5) through (u_0,v_0), where T is the semigroup generated by A. This fact, as well as the asymptotic properties of the solution, will be treated in Section 6.

5. Solution of the Cauchy Problem

The matter of solving the evolution equation

$$(5.1) \qquad \frac{du}{dt} + Au = 0, \quad u(0) = u_0 \in \mathscr{D}(A)$$

is referred to as the *Cauchy problem*. So to begin with, we must have a notion of what is means for a function u to be a solution of Equation (5.1). When A is accretive and satisfies $\mathscr{R}(I+A) \supset \overline{\mathscr{D}(A)}$, $T(\cdot)u_0$ appears to be an obvious candidate for a solution. This requires (strong) differentiability of the mapping $t \to T(t)u_0$. For the sake of generality (which we will require later), we will formulate the Cauchy problem for the nonautonomous evolution equation with

$$\frac{du}{dt} + Au = f, \quad u(0) = u_0 \in \mathscr{D}(A), \tag{5.2}$$

where $f \in L^1_{loc}(\mathbb{R}^+;X)$.

Definition 5.1. A function $u: \mathbb{R}^+ \to X$ is called a *strong solution* of the nonautonomous Cauchy problem, Equation (5.2), provided

(i) $u(t) \in \mathscr{D}(A)$ a.e.,

(ii) $u \in C(\mathbb{R}^+;X)$,

(iii) there exists $v \in L^1_{loc}(\mathbb{R}^+;X)$ so that

$$u(t) - u(s) = \int_s^t v(\tau)d\tau, \quad 0 \leq s \leq t < \infty, \text{ and}$$

(iv) u satisfies Equation (5.2) a.e.

Fortunately, if the Cauchy problem has a strong solution for a closed operator satisfying the hypotheses of Theorem 2.4, then the strong solution is indeed the semigroup generated by A. This is a consequence of the following result. As in Section 3, there is no loss of generality in assuming $\mathscr{R}(I+\lambda A) = \overline{\mathscr{D}(A)} = X$ for all sufficiently small $\lambda > 0$.

Theorem 5.2. Suppose $A: \mathscr{D}(A) \subset X \to X$ is a densely defined, closed operator with $A + \omega I$ m-accretive for some $\omega \in \mathbb{R}$. If $u_0 \in \mathscr{D}(A)$ and $u: \mathbb{R}^+ \to X$, then u is a strong solution of Equation (5.1) if and only if $u(t) = T(t)u_0$ for all $t \in \mathbb{R}^+$ and $T(t)u_0$ is differentiable a.e. in $t \in \mathbb{R}^+$.

Proof: Suppose u is a strong solution of Equation (5.1) with $u_0 \in \mathscr{D}(A)$. Let $\lambda > 0$ and consider the discrete problem expressed by Equation (1.5) and its solution u_λ given by

Equation (1.6). According to Inequality (8.3) (Section 8, Appendix of this chapter) we have that $\lim_{\lambda \downarrow 0} u_\lambda(s) = T(s)u_0$ uniformly for s in compact subsets of $\mathbb{R}^+$. Extend the strong solution u to $\mathbb{R}$ by defining $u(s) = x$ for all $s \leq 0$ and set

$$g_\lambda(s) = \frac{u(s)-u(s-\lambda)}{\lambda} - \dot{u}(s), \quad s > 0.$$

For any $t > 0$ we must have

$$\lim_{\lambda \downarrow 0} \int_0^t \|g_\lambda(s)\| ds = 0.$$

From the fact that $(I+\lambda A)u = u-\lambda\dot{u}$, we may write $u = J_\lambda[u-\lambda\dot{u}]$, so $u(s) = J_\lambda[u(s-\lambda)+\lambda g_\lambda(s)]$ a.e.. As $u_\lambda(s) = J_\lambda u_\lambda(s-\lambda)$ and J_λ is $(1-\lambda\omega)^{-1}$-Lipschitz, then

$$\|u_\lambda(s)-u(s)\| \leq (1-\lambda\omega)^{-1}\{\|u_\lambda(s-\lambda)-u(s-\lambda)\| + \lambda\|g_\lambda(s)\|\}.$$

Integrate over $[0,t]$ and rearrange to obtain

$$\frac{1}{\lambda}\int_{t-\lambda}^t \|u_\lambda(s)-u(s)\| ds$$

$$\leq \frac{(1-\lambda\omega)^{-1}-1}{\lambda}\int_0^{t-\lambda} \|u_\lambda(s)-u(s)\| ds$$

$$+ (1-\lambda\omega)^{-1}\int_0^t \|g_\lambda(s)\| ds.$$

Let $\lambda \downarrow 0$. Then

$$\|T(t)u_0-u(t)\| \leq \omega\int_0^t \|T(s)u_0-u(s)\| ds.$$

An application of Gronwall's inequality (Appendix A) establishes $\|T(t)u_0-u(t)\| = 0$. As this holds for every $t \in \mathbb{R}^+$, we must have $u(t) = T(t)u_0$.

Conversely, suppose $T(t)u_0$ is strongly differentiable

at $t_0 > 0$. Choose $\lambda \in (0,t_0)$ and define $u_\lambda = J_\lambda T(t_0-\lambda)u_0$. Then

$$u_\lambda + \lambda A u_\lambda = T(t_0-\lambda)u_0 = T(t_0)u_0 - \lambda v_0 + o(\lambda), \tag{5.3}$$

where $v_0 = \frac{d}{dt} T(t)u_0 \Big|_{t=t_0}$, and $\lim\limits_{\lambda \downarrow 0} \frac{o(\lambda)}{\lambda} = 0$. Now replace x_0 with u_λ and x with $T(t_0)u_0$ in the Inequality (3.3). There must exist $\eta^* \in F(u_\lambda - T(t_0)u_0)$ so that

$$(v_0,\xi^*) \leq (Au_\lambda,\eta^*) + \omega\|u_\lambda - T(t_0)u_0\|^2$$

for all $\xi^* \in F(T(t_0)u_0 - u_\lambda)$. Choose $\xi^* = -\eta^*$. We obtain

$$(Au_\lambda + v_0,\eta^*) + \omega\|u_\lambda - T(t_0)u_0\|^2 \leq 0.$$

Consequently, substituting for v_0 from Equation (5.3) we have

$$(T(t)u_0 - u_\lambda + o(\lambda),\eta^*) + \lambda\omega\|u_\lambda - T(t_0)u_0\|^2 \geq 0.$$

As $\eta^* \in F(u_\lambda - T(t_0)u_0)$, we find that $(1-\lambda\omega)\|T(t_0)u_0 - u_\lambda\| \leq o(\lambda)$. Thus, $\lim\limits_{\lambda \downarrow 0} u_\lambda = T(t_0)u_0$. We also have

$$\|Au_\lambda + v_0\| = \left\|\frac{T(t_0)u_0 - u_\lambda}{\lambda} + \frac{o(\lambda)}{\lambda}\right\| \leq [(1-\lambda\omega)^{-1}+1]\frac{o(\lambda)}{\lambda},$$

so $\lim\limits_{\lambda \downarrow 0} Au_\lambda = -v_0$. As A is a closed operator, it follows that $AT(t_0)u_0 + v_0 = 0$. Thus $u(t) = T(t)u_0$ satisfies Equation (5.1) at $t = t_0$ and indeed, is a strong solution. □

Corollary 5.3. Assume the hypotheses of Theorem 5.2. If the semigroup T generated by A is such that $T(t)u_0$ for $u_0 \in \mathcal{D}(A)$ is differentiable a.e. in $t \in \mathbb{R}^+$, then $T(t)u_0$ is the unique strong solution of Equation (5.1).

Regrettably, there exist operators A satisfying the hypotheses of Theorem 5.2 for which Equation (5.1) does not possess a strong solution on any interval $[0,\tau] \subset \mathbb{R}^+$. In the event X is reflexive though, Equation (5.1) always possesses a unique strong solution on $\mathbb{R}^+$ through any $u_0 \in \mathcal{D}(A)$. This is because the mapping $t \to T(t)u_0$ is Lipschitz continuous in t belonging to compact subsets of $\mathbb{R}^+$ (Inequality (8.4)). If X is reflexive, it is a fact (Kōmura [1]) that every Lipschitz continuous X-valued function is differentiable a.e. in $t \in \mathbb{R}^+$. Hence we obtain

Theorem 5.4. Let X be reflexive and suppose $A: \mathcal{D}(A) \subset X \to X$ is a densely defined closed operator with $A + \omega I$ m-accretive for some $\omega \in \mathbb{R}$. If $u_0 \in \mathcal{D}(A)$, then $T(t)u_0$ is the unique strong solution of Equation (5.1) where $t \in Q_\omega$ is generated by A.

Example 5.5. (Continuation of Example 4.2) It was established in Example 4.2 that the operator A was densely defined m-accretive on the Hilbert space $X = H_0^1 \times L^2$ with domain $\mathcal{D}(A) = (H_0^1 \cap H^2) \times H_0^1$. It is only left to verify that A is a closed operator. So let $\{(u_n,v_n)\} \subset \mathcal{D}(A)$ converge to (u,v) in X, and let $\{A(u_n,v_n)\}$ converge to $(f,g) \in X$. We must show that $(u,v) \in \mathcal{D}(A)$ and $A(u,v) = (f,g)$. It will be sufficient to show $u \in \mathcal{D}(\Delta) = H_0^1 \cap H^2$ and $\Delta u = -g$. But recall from Example 4.2 that Δ^{-1} is a continuous operator. Since $A(u_n,v_n) \to (f,g)$ implies $-\Delta u_n \to g$, then $-u_n \to \Delta^{-1}g$. As $u_n \to u$, it follows that $-u = \Delta^{-1}g$; that is, $\Delta u = -g$. Now Theorem 5.4 is applicable and so the wave equation has a unique strong solution.

There are other types of solutions of Equation (5.1) which do not require strong differentiability. For example, the derivative $\frac{du}{dt}$ may be taken in the weak topology of X. An entirely different approach is to approximate T by a (strongly) differentiable semigroup. This was done in Theorem 2.13. For the more general case of Equation (5.2), another approach is to take limits of strong solutions of Equation (5.2) in the following sense.

Definition 5.6. A function $u\colon \mathbb{R}^+ \to X$ is called a *weak solution* of the nonautonomous Cauchy problem, Equation (5.2), provided there exist sequences $\{u_n\} \subset C(\mathbb{R}^+;X)$ and $\{f_n\} \subset L^1_{loc}(\mathbb{R}^+;X)$ so that

(i) $\lim_{n\to\infty} f_n = f$ in $L^1_{loc}(\mathbb{R}^+;X)$,

(ii) u_n is a strong solution of $\frac{du}{dt} + Au = f_n$, and

(iii) $\lim_{n\to\infty} u_n = u$, uniformly on compact subsets of $\mathbb{R}^+$.

We focus our attention on weak solutions of Equation (5.2) in the event X is a Hilbert space. The inner product there will be denoted by $\langle\cdot,\cdot\rangle$ and associated norm $\|\cdot\|$.

Lemma 5.7. Suppose A is accretive, and let $f,g \in L^1_{loc}(\mathbb{R}^+;X)$. If u and v are weak solutions of $\frac{du}{dt} + Au = f$ and $\frac{dv}{dt} + Av = g$ respectively, then

$$\|u(t)-v(t)\| \le \|u(s)-v(s)\| + \int_s^t \|f(\xi)-g(\xi)\|\,d\xi \tag{5.4}$$

for all $0 \le s \le t < \infty$.

Proof: Compute

$$\begin{aligned}\frac{1}{2}\frac{d}{dt}\|u(t)-v(t)\|^2 &= \langle \tfrac{du}{dt}(t) - \tfrac{dv}{dt}(t), u(t) - v(t)\rangle \\ &= \langle f(t) - g(t), u(t) - v(t)\rangle,\end{aligned}$$

as $\langle Au(t)-Av(t),u(t)-v(t)\rangle \geq 0$ by hypothesis. Integrate to obtain

$$\frac{1}{2}\|u(t)-v(t)\|^2 - \frac{1}{2}\|u(s)-v(s)\|^2 \leq \int_s^t \|f(\xi)-g(\xi)\| \|u(\xi)-v(\xi)\| d\xi.$$

Apply the generalized Gronwall inequality (Appendix A) to conclude the proof. □

The last lemma establishes uniqueness of weak solutions to the nonautonomous Cauchy problem. Next we turn to the question of existence in Hilbert space. The following lemma characterizes m-accretive operators in a Hilbert space. The proof is a direct consequence of Lemma 3.4.

<u>Lemma 5.8</u>. Let X be a Hilbert space. Then $A: \mathscr{D}(A) \subset X \to X$ is m-accretive if and only if A is accretive and for each $x \in X$ such that $\langle y-A\xi,x-\xi\rangle \geq 0$ for all $\xi \in \mathscr{D}(A)$ then $y = Ax$.

<u>Corollary 5.9</u>. Suppose X is a Hilbert space and $A: \mathscr{D}(A) \subset X \to X$ is m-accretive. Then A is closed.

<u>Proof</u>: Let $\{x_n\} \subset \mathscr{D}(A)$ with $x_n \to x$ and $Ax_n \to y$. Then $\langle Ax_n-A\xi,x_n-\xi\rangle \geq 0$ for all x_n and $\xi \in \mathscr{D}(A)$. Take limits to obtain $\langle y-A\xi,x-\xi\rangle \geq 0$ for all $\xi \in \mathscr{D}(A)$. Thus $y = Ax$ by Lemma 5.8. □

<u>Theorem 5.10</u>. Suppose $A: \mathscr{D}(A) \in X \to X$ is a densely defined m-accretive operator in the Hilbert space X. Then for every $f \in L^1_{loc}(\mathbb{R}^+;X)$ and $u_0 \in X$, there exists a unique weak solution of $\frac{du}{dt} + Au = f$ which satisfies $u(0) = u_0$.

<u>Proof</u>: Let $u_0 \in \mathscr{D}(A)$ and fix a compact interval $[0,\tau] \subset \mathbb{R}^+$. We will first show that Equation (5.2) admits a strong solution

whenever f is a step function on $[0,\tau]$ with $f(t) = x_i$ for $t \in [a_{i-1}, a_i)$. The operator $A - x_i: \mathcal{D}(A) \to X$ defined by $(A-x_i)x = Ax - x_i$ is also densely defined, closed, and accretive, so it generates a contraction semigroup T_i. Define

$$u(t) = \begin{cases} T_i(t-a_{i-1})u(a_{i-1}), & t \in [a_{i-1}, a_i),\ i = 1,2,\ldots,n. \\ u_0 \quad , & t = 0. \end{cases}$$

In view of Theorem 5.4, u is a strong solution of Equation (5.2) for the step function f. Now suppose $f \in L^1([0,\tau];X)$ and $u_0 \in X$. There exists a sequence $\{f_n\}$ of step functions on $[0,\tau]$ which converge to f in $L^1([0,\tau];X)$, and a sequence $\{x_n\} \subset \mathcal{D}(A)$ so that $x_n \to u_0$ in X. Denote by u_n the strong solution of $\frac{du}{dt} + Au = f_n$, $u(0) = x_n$. By Lemma 5.7 we obtain

$$\|u_n(t) - u_m(t)\| \leq \|x_n - x_m\| + \int_0^t \|f_n(\xi) - f_m(\xi)\| d\xi, \quad t \in [0,\tau].$$

As $\{x_n\}$ and $\{f_n\}$ are Cauchy, then so is $\{u_n\}$. Hence $\{u_n\}$ converges uniformly to a weak solution of Equation (5.2) on $[0,\tau]$. Uniqueness of the weak solution follows immediately from Lemma 5.7. □

We conclude this section with a theorem and an example on the existence of strong solutions to the nonautonomous Cauchy problem. See Brezis [2] for a proof. We remind the reader (see Appendix A) that $W^{1,p}(\mathbb{R}^+;X)$ is the set of all functions $f: \mathbb{R}^+ \to X$ so that f and its distributional derivative f' all belong to $L^p(\mathbb{R}^+;X)$.

<u>Theorem 5.11</u>. Suppose $A: \mathcal{D}(A) \subset X \to X$ is a densely defined m-accretive operator in a Hilbert space X. For every $f \in W^{1,p}_{loc}(\mathbb{R}^+;X)$ and every $u_0 \in X$, there exists a unique strong

solution to $\frac{du}{dt} + Au = f$ which satisfies $u(0) = u_0$.

Example 5.12. We consider a type of wave equation with weak damping of the form

$$\frac{\partial^2 u}{\partial t^2} = \frac{\partial^2 u}{\partial x^2} - a(x)q(\frac{\partial u}{\partial t}) + f(x,t), \quad x \in [0,1] \tag{5.5}$$

where $a \in C^1([0,1]; \mathbb{R})$, $a(x) \geq 0$ for all $x \in [0,1]$ with $a(x_0) > 0$ for some $x_0 \in (0,1)$; $q \in C^1(\mathbb{R}; \mathbb{R})$ is strictly increasing with bounded derivative and $q(0) = 0$. For boundary conditions we take

$$u(0,t) = u(1,t) = 0 \quad \text{for all} \quad t > 0$$

and initial values

$$u(x,0) = u_0(x), \quad x \in [0,1]$$

$$\frac{\partial u}{\partial t}(x,0) = v_0(x), \quad x \in [0,1].$$

With Example 4.2 as a guide we can write this equation as the first order system

$$\frac{du}{dt} = v$$

$$\frac{dv}{dt} = \Delta u - aq(v) + f$$

in the phase space $X = H_0^1 \times L^2$. Define the operator $A: X \to X$ by

$$A(u,v) = (-v, -\Delta u + aq(v)).$$

As in Example 4.2 we see that $\mathcal{D}(A) = H_0^1 \cap H^2 \times H_0^1$ which is dense in X. The following calculation establishes the accretiveness of A; for (u,v), $(\tilde{u},\tilde{v}) \in \mathcal{D}(A)$

$$\langle A(u,v) - A(\tilde{u},\tilde{v}),\ (u,v) - (\tilde{u},\tilde{v})\rangle$$

$$= \langle(-v,-u'' + aq(v)) - (-\tilde{v},-\tilde{u}'' + aq(\tilde{v})),\ (u-\tilde{u},v-\tilde{v})\rangle$$

$$= \langle(-v+\tilde{v},\ a[q(v)-q(\tilde{v})] - u'' + \tilde{u}''),\ (u-\tilde{u},v-\tilde{v})\rangle$$

$$= \int_0^1 \{(-v'+\tilde{v}')(u'-\tilde{u}') + a[q(v)-q(\tilde{v})](v-\tilde{v}) - (u''-\tilde{u}'')(v-\tilde{v})\}dx$$

$$= \int_0^1 a(x)[q(v(x))-q(\tilde{v}(x))][v(x)-\tilde{v}(x)]dx \geq 0,$$

where the last equality comes from integration by parts and the final inequality comes from the hypotheses on q.

In order to prove that A is m-accretive we must show that for every $(g,h) \in X$, there exist $(u,v) \in \mathscr{D}(A)$ so that $(I+A)(u,v) = (g,h)$. This is equivalent to solving the second order differential equation for v,

$$v(x) - a(x)q(v(x)) - v''(x) = h(x) - g''(x).$$

The hypotheses on q ensure that this equation admits a solution $v \in H_0^1$ via the technique of Example 2.7.

6. Structure of Positive Limit Sets for Contraction Semigroups

Because the phase space of the semidynamical systems considered in this chapter is a Banach space, its rich properties enable us to go much further in studying the structure of positive limit sets than would otherwise be possible. These results are of interest in themselves as well as being useful in the analysis of asymptotic behavior of equations of evolution.

Theorem 6.1. Suppose T is a contraction semigroup on a closed subset $\mathscr{C}$ of a real Banach space X so that $T(t)\mathscr{C}\subset\mathscr{C}$

for every $t \in \mathbb{R}^+$. If for some $x \in \mathscr{L}$, the positive limit set $L^+(x) \neq \emptyset$, then

(i) $L^+(x)$ is positively minimal.

(ii) $T(t)$ is an isometry on $L^+(x)$ for every $t \in \mathbb{R}$.

(iii) If $L^+(x)$ is compact then T is equi-almost periodic on $L^+(x)$; that is, for each $\varepsilon > 0$ the set $\{\tau \in \mathbb{R}^+ : \sup \|T(\tau)y-y\| < \varepsilon\}$ is relatively dense in $\mathbb{R}^+$.

(iv) $L^+(x)$ is invariant, hence T extends to a group on $L^+(x)$.

(v) If T has a critical point $x_0 \in \mathscr{L}$, then $L^+(x) \subset \{y \in \mathscr{L}: \|y-x_0\| = r\}$ for some $r \leq \|x-x_0\|$.

Proof: (i) As T is a contraction semigroup, then $(\mathscr{L},T)$ is a Lyapunov stable system. So by Theorem 2.4 of Chapter III, $L^+(x)$ is positively minimal.

(ii) Let $y \in L^+(x) = L^+(y)$. There exists $\{\tau_n\} \subset \mathbb{R}^+$ with $\tau_n \to \infty$ so that $T(\tau_n)y \to y$. We claim that $T(\tau_n)z \to z$ for every $z \in L^+(x)$. Indeed, if $T(t_n)y \to z$ for some sequence $\{t_n\} \subset \mathbb{R}^+$ with $t_n \to \infty$, then $\|T(\tau_n)z-z\| \leq \|T(\tau_n)z-T(\tau_n+t_n)y\| + \|T(\tau_n+t_n)y-T(t_n)y\| + \|T(t_n)y-z\| \leq 2\|z-T(t_n)y\| + \|T(\tau_n)y-y\| \to 0$. Thus for any $t \in \mathbb{R}^+$ we have

$$\|y-z\| = \lim_{n\to\infty} \|T(\tau_n)y-T(\tau_n)z\| = \lim_{n\to\infty} \|T(\tau_n-t)T(t)y-T(\tau_n-t)T(t)z\|$$
$$\leq \|T(t)y-T(t)z\| \leq \|y-z\| .$$

Thus $T(t)$ is an isometry for $t \in \mathbb{R}^+$, so (ii) is proved.

(iii) We can use Theorem 4.19 of Chapter III. We offer though, a simple proof. In view of what we have already proved, $L^+(x)$ is positively minimal. By Theorem 3.5 of

Chapter III, the positive motion through any $y \in L^+(x)$ is recurrent. Thus $\{\tau \in \mathbb{R}^+ : \|T(\tau)y-y\| < \varepsilon\}$ is relatively dense in $\mathbb{R}^+$. As $T(t)$ is an isometry, we must have $\{\tau \in \mathbb{R}^+ : \sup_{z\in\gamma^+(y)} \|T(\tau)z-z\| < \varepsilon\} = \{\tau \in \mathbb{R}^+ : \|T(\tau)y-y\| < \varepsilon\}$. But $\overline{\gamma^+(y)} = L^+(x)$, so (iii) is proved.

(iv) This follows directly from Corollary 4.18 of Chapter III.

(v) Let $T(t)x_0 = x_0$ for all $t \in \mathbb{R}^+$. $\|T(t)x-x_0\|$ is nonincreasing in t. Let r denote the corresponding limit. Clearly, $r \le \|x-x_0\|$. If $y \in L^+(x)$ with $T(t_n)x \to y$ for some $\{t_n\} \subset \mathbb{R}^+$ with $t_n \to \infty$, then $\|y-x_0\| = \lim_{n\to\infty} \|T(t_n)x-x_0\| = r$. □

With additional restrictions on the Banach space X we may say more about the positive limit sets. Recall that X is *strictly convex* if $\|x+y\| = \|x\| + \|y\|$, $x,y \in X$ implies x and y are linearly dependent; it follows that $x = \alpha y$, where $\alpha = \|x\|/\|y\|$, $y \ne 0$. Also the mapping $F: X \to X$ is called *affine* if it preserves convex combinations.

<u>Theorem 6.2</u>. Let T be a strongly continuous contraction semigroup on a strictly convex real Banach space X. Suppose for some $x \in X$ we have $L^+(x) \ne \emptyset$,

(i) There is a strongly continuous affine contraction group $\hat{T}$ on the closed linear variety spanned by $L^+(x)$ which agrees with T on $\overline{\text{co}}\, L^+(x)$.

(ii) $\overline{\text{co}}\, L^+(x)$ is invariant with respect to T.

(iii) If $x_0 \in \overline{\text{co}}\, L^+(x)$ is a critical point of T, then

$$x_0 = \lim_{t\to\infty} t^{-1} \int_0^t T(s)y\, ds \tag{6.1}$$

for every $y \in \overline{co}\, L^+(x)$. If $x_0 = 0$, $\hat{T}$ is linear.

(iv) If $\overline{co}\, L^+(x)$ is weakly compact, it contains a critical point.

Proof: (i) We begin by showing the restriction of T to $co\, L^+(x)$ is affine. Let $y = \alpha x_1 + (1-\alpha)x_2$ with $x_1, x_2 \in L^+(x)$ and $0 \leq \alpha \leq 1$. We must show $T(t)y = \alpha T(t)x_1 + (1-\alpha)T(t)x_2$. Compute $\|y-x_1\| + \|y-x_2\| = \|x_1-x_2\|$. As T is an isometry on $L^+(x)$, we get

$$\begin{aligned}\|y-x_1\| &= \|x_1-x_2\| + \|y-x_2\| \\ &\leq \|T(t)x_1-T(t)x_2\| - \|T(t)y-T(t)x_2\| \\ &\leq \|T(t)y-T(t)x_1\| \leq \|y-x_1\| .\end{aligned}$$

Thus $\|T(t)y-T(t)x_1\| = \|y-x_1\|$. Similarly, $\|T(t)y-T(t)x_2\| = \|y-x_2\|$. Therefore

$$\begin{aligned}\|T(t)x_1-T(t)x_2\| &= \|x_1-x_2\| = \|y-x_1\| + \|y-x_2\| \\ &= \|T(t)y-T(t)x_1\| + \|T(t)y-T(t)x_2\|.\end{aligned}$$

By the strict convexity of X we get

$$\begin{aligned}T(t)y-T(t)x_1 &= \frac{\|T(t)y-T(t)x_1\|}{\|T(t)y-T(t)x_2\|}(T(t)y-T(t)x_2) \\ &= \frac{\|y-x_1\|}{\|y-x_2\|}(T(t)y-T(t)x_2) \\ &= \frac{1-\alpha}{\alpha}(T(t)y-T(t)x_2).\end{aligned}$$

Thus, $T(t)y = \alpha T(t)x_1 + (1-\alpha)T(t)x_2$. We extend to $y = \sum_{i=1}^{n} \alpha_i x_i$ with $\sum_{i=1}^{n} \alpha_i = 1$ and $x_1, x_2, \ldots, x_n \in L^+(x)$ by induction on

$$T(t)y = \sum_{i=1}^{n} \alpha_i T(t)x_i. \tag{6.2}$$

Thus T is affine on $\text{co } L^+(x)$. Since $L^+(x)$ is invariant under T, then just use Equation (6.2) to define $T(t)$ on $\text{co } L^+(x)$ for each $t \in \mathbb{R}^-$. Now let V be the linear variety spanned by $L^+(x)$. Then $V \supset \text{co } L^+(x)$. We can extend $T(t)$ to V for all $t \in \mathbb{R}^-$ as follows: for $y \in V$, write $y = \beta_1 y_1 + \beta_2 y_2$ where $\alpha_1, \alpha_2 \in \mathbb{R}$ and $y_1, y_2 \in \text{co } L^+(x)$. Set $\hat{T}(t)y = \beta_1 T(t)y_1 + \beta_2 T(t)y_2$ for all $t \in \mathbb{R}^-$. A direct computation shows that $\hat{T}(t)y$ depends solely on y and not on its particular representation $\beta_1 y_1 + \beta_2 y_2$. So $\hat{T}$ is a strongly continuous affine group on V. We may extend $\hat{T}$ to the closure of V by continuity of $\hat{T}(t)$. Moreover, $\hat{T}|_{\text{co } L^+(x)} = T$.

(ii) As $\overline{\text{co}}\, L^+(x) \subset \overline{V}$, then $\overline{\text{co}}\, L^+(x)$ is invariant under T.

(iii) Let $x_0 \in \overline{\text{co}}\, L^+(x)$ be a critical point. Fix some $y \in L^+(x)$ and let $z \in \text{co } \gamma^+(y)$. Say that

$$z = \sum_{i=1}^{n} \lambda_i T(t_i)y, \; t_i, \lambda_i \in \mathbb{R}^+, \; i = 1,2,\ldots,n, \; \sum_{i=1}^{n} \lambda_i = 1.$$

Then

$$t^{-1} \int_0^t T(s)z\, ds = t^{-1} \sum_{i=1}^{n} \lambda_i \int_{t_i}^{t_i+t} T(s)y\, ds$$

$$= t^{-1} \int_0^t T(s)y\, ds + \sum_{i=1}^{n} \lambda_i t^{-1} \left\{ \int_{t_i}^{0} T(s)y\, ds + \int_t^{t+t_i} T(s)y\, ds \right\}.$$

The last sum tends to zero as $t \to \infty$ since $\gamma^+(y)$ is bounded. Thus

$$\limsup_{t\to\infty} \left\| t^{-1}\int_0^t T(s)y\,ds - x_0 \right\| = \limsup_{t\to\infty} \left\| t^{-1}\int_0^t T(s)z\,ds - x_0 \right\|$$

(6.3)

$$= \limsup_{t\to\infty} \left\| t^{-1}\int_0^t [T(s)z - T(s)x_0]\,ds \right\| \leq \|z - x_0\|.$$

As $L^+(x)$ is positively minimal so $\overline{co}\, L^+(x) = \overline{co\, \gamma^+(y)}$, and $\inf\{\|z-x_0\|: z \in co\, \gamma^+(y)\} = 0$. Since (6.3) holds for any $z \in co\, \gamma^+(y)$, then we obtain Equation (6.1) for any $y \in L^+(x)$, and hence for any $y \in \overline{co}\, L^+(x)$. In the event $x_0 = 0$, we leave it as an exercise to prove that T is linear.

(iv) For any $y \in \overline{co}\, L^+(x)$ we have $T(t)y \in \overline{co}\, L^+(x)$, and hence $p(t) \overset{def}{=} t^{-1}\int_0^t T(s)y\,ds$ belongs to $\overline{co}\, L^+(x)$ for all $t > 0$. This is true since the integral is a weak limit of linear combinations of elements of $\overline{co}\, L^+(x)$. Multiplying by t^{-1} ensures that the combinations are convex. Thus its weak limit, $p(t)$, must lie in $\overline{co}\, L^+(x)$.

As $\overline{co}\, L^+(x)$ is weakly compact there exists a sequence $\{t_n\} \subset \mathbb{R}^+$ with $t_n \to \infty$ so that $p(t_n)$ converges weakly to some $x_0 \in \overline{co}\, L^+(x)$. T is affine (hence weakly continuous) on $\overline{co}\, L^+(x)$, so in view of the remarks about $p(t)$ we have for every $t > 0$

$$T(t)x_0 = \operatorname*{w-lim}_{n\to\infty} t_n^{-1} \int_t^{t+t_n} T(s)y\,ds$$

$$= \operatorname*{w-lim}_{n\to\infty} t_n^{-1} \int_0^{t_n} T(s)y\,ds + \operatorname*{w-lim}_{n\to\infty} t_n^{-1} \left\{ \int_t^0 T(s)y\,ds + \int_{t_n}^{t_n+t} T(s)y\,ds \right\} = x_0. \qquad \square$$

We specialize some of the results of the last two theorems to the asymptotic behavior of weak solutions to the nonautonomous Cauchy problem in Hilbert space.

Theorem 6.3. Let X be a real Hilbert space and $A: \mathcal{D}(A) \subset X \to X$ a densely defined m-accretive operator. Assume $0 \in \mathcal{R}(A)$ and $(I + \lambda A)^{-1}$ is compact for some $\lambda > 0$. Then for any $u_0 \in X$ and $f \in L^1(\mathbb{R}^+;X)$, the weak solution u of

$$\frac{du}{dt} + Au = f, \quad u(0) = u_0 \in \mathcal{D}(A) \tag{6.4}$$

approaches a compact subset Ω of the sphere $S_r(x_0)$, where $x_0 \in A^{-1}(0)$ and $r \leq \|u_0 - x_0\| + \int_0^\infty \|f(\xi)\| d\xi$. Moreover,

(i) Ω is positively minimal, invariant, and equi-almost periodic under T, the semigroup generated by A,

(ii) the restriction of T to $\overline{\text{co}}\ \Omega$ is an affine group of isometries,

(iii) $A\Omega$ is compact and lies on a sphere centered at 0.

(iv) $\overline{\text{co}}\ \Omega \subset \mathcal{D}(A)$, and

(v) the restriction of A to $\overline{\text{co}}\ \Omega$ is affine.

Proof: Set $Y = X \times L^1(\mathbb{R}^+;X)$ with norm

$$\|(u_0,f)\|_Y = \|u_0\| + \int_0^\infty \|f(\tau)\| d\tau.$$

For $t \in \mathbb{R}^+$, define $S(t): Y \to Y$ by

$$S(t)(u_0,f) = (u(t),f_t)$$

where u is the weak solution of Equation (6.4) and f_t is the t-translation of f. Now S is a strongly continuous semigroup of contractions on Y. Indeed, property (i) of Definition (2.1) is obvious. Property (ii) follows from the uniqueness of solutions to Equation (6.4). In fact, the proof is identical to the case for nonautonomous ordinary differential equations. (See the discussion following Definition 2.1 of Chapter IV.) Finally, property (iii) is a consequence of the continuity of the mappings $t \to u(t)$ and $t \to f_t$. An application of Lemma 5.7 readily establishes that $S(t)$ is a contraction in the norm $\|\cdot\|_Y$.

Next we show that S has a critical point $(x_0,0) \in Y$. So consider the strongly continuous semigroup of contractions T generated by the densely defined m-accretive operator A on the Hilbert space X. As $0 \in \mathscr{R}(A)$ there must exist $x_0 \in \mathscr{D}(A)$ with $Ax_0 = 0$. It was shown in the proof of Theorem 4.5 that x_0 is a critical point of T. Moreover, $T(\cdot)u_0$ is the unique strong solution of $\frac{du}{dt} + Au = 0$, $u_0 \in \mathscr{D}(A)$ according to Theorem 5.4. As a strong solution is necessarily a weak solution, then the definition of S implies $S(t)(x_0,0) = (x_0,0)$. Thus $(x_0,0)$ is a critical point of S.

We now will produce a generator B of S. Define $B\colon \mathscr{D}(B) \subset Y \to Y$ by

$$B(u_0,f) = (Au_0 - f(0),-\dot{f}).$$

If u is a strong solution of Equation (6.4), then

$$\frac{(u_0,f) - (u(t),f_t)}{t} = \left(-\frac{u(t) - u(0)}{t}, -\frac{f_t - f}{t}\right)$$

$$\to \left(-\frac{du}{dt}(0),-\dot{f}\right) = (Au_0 - f(0),-\dot{f}) \quad \text{as} \quad t \downarrow 0.$$

In view of Theorem 5.11 we may take for $\mathscr{D}(B)$ the set $\mathscr{D}(A) \times W^{1,1}(\mathbb{R}^+;X)$. Note that $\overline{\mathscr{D}(B)} = Y$.

The next step is to show that the positive motion $t \to S(t)(u_0,f)$ is compact for every $(u_0,f) \in \mathscr{D}(B)$. First observe that

$$\lim_{t\to\infty} \|f_t\| = \lim_{t\to\infty} \int_0^\infty \|f_t(\tau)\| d\tau = \lim_{t\to\infty} \int_t^\infty \|f(\tau)\| d\tau = 0.$$

Then the positive motion $t \to f_t$ is precompact. Without loss of generality we'll proceed by assuming the critical point x_0 of T satisfies $x_0 = 0$. Calculate for $\lambda > 0$, $t \in \mathbb{R}^+$,

$$\|(I + \lambda B)S(t)(u_0,f)\|_Y = \|(u(t),f_t) + \lambda B(u(t),f_t)\|_Y$$

$$= \|((I + \lambda A)u(t) - f(t), f_t - \dot{f}_t)\|_Y \geq \|(I + \lambda A)u(t) - f(t)\|$$

$$\geq \|(I + \lambda A)u(t)\| - \|f(t)\|.$$

Therefore

$$\|(I + \lambda A)u(t)\| \leq \|f(t)\| + \|(I + \lambda B)S(t)(u_0,f)\|_Y$$

$$\leq \|f(t)\| + \|S(t)(u_0,f)\|_Y + \lambda\|BS(t)(u_0,f)\|_Y.$$

We show the right side of the last inequality is bounded in $t \in \mathbb{R}^+$. Since $f \in W^{1,1}(\mathbb{R}^+;X)$, then

$$\|f(t)\| \leq \|f(t)-f(0)\| + \|f(0)\| \leq \int_0^\infty \|\dot{f}(\tau)\| d\tau + \|f(0)\|.$$

Next we have

$$\|S(t)(u_0,f)\|_Y \leq \|(u_0,f)\|_Y, \quad \|BS(t)(u_0,f)\|_Y \leq \|B(u_0,f)\|_Y.$$

Thus $(I + \lambda A)u(t)$ is bounded in X. As $(I + \lambda A)^{-1}$ is compact, then the positive motion $t \to u(t)$ must be compact. Consequently, the positive orbit $\gamma^+(u_0,f)$ of the semidynamical system (Y,S) is precompact.

Now we prove that $\gamma^+(u_0,f)$ is precompact for every $(u_0,f) \in Y$. Define $C = \{(u_0,f) \in Y\colon \gamma^+(u_0,f)$ is precompact$\}$. We show that C is closed. Let $(u_0,f) \in \overline{C}$. There exists a sequence $\{(u_{0,n},f_n)\} \subset C$ with $(u_{0,n},f_n) \to (u_0,f)$ in Y. Denote by u_n the weak solution of $\frac{du_n}{dt} + Au_n = f_n$,

$u_n(0) = u_{0,n}$, and by u the weak solution $\frac{du}{dt} + Au = f$, $u(0) = u_0$. Suppose $\{\tau_m\}$ is any sequence in $\mathbb{R}^+$. We will demonstrate that $\{u(\tau_m)\}$ admits a convergent subsequence. By choosing subsequences if necessary, we may assume $\{u_n(\tau_m)\}$ is convergent for each fixed $n \in \mathbb{N}$. This is because each $\gamma^+(u_{0,n},f_n)$ is precompact. From Lemma 5.7 we obtain

$$\|u_n(t)-u(t)\| \leq \|u_{0,n}-u_0\| + \int_0^\infty \|f_n(\tau)-f(\tau)\| d\tau.$$

Fix $\varepsilon > 0$. There exists $n_0 \in \mathbb{N}$ so that

$$\|u_{n_0}(t)-u(t)\| < \tfrac{1}{2}\varepsilon \quad \text{for all} \quad t \in \mathbb{R}^+.$$

Also there exists a positive integer N so that $k,m \geq N$ implies

$$\|u_{n_0}(\tau_k)-u_{n_0}(\tau_m)\| < \tfrac{1}{4}\varepsilon.$$

So for $k,m \geq N$,

$$\begin{aligned}\|u(\tau_k)-u(\tau_m)\| &\leq \|u(\tau_k)-u_{n_0}(\tau_k)\| + \|u_{n_0}(\tau_k)-u_{n_0}(\tau_m)\| \\ &\quad + \|u_{n_0}(\tau_m)-u(\tau_m)\| < \varepsilon.\end{aligned}$$

Therefore $\{u(\tau_m)\}$ is Cauchy, and so it converges. This means $\gamma^+(u_0,f)$ is precompact, whereby $(u_0,f) \in C$. We have just shown $\overline{C} = C$. As $\mathscr{D}(B) \subset C$ was shown earlier, then $Y = \overline{\mathscr{D}(B)} \subset \overline{C} = C$.

We have seen that $\lim_{t\to\infty} f_t = 0$ in $L^1(\mathbb{R}^+;X)$. Thus the positive limit set of (u_0,f) is of the form $L^+(u_0,f) = \Omega \times \{0\}$ where $\Omega = \{v_0 \in X: v_0 = \lim_{n\to\infty} u(t_n)$ for some sequence $\{t_n\} \subset \mathbb{R}^+$ with $t_n \to \infty\}$ and u is the weak solution of $\frac{du}{dt} + Au = f$, $u(0) = u_0$. But $L^+(u_0,f)$ is compact, hence Ω is compact in X. Now apply Theorem 6.1. $L^+(u_0,f)$ is positively minimal, invariant, and equi-almost periodic with respect to S. As points in $L^+(u_0,f)$ take the form $(v_0,0)$, then $S(t)(v_0,0) = (T(t)v_0,0)$ for all $t \in \mathbb{R}^+$. This means that Ω is also positively minimal, invariant, and equi-almost periodic with respect to T. In addition, $S(t)$ is an isometry on $L^+(u_0,f)$, so in view of the norm $\|\cdot\|_Y$, T must be an isometry on Ω. Finally, as $L^+(u_0,f) \subset S_r(0,0)$ where $r \leq \|(u_0,f)\|_Y$, then $\Omega \subset S_r(0)$ in X.

The next step is to show that the restriction of T to $\overline{\text{co}}\,\Omega$ is an affine group of isometries. Consider $L^+(u_0,f)$. The proof of Theorem 6.1 (ii) showed that there exists a sequence $\{t_n\} \subset \mathbb{R}^+$ with $t_n \to \infty$ so that $\lim_{n\to\infty} S(t_n)y_0 = y_0$ for every $y_0 \in L^+(u_0,f)$. As $y_0 = (v_0,0)$ for some $v_0 \in \Omega$, we may write $\lim_{n\to\infty} T(t_n)v_0 = v_0$ for every $v_0 \in \Omega$. Define the set

$$\mathscr{S} = \{v \in X: \lim_{n\to\infty} T(t_n)v = v\}$$

$\mathscr{S}$ is closed and positively invariant with respect to T. Also, $\Omega \subset \mathscr{S}$. Now T is an isometry on $\mathscr{S}$. If not, there exists $v_1,v_2 \in \mathscr{S}$ with $\varepsilon = \|v_1-v_2\| - \|T(t)v_1-T(t)v_2\| > 0$ for some $t \in \mathbb{R}^+$. Choose $t_n \geq t$ so that $\|T(t_n)v_i-v_i\| < \frac{1}{2}\varepsilon$, $i = 1,2$. Then

$$\varepsilon \leq \|v_1-v_2\| - \|T(t_n)v_1-T(t_n)v_2\| \leq \|v_1-T(t_n)v_1\| + \|v_2-T(t_n)v_2\| < \varepsilon \quad ,$$

a contradiction. Thus T is an isometry on $\mathscr{S}$. The proof of Theorem 6.2 (i) shows that the restriction of T to $\overline{\text{co}}\,\mathscr{S}$ is an affine group. We use this fact to show $\mathscr{S}$ is convex. So let $v = \alpha v_1 + (1-\alpha)v_2$, $v_1,v_2 \in \mathscr{S}$. Then

$$T(t_n)v = \alpha T(t_n)v_1 + (1-\alpha)T(t_n)v_2.$$

Letting $t_n \to \infty$ we obtain

$$\lim_{n\to\infty} T(t_n)v = \alpha v_1 + (1-\alpha)v_2 = v,$$

which implies $v \in \mathscr{S}$. As $\mathscr{S}$ is a closed convex set, then $\overline{co}\ \Omega \subset \mathscr{S}$. So T is an isometry on $\overline{co}\ \Omega$.

Now we demonstrate that $A\Omega$ lies on a sphere centered at 0. It will be sufficient to prove that $\|Au\| = \|Av\|$ for all $u,v \in \Omega$. So suppose $\|Au\| \leq \|Av\|$. As Ω is positively minimal under T, there exists a sequence $\{t_n\} \subset \mathbb{R}^+$ with $v = \lim_{n\to\infty} T(t_n)u$. As $\|AT(t_n)u\| \leq \|Au\|$, we may assume (by choosing a subsequence if necessary) that $\{AT(t_n)u\}$ converges weakly to some point $y \in X$. But A is m-accretive, so

$$\langle AT(t_n)u-A\xi, T(t_n)u-\xi\rangle \geq 0$$

for all $\xi \in \mathscr{D}(A)$. The inequality is preserved under limits, so

$$\langle y-A\xi, v-\xi\rangle \geq 0 \quad \text{for all} \quad \xi \in \mathscr{D}(A).$$

It follows from Lemma 5.8 that $y = Av$. Therefore

$$\|Av\| = \|y\| \leq \liminf_{n\to\infty} \|AT(t_n)u\| \leq \|Au\| \leq \|Av\|.$$

Hence $\|Au\| = \|Av\|$.

To show compactness of $A\Omega$, let $\{y_n\} \subset A\Omega$, $y_n = Ax_n$ with $\{x_n\} \subset \Omega$. Since Ω is compact and $\{y_n\}$ is bounded (lies on a sphere), we may assume (by choosing a subsequence if necessary) that $\{x_n\}$ converges (strongly) to some point $x \in \Omega$ and that $\{y_n\}$ converges weakly to some $y \in X$. As before we conclude that $Ax = y$ and

$$\|Ax\| \leq \liminf_{n\to\infty} \|Ax_n\| = \|Ax\|$$

as $A\Omega$ lies on a sphere centered at 0. Thus $\lim_{n\to\infty} \|y_n\| = \|y\|$, so $\{y_n\}$ converges (strongly) to y. Therefore, $A\Omega$ is compact.

Now suppose $u_0 \in \mathscr{D}(A)$. We'll show that $\overline{\text{co}}\ \Omega \subset \mathscr{D}(A)$. First we prove $\Omega \subset \mathscr{D}(A)$. Let $v_0 \in \Omega$, say $v_0 = \lim_{n\to\infty} u(t_n)$ for some sequence $\{t_n\}$ with $t_n \to \infty$. From the fact established earlier that $(I + \lambda A)u(t)$ is bounded in X and that $\{u(t)\colon t \in \mathbb{R}^+\}$ is precompact in X, it follows that $\{\|Au(t_n)\|\}$ is bounded. We may assume (by choosing a subsequence if necessary) that $\{Au(t_n)\}$ converges weakly to some point $y_0 \in X$. Once again we conclude that $y_0 = Av_0$. Therefore $v_0 \in \mathscr{D}(A)$. Since the restriction of T to $\overline{\text{co}}\ \Omega$ is affine, we must have $\text{co}\ \Omega \subset \mathscr{D}(A)$ and the restriction of A to $\text{co}\ \Omega$ is also affine. We have already shown that $A\Omega$ is compact, so $A(\text{co}\ \Omega)$ is bounded. Finally we use Lemma 5.7 another time to get $\overline{\text{co}}\ \Omega \subset \mathscr{D}(A)$ and that the restriction of A to $\overline{\text{co}}\ \Omega$ is affine. □

We return to Example 5.12 and demonstrate that the positive orbits are precompact.

Example 6.4. We have already proved in Example 5.12 that the operator $A\colon X = H_0^1 \times L^2 \to X$ with domain $\mathscr{D}(A) = (H_0^1 \cap H^2) \times H_0^1$ given by $A(u,v) = (-v, -\Delta u + aq(v))$ is densely defined and m-accretive. As $q(0) = 0$, then $A(0,0) = (0,0)$. So in order to use Theorem 6.3 we need only prove $(I+A)^{-1}$ is compact. In view of Equation (4.6) it will be sufficient to show A^{-1} is compact.

First observe that for any $(g,h) \in X$ we may write

$$A^{-1}(g,h) = (\Delta^{-1}[aq(g)-h], -g).$$

Now let $\{(g_n,h_n)\}$ be a bounded sequence in $H_0^1 \times L^2$. We claim $\{aq(g_n)-h_n\}$ is a bounded sequence in L^2. Indeed, in view of the hypotheses of Example 5.12, we may write

$$|q(y)| = |q(y)-q(0)| \leq |q'(\xi)||y-0| \leq Q|y|,$$

where q' is bounded on $\mathbb{R}$ by $Q > 0$. Consequently, $|a(x)q(g_n(x))| \leq |a(x)g_n(x)|$ for all $x \in [0,1]$. It follows that $\{aq(g_n)-h_n\}$ is bounded in L^2 as both $\{aq(g_n)\}$ and $\{h_n\}$ are.

It was shown in Example 4.2 that Δ^{-1} is compact. Thus $\{\Delta^{-1}[aq(g_n)-h_n]\}$ must converge (strongly) in L^2. By assumption $\{g_n\} \subset H_0^1$, so $\{g_n\}$ must converge strongly in L^2 by the Sobolev embedding theorem. Therefore $A^{-1}(g_n,h_n)$ converges (strongly) in $X = H_0^1 \times L^2$, hence A^{-1} is compact. Thus we may apply Theorem 6.3 to characterize the limit sets.

Let us consider only the special case when $f = 0$ and use a Liapunov function to locate the limit set. Choose for the Liapunov function the (square of the) norm of the space X:

$$V(u,v) = \frac{1}{2}\int_0^1 [(u')^2+v^2]dx.$$

Now compute

$$\dot{V}(u,v) = \frac{d}{dt}V(u,v) = -\int_0^1 a(x)q(v(x))v(x)dx \leq 0. \tag{6.5}$$

Fix any $w_0 = (u_0,v_0) \in X$. We know from Chapter II, Section 8 that V is constant on the positive limit set $L^+(w_0)$. Consequently if $\hat{w}_0 = (\hat{u}_0,\hat{v}_0) \in L^+(w_0)$, then $\dot{V}(T(t)\hat{w}_0) = 0$ for all $t \in \mathbb{R}^+$, where T is the semigroup generated by A. Write $\hat{w}(t) = (\hat{u}(t),\hat{v}(t)) = T(t)\hat{w}_0$. In view of the Inequality (6.5) and the assumptions on the function a (see Example 5.12), we must have

$$\hat{v}(t) = 0 \quad \text{for all} \quad t \in \mathbb{R}^+ \quad \text{and} \quad x \in \text{supp } a.$$

Here, supp a is the closure of the points $x \in [0,1]$ for which $a(x) > 0$. For all such x, $\hat{u}(t)$ becomes a solution to the undamped wave equation $\frac{\partial^2 u}{\partial t^2} = \Delta u$. The technique of separation of variables provides a solution to this equation in L^2, namely

$$\hat{u}(t) = \text{Re} \sum_{k=1}^{\infty} e^{i\theta_k t} \phi_k, \tag{6.6}$$

where ϕ_k is a solution of the eigenvalue equation in L^2

$$\begin{aligned} &\Delta\phi + \theta^2\phi = 0, \quad x \in [0,1] \\ &\phi(0) = \phi(1) = 0. \end{aligned} \tag{6.7}$$

Recall that $v = \frac{du}{dt}$, so from Equation (6.6)

$$\hat{v}(t) = \text{Re} \sum_{k=1}^{\infty} i\theta_k e^{i\theta_k t} \phi_k, \quad x \in \text{supp } a.$$

According to the conclusion of Theorem 6.3, $\hat{w}(t) = (\hat{u}(t),\hat{v}(t))$ is almost periodic in t. We then have (see Appendix A for some properties of almost periodic functions) $\phi_k(x) = 0$ for all $x \in$ supp a. Since solutions to Equation (6.7) are analytic in $(0,1)$ and supp a contains an open set containing x_0 where $a(x_0) > 0$, it follows that $\phi_k(x) = 0$ for all $x \in [0,1]$. Thus

$$\hat{w}(t) = (\hat{u}(t),\hat{v}(t)) = 0 \quad \text{for} \quad t \in \mathbb{R}^+, \ x \in [0,1].$$

So all solutions of the damped wave equation (5.5) tend to zero as $t \to \infty$. Note that $L^+(u_0,v_0) = (0,0)$.

This example suggests a representation of the form of Equation (6.6) for solutions of the Cauchy problem in positive limit sets. The almost periodicity guaranteed by

Theorem 6.3 makes such a representation reasonable. We refer the reader to Appendix A for a summary of results concerning the Fourier series of an almost periodic motion.

Theorem 6.5. Let X be a real Hilbert space and $A: \mathscr{D}(A) \subset X \to X$ a densely defined m-accretive operator. Assume $A^{-1}0 = 0$ and $(I + \lambda A)^{-1}$ is compact for some $\lambda > 0$. If T is the contraction semigroup generated by A, and $\hat{T}$ the restriction of T to $\overline{co}\, L^+(u_0) \in \mathscr{D}(A)$, then

(i) $\hat{T}$ is linear and its infinitesimal generator $\hat{A}$ coincides with A on $\overline{co}\, L^+(u_0)$,

(ii) 0 is the only critical point of T, and $0 \in \overline{co}\, L^+(u_0)$,

(iii) if $v \in L^+(u_0)$, then $T(t)$ admits the representation

$$T(t)v = \hat{T}(t)v \sim \sum_{k=1}^{\infty} e^{i\theta_k t} v_k, \tag{6.8}$$

where

$$v_k \in \overline{co}\, L^+(u_0), \quad \hat{A}v_k = i\theta_k v_k \quad \text{for each } k \in \mathbb{N}.$$

If $y \in \overline{co}\, L^+(u_0)$, then

(iv) $\langle \hat{A}y, y\rangle = \frac{1}{2}\frac{d}{dt}\|\hat{T}(t)y\|^2\Big|_{t=0} = 0$, and

(v) if the equation $\hat{A}v_k = i\theta_k v_k$ admits only the trivi- trivial solution $v_k = 0$ for every $k \in \mathbb{N}$, then $\lim_{t\to\infty} T(t)v = 0$.

Proof: (i) $A^{-1}0 = 0$ implies that $x_0 = 0$ is the only critical point of T. According to Theorem 6.2, then $\hat{T}$ is linear. A direct computation of the infinitesimal generator for $\hat{T}$ shows

$$\hat{A}x = \lim_{t \downarrow 0} \frac{x - \hat{T}(t)x}{t} = \lim_{t \downarrow 0} \frac{x - T(t)x}{t} = Ax$$

for every $x \in \mathscr{D}(\hat{A})$, since $T(t)x$ is the strong solution of $\frac{du}{dt} + Au = 0$, $u(0) = x$.

(ii) Since $L^+(u_0)$ is compact, then $\overline{co}\, L^+(u_0)$ is closed and bounded, hence $\overline{co}\, L^+(u_0)$ is weakly compact. Again, according to Theorem 6.2, $\overline{co}\, L^+(u_0)$ must contain the unique critical point 0.

(iii) $L^+(u_0)$ is positively minimal and equi-almost periodic. Accordingly, if $v \in L^+(u_0)$, then the motion $T(\cdot)v$ is almost periodic. Moreover, $T(t)v$ admits the representation given by Equation (6.8). (See Appendix A.) The linearity of $\hat{A}$ implies

$$\hat{A}v_k = \lim_{t \to \infty} \frac{1}{t} \int_0^t \hat{A}e^{i\theta_k s}\, \hat{T}(s)v\,ds = i\theta_k v_k.$$

The proof of Theorem 6.2 (iv) shows that v_k must belong to $\overline{co}\, L^+(u_0)$.

(iv) Because of the linearity of $\hat{T}$ we may compute

$$\frac{1}{2}\frac{d}{dt}\|\hat{T}(t)y\|^2 = \frac{1}{2}\frac{d}{dt}\langle \hat{T}(t)y, \hat{T}(t)y\rangle = \langle \frac{d}{dt}\hat{T}(t)y, \hat{T}(t)y\rangle$$
$$= -\langle \hat{A}\hat{T}(t)y, \hat{T}(t)y\rangle.$$

Evaluate at $t = 0$ to get $\langle \hat{A}y, y\rangle = -\frac{1}{2}\frac{d}{dt}\|\hat{T}(t)y\|^2\Big|_{t=0}$. For $y \in \overline{co}\, L^+(u_0)$, $\hat{T}(t)$ is an isometry, so $\|\hat{T}(t)y\|$ is constant for all $t \in \mathbb{R}^+$. Thus property (iv) obtains.

(v) $L^+(u_0) = \{0\}$ as $Av = i\theta v$ admits only the trivial solution. Consequently $\lim_{t \to \infty} T(t)u_0 = 0$ by Theorem 3.11 of Chapter II. □

Remark 6.6. The relation $\langle \hat{A}y, y\rangle = 0$ can be used to delimit $\mathscr{D}(\hat{A})$. Even though T may be nonlinear, its asymptotic behavior is characterized by the linear operator $\hat{A}$.

Example 6.7. We return to Example 4.6 and characterize the limiting behavior of the solutions. The eigenvalue equation takes the form

$$\left.\begin{aligned} v &= -i\theta u \\ \Delta u - u &= -i\theta v \end{aligned}\right\} \quad \text{on} \quad \Lambda$$

and

$$v = 0, \quad \frac{\partial u}{\partial n} = 0 \quad \text{on} \quad \partial\Lambda.$$

Eliminating v we obtain

$$\begin{aligned} &\Delta u + (\theta^2 - 1)u = 0 \quad \text{on} \quad \Lambda \\ (6.9)\qquad &u = 0, \quad \frac{\partial u}{\partial n} = 0 \quad \text{on} \quad \partial\Lambda. \end{aligned}$$

Due to the uniqueness of solutions of the Cauchy problem, Equations (6.9) admit only the trivial solution. Thus, all solutions of Equations (4.3) and (4.4) tend to zero as $t \to \infty$.

7. Exercises

7.1. Prove that if $T \in Q_\omega$ then $A_T + \omega I$ is accretive (A_T is the infinitesimal generator of T).

7.2. Suppose B is accretive on X. Show that B is m-accretive if and only if there exists $\lambda > 0$ so that $\mathscr{R}(I + \lambda B) = X$.

7.3. Prove that if $B\colon X \to X$ is a contraction, then $I + B$ is accretive.

7.4. Let $X = C([-1,1]; \mathbb{R})$ and $\mathscr{K} = \{f \in X\colon -1 \le f(x) \le x, x \in [-1,1]\}$. Define $A\colon \mathscr{K} \to \mathscr{K}$ by

$$(Af)(x) = \begin{cases} -1, & f(x) < x, \\ 0, & f(x) = x \end{cases}$$

and let $T(t): X \to X$ by $(T(t)f)(x) = \min\{t+f(x),x\}$. Prove that

(i) $\mathscr{K}$ is convex,

(ii) T is a contraction on $\mathscr{K}$,

(iii) $T(t)f$ is differentiable at $t = 0$ if and only if $f(x) \equiv x$ on $[-1,1]$.

7.5. Let $A + \omega I$ be accretive with $\mathscr{R}(I + \lambda A) \supset \overline{\mathscr{D}(A)}$ for all $\lambda > 0$ so that $\lambda\omega < 1$. Suppose $\{x_n\} \subset \mathscr{D}(A)$ where $\lim_{n\to\infty} x_n = x \in \mathscr{R}(I + \lambda A)$ and $\{|Ax_n|\}$ is bounded. Show that $x \in \hat{\mathscr{D}}(A)$.

7.6. Verify that if $x,y,z \in X$, $\alpha \in \mathbb{R}$, then

(i) $\langle x+\alpha y,y\rangle_s = \langle x,y\rangle_s + \alpha\|y\|^2$,

(ii) $\langle x+z,y\rangle_s \le \langle x,y\rangle_s + \|z\|\,\|y\|$.

7.7. If $A + \omega I$ is m-accretive for some $\omega \in \mathbb{R}$, prove that

(i) $|Ax| \le \|Ax\|$,

(ii) $\mathscr{D}(A) \subset \hat{\mathscr{D}}(A)$,

(iii) $|Ax| = \sup\{(1-\lambda\omega)\,\|A_\lambda x\|: 0 < \lambda < \omega^{-1}\}$,

(iv) The map $x \to |Ax|$ is lower semicontinuous.

7.8. Let $X = C_0(\mathbb{R}^+;\mathbb{R})$ be the set of all continuous real valued functions on $\mathbb{R}^+$ for which $|f(x)| \to 0$ as $x \to \infty$. X is a Banach space with norm $\|f\| = \sup\{|f(x)|: x \in \mathbb{R}^+\}$. Define the operator A on X by $Af = -f'$ where $\mathscr{D}(A) = \{f \in X: f' \in X\}$. Let T be the semigroup generated by A. Show

(i) $T(t)f(x) = f(x+t)$ for all $x,t \in \mathbb{R}^+$, $f \in X$,

(ii) $A_\lambda f(x) = \lambda^{-2}\int_x^\infty \exp[(x-s)/\lambda](f(x)-f(s))ds$,

(iii) $\hat{\mathscr{D}}(A) = \{f \in X: f$ is Lipschitz continuous$\}$,

(iv) $|Af|$ is the least Lipschitz constant for f.

7.9. Let T be a quasi-contracting semigroup on a closed subset C of X with $T(t)C \subset C$ for every $t \in \mathbb{R}^+$. Suppose the infinitesimal generator A_T satisfies $\mathcal{D}(A_T) \subset C \subset \overline{\mathcal{D}(A)}$, and that $u(t) = T(t)x$ is the unique solution to the Cauchy problem:

$$\frac{du}{dt} + A_T u(t) = 0, \quad u(0) = x \in \mathcal{D}(A).$$

If $P: C \to X$ is a homeomorphism with $P(C) \subset C$ and $P\mathcal{D}(A) \subset \mathcal{D}(A)$, and if P admits a linear extension to all of X, then the condition $PA_T x = A_T Px$ for all $x \in \mathcal{D}(A)$ is equivalent to $PT(t)x = T(t)Px$ for all $(x,t) \in C \times \mathbb{R}^+$.

7.10. Let T, A_T, and u be as in problem 7.9. Suppose $A_T = L + N$, $\mathcal{D}(A) = \mathcal{D}(N) = \mathcal{D}(L) \cap C$, N is bounded on bounded sets and L is m-accretive. Moreover assume $J_\lambda = (I + \lambda L)^{-1}$ is compact for some small $\lambda > 0$. If $\gamma^+(x)$ is bounded for $x \in \mathcal{D}(A)$, show that $\gamma^+(x)$ is precompact and $T(\cdot)x$ is uniformly Lipschitz continuous on $\mathbb{R}^+$.

7.11. Suppose T is a linear semigroup on X with a densely defined infinitesimal generator A_T. Suppose there exists a polynomial p so that $p(A)$ is compact, and $\mathcal{D}(p(A)) = X$, $\overline{\mathcal{R}(p(A))} = X$. Show that if $u = 0$ is a stable solution of $\frac{du}{dt} + A_T u = 0$, then all positive orbits are precompact.

7.12. Let X be a Hilbert space, $A: \mathcal{D}(A) \subset X \to X$ a densely defined, self adjoint linear operator with $A + \omega I$ m-accretive for some $\omega < 0$. If $f_0 \in X$ and u is a strong solution of $\frac{du}{dt} + Au = f_0$, show there exists $v_0 \in X$ so that $\lim_{t\to\infty} u(t) = v_0$.

7.13. Let $T \in Q_\omega$ have domain X. If every positive orbit $\gamma^+(x)$, $x \in X$, is precompact, show that $\cup\{L^+(x): x \in X\}$ is a closed convex set.

8. Appendix: Proofs of Theorems 2.4 and 2.16

The proof of Theorem 2.4 requires the following preparatory lemma.

Lemma 8.1. Let $\lambda \in \mathbb{R}^+$ with $\lambda\omega < 1$. Then J_λ is $(1-\lambda\omega)^{-1}$-Lipschitz on $\mathscr{D}(J_\lambda)$.

Proof: The proof of Lemma 2.5 shows that $\|(I+\lambda A)x-(I+\lambda A)y\| \geq (1-\lambda\omega)\|x-y\|$ for all $x,y \in \mathscr{D}(A)$. Set $x = J_\lambda u$, $y = J_\lambda v$ for $u,v \in \mathscr{D}(J_\lambda)$. Then

$$\|J_\lambda u-J_\lambda v\| \leq (1-\lambda\omega)^{-1} \|u-v\| . \qquad \square$$

Lemma 8.2. Let $\lambda \geq \mu > 0$, $\omega\lambda < 1$, and $n \geq m$ be positive integers. Then for each $x \in \mathscr{D}(J_\lambda^m) \cap \mathscr{D}(J_\mu^n)$,

$$(8.1) \quad \|J_\mu^n x - J_\lambda^m x\| \leq (1-\mu\omega)^{-n} \sum_{j=0}^{m-1} \alpha^j \beta^{n-j} \binom{n}{j} \|J_\lambda^{m-j} x-x\| + \sum_{j=m}^{n} (1-\mu\omega)^{-j} \alpha^m \beta^{j-m} \binom{j-1}{m-1} \|J_\mu^{n-j} x-x\|,$$

where $\alpha = \mu/\lambda$ and $\beta = (\lambda-\mu)/\lambda$.

Proof: For $0 \leq j \leq n$, $0 \leq k \leq m$, set $a_{k,j} = \|J_\mu^j x-J_\lambda^k x\|$. If $j > 0$, $k > 0$, then

$$\begin{aligned} a_{k,j} &= \|J_\mu^j x - J_\mu(\tfrac{\mu}{\lambda} J_\lambda^{k-1} x + \tfrac{\lambda-\mu}{\lambda} J_\lambda^k x)\| \\ &\leq (1-\mu\omega)^{-1}\|J_\mu^{j-1} x - \tfrac{\mu}{\lambda} J_\lambda^{k-1} x + \tfrac{\lambda-\mu}{\lambda} J_\lambda^k x)\| \\ &\leq (1-\mu\omega)^{-1} [\tfrac{\mu}{\lambda} \|J_\mu^{j-1} x - J_\lambda^{k-1} x\| + \tfrac{\lambda-\mu}{\lambda}\|J_\mu^{j-1} x-J_\lambda^k x\|]. \end{aligned}$$

Set $\alpha_1 = (1-\mu\omega)^{-1}(\mu/\lambda)$ and $\beta_1 = (1-\mu\omega)^{-1}(\lambda-\mu)/\lambda$. Solve the inequalities

$$a_{k,j} \leq \alpha_1 a_{k-1,j-1} + \beta_1 a_{k,j-1}$$

to estimate $a_{m,n}$ in terms of $a_{k,0}$ and $a_{0,j}$ in precisely the form indicated in Inequality (8.1) when $n \geq m$. If $n \leq m$ the estimate becomes

$$(8.2) \qquad \|J_\mu^n x - J_\lambda^m x\| \leq (1-\mu\omega)^{-n} \sum_{j=0}^{n} \alpha^j \beta^{n-j} \binom{n}{j} \|J_\lambda^{m-j} x - x\| . \qquad \square$$

Our last lemma before proving Theorem 2.4 is a technical result whose proof may be found in Crandall and Liggett [1].

<u>Lemma 8.3</u>. Let $n \geq m > 0$ be integers, and $\alpha + \beta = 1$. Then

$$\text{(i)} \qquad \sum_{j=0}^{m} \binom{n}{j} \alpha^j \beta^{n-j} (m-j) \leq [(n\alpha-m)^2 + n\alpha\beta]^{\frac{1}{2}}$$

$$\text{(ii)} \qquad \sum_{j=m}^{n} \binom{j-1}{m-1} \alpha^m \beta^{j-m} (n-j) \leq [m\beta/\alpha^2 + (m\beta/\alpha + m-n)^2]^{\frac{1}{2}}$$

<u>Proof of Theorem 2.4</u>: Let $x \in \mathscr{D}(A)$ and $n \geq m > 0$ be integers, $\lambda \geq \mu > 0$ be real numbers with $\lambda\omega < 1$. We'll first show the sequence $\{J_{t/n}^n x\}$, $t \in \mathbb{R}^+$ is Cauchy.

$$\|J_\mu^n x - J_\lambda^m x\| \leq (1-\mu\omega)^{-m} \lambda \sum_{j=0}^{m-1} \binom{n}{j} \alpha^j \beta^{n-j} (m-j)$$
$$+ (1-\mu\omega)^{-2n} \sum_{j=m}^{n} \binom{j-1}{m-1} \alpha^m \beta^{j-m} (n-j) \|Ax\| .$$

If $t \in [0,\frac{1}{2}]$, it is easy to verify that $(1-t)^{-n} \leq e^{2nt}$. If $\lambda\omega < \frac{1}{2}$, then using Inequality (8.2) and (i) of Lemma 8.3 we get

$$(8.3) \quad \|J_\mu^n x - J_\lambda^m x\| \leq \{[(n\mu-\lambda m)^2 + n\mu(\lambda-\mu)]^{\frac{1}{2}} e^{2\omega(n\mu+m\lambda)}$$
$$+ [m\lambda(\lambda-\mu) + (m\lambda-n\mu)^2]^{\frac{1}{2}} e^{4\omega n\mu}\} \|Ax\| .$$

Take $\mu = t/n$, $\lambda = t/m$. Then

$$\|J^n_{t/n}x - J^m_{t/m}x\| \le 2te^{4\omega t}[m^{-1}-n^{-1}]^{\frac{1}{2}}\|Ax\|.$$

If $0 < n \le m$, we use the Inequality (8.2) to obtain a similar bound. Therefore, $\{J^n_{t/n}x\}$ is Cauchy for each $t \in \mathbb{R}^+$. Thus $\lim_{n\to\infty} J^n_{t/n}x$ exists for $x \in \mathcal{D}(A)$, $t \in \mathbb{R}^+$.

$J^n_{t/n}x$ is also $(1-\omega t/n)^{-1}$-Lipschitz. As $\lim_{n\to\infty}(1-\omega t/n)^{-n} = e^{\omega t}$, we see that $T(t)x$ as defined by Equation (2.2) exists for $x \in \overline{\mathcal{D}(A)}$ and $t \in \mathbb{R}^+$. Moreover $T(t)$ is $e^{\omega t}$-Lipschitz.

We now verify properties (i), (ii), and (iii) of Definition 2.1. Property (i) is trivial. To obtain property (iii) let $x \in \mathcal{D}(A)$, $s > t \ge 0$ be real numbers, and take the limit in the Inequality (8.3) as $n \to \infty$ with $n = m$, $\mu = t/n$, $\lambda = s/n$. We get

$$\|T(s)x - T(t)x\| \le 2e^{4\omega s}\|Ax\|(s-t). \tag{8.4}$$

Thus the mapping $t \to T(t)x$ is Lipschitz continuous in t on bounded subsets of $\mathbb{R}^+$. The continuity naturally extends to $x \in \overline{\mathcal{D}(A)}$.

Lastly, we verify the semigroup property (ii). We demonstrate that property (ii) holds for rationals s and t. We extend (ii) to all $s,t \in \mathbb{R}^+$ by the continuity of T in t and x. Now if $t \in \mathbb{R}^+$

$$[T(t)]^m = \lim_{n\to\infty}[J^n_{t/n}]^m = \lim_{n\to\infty}[J^m_{t/n}]^n.$$

Also

$$[T(mt)] = \lim_{n\to\infty} J^n_{mt/n} = \lim_{k\to\infty} J^{mk}_{mt/mk} = \lim_{k\to\infty}[J^m_{t/k}]^k = [T(t)]^m.$$

If k, m, r, and s are positive integers, then

$$T(\frac{r}{k} + \frac{m}{s}) = T(\frac{rn+mk}{kn}) = [T(\frac{1}{kn})]^{rn+mk}$$

$$= [T(\frac{1}{kn})]^{rn}[T(\frac{1}{kn})]^{mk} = T(\frac{r}{k})T(\frac{m}{n}).$$

This concludes the proof of Theorem 2.4. □

The proof of the concluding statement of Theorem 2.16 namely that $T_\lambda(t)x \to T(t)x$ as $\lambda \downarrow 0$ is rather involved. It is accomplished through the next four lemmas. First define the semigroup S by

(8.5) $$S(t) = T_\lambda(\lambda t).$$

Then $S \in Q_{\alpha-1}$ where $\alpha = (1 - \lambda\omega)^{-1}$ is the Lipschitz constant for J_λ. Furthermore, for each $x \in X$ we have

(8.6) $$\frac{d}{dt} S(t)x = (J_\lambda - I)S(t)x.$$

<u>Lemma 8.4</u>. $\|S(t)x-x\| \le te^{(\alpha-1)t} \|J_\lambda x-x\|$.

<u>Proof</u>: Upon integrating Equation (8.6) we obtain

(8.7) $$S(t)x = e^{-t}x + \int_0^t e^{-(t-s)} J_\lambda S(s)x\, ds.$$

Also observe that as $J_\lambda - I$ is the infinitesimal generator of $S(t)$, then for any $h > 0$ we have

$$\left\|\frac{S(t+h)x-S(t)x}{h}\right\| \le e^{(\alpha-1)t} \left\|\frac{S(h)x-x}{h}\right\| \to e^{(\alpha-1)t} \|J_\lambda x-x\|$$

as $h \downarrow 0$.

Consequently in view of Equation (8.6) we have

(8.8) $$\|(J_\lambda - I)S(t)x\| \le e^{(\alpha-1)t} \|J_\lambda x-x\|.$$

Then

$$\|S(t)x-x\| = \|\int_0^t e^{-(t-s)}[J_\lambda S(s)x-x]ds\|$$

$$\leq \|\int_0^t e^{-(t-s)}[(J_\lambda - I)S(s)x]ds\| + \|\int_0^t e^{-(t-s)}[S(s)x-x]ds$$

$$\leq e^{-t}\Big[\int_0^t e^{\alpha s}ds\Big]\|J_\lambda x-x\| + e^{-t}\int_0^t e^{s}\|S(s)x-x\|\, ds.$$

It is left to the reader to verify that

$$\|S(t)x-x\| \leq e^{-t}\Big[\int_0^t \sum_{k=0}^{n} \frac{(t-s)^k}{k!} e^{\alpha s}ds\Big]\|J_\lambda x-x\|$$

$$+ \frac{e^{-t}}{n!}\int_0^t (t-s)^n\|S(s)x-x\|\, ds.$$

Let $n \to \infty$ and we obtain the desired inequality. □

Lemma 8.5.

$$(8.9) \qquad \|S(t)x-J_\lambda^m x\| \leq e^{-t}\alpha^{m-1} \sum_{j=0}^{\infty} \frac{|j-m|t^j\alpha^j}{j!}\|J_\lambda x-x\|.$$

Proof: If we use Equation (8.7) we obtain

$$S(t)x-J_\lambda^m x = e^{-t}(x-J_\lambda^m x) + \int_0^t e^{-(t-s)}[J_\lambda S(s)x-J_\lambda^m x]ds.$$

So

$$\|S(t)x-J_\lambda^m x\| \leq e^{-t}\, \|x-J_\lambda^m x\| + \alpha\int_0^t e^{-(t-s)}\|S(s)x-J_\lambda^{m-1}x\|\, ds.$$

Repeat the argument m times and obtain

$$(8.10) \qquad \|S(t)x-J_\lambda^m x\| \leq \sum_{j=0}^{m-1} \frac{t^j\alpha^j}{j!}\|x-J_\lambda^{m-j}x\|$$

$$+ \frac{\alpha^m}{(m-1)!}\int_0^t e^{-(t-s)}(t-s)^{m-1}\, \|S(s)x-x\|\, ds.$$

Moreover using Lemma 2.15(ii) we have

$$(8.11) \qquad \sum_{j=0}^{m-1} \frac{t^j\alpha^j}{j!}\|x-J_\lambda^{m-j}x\| \leq \alpha^{m-1}\sum_{j=0}^{m-1} \frac{(m-j)t^j\alpha^j}{j!}\, \|x-J_\lambda x\|\, ,$$

and from Lemma 8.4 that

$$\int_0^t e^{-(t-s)}(t-s)^{m-1}\|S(s)x-x\|ds$$

$$\le \int_0^t e^{-t+\alpha s}(t-s)^{m-1}s\|x-J_\lambda x\|\,ds$$

$$= e^{-t}\sum_{j=0}^{\infty}\frac{\alpha^j}{j!}\int_0^t(t-s)^{m-1}s^{j+1}\,\|x-J_\lambda x\|\,ds$$

$$(8.12)\qquad = (m-1)!e^{-t}\sum_{j=0}^{\infty}\frac{(j+1)\alpha^j t^{m+j+1}}{(m+j+1)!}\,\|x-J_\lambda x\|$$

$$\le \frac{(m-1)!}{\alpha}\,e^{-t}\sum_{j=m+1}^{\infty}\frac{(j-m)t^j\alpha^j}{j!}\|x-J_\lambda x\|\,.$$

Combine Inequalities (8.10), (8.11), and (8.12) to obtain the desired Inequality (8.9). □

Lemma 8.6.

$$\sum_{j=0}^{\infty}\frac{|j-m|m^j\alpha^j}{j!} \le e^{m\alpha}[m^2(\alpha-1)^2 + m\alpha]^{1/2}.$$

Proof: Apply the Schwarz inequality. □

We conclude the proof of Theorem 2.16, part (iii).

Proof: Choose a positive integer m so that $t = m\lambda + \delta$, $0 \le \delta < \lambda$. Then

$$\|T_\lambda(t)x - T(t)x\| \le \|T_\lambda(t)x - T_\lambda(m\lambda)x\| + \|T_\lambda(m\lambda)x - J_\lambda^m x\|$$
$$+ \|J_\lambda^m x - T(m\lambda)x\| + \|T(m\lambda)x - T(t)x\|\,.$$

We'll show that each of the terms on the right hand side admits an estimate which tends to zero as $\lambda \downarrow 0$.

(i) $\|T_\lambda(t)x - T_\lambda(m\lambda)x\| \le \delta\alpha\|Ax\| \le \lambda(1-\lambda\omega)^{-1}\|Ax\|$

in view of the fact that $T_\lambda(t)x$ is $\alpha\|Ax\|$-Lipschitz in t. Indeed, we already know from Theorem 2.16(ii) that

$$\|\frac{d}{dt} T_\lambda(t)x|_{t=0}\| = \|A_\lambda x\| \le \|Ax\|.$$

(ii) Set $t = m$ in the Inequality (8.9) and apply Lemma 8.6. We get

$$\|T_\lambda(m\lambda)x - J_\lambda^m x\| \le \alpha^{m-1} e^{m(\alpha-1)}[m^2(\alpha-1)^2+m\alpha]^{\frac{1}{2}} \|x-J_\lambda x\|$$
$$\le \alpha^{m-1}e^{\alpha\omega t}m[(\alpha-1)^2+\alpha^2]^{\frac{1}{2}}\lambda\alpha \|Ax\|$$
$$\le (1-\lambda\omega)^{-m}e^{\alpha\omega t}t[(\alpha-1)^2+\alpha^2]^{\frac{1}{2}} \|Ax\|.$$

(iii) Set $\mu = m\lambda/n$ in Inequality (8.3) and obtain

$$\|J_\lambda^m x - T(m\lambda)x\| \le 2\lambda\sqrt{m}\; e^{4\omega m\lambda} \|Ax\| \le 2\sqrt{\lambda t}\; e^{4\omega t} \|Ax\|.$$

(iv) As $T \in Q_\omega$ we have

$$\|T(m\lambda)x - T(t)x\| \le e^{\omega m\lambda} \|x-T(\delta)x\| \le e^{\omega t} \sup_{0\le s\le\lambda} \|x-T(s)x\|. \quad \square$$

9. Notes and Comments

Section 1. The heuristic approach to the generation of the solution of Equation (1.1) is borrowed from Crandall [5]. Indeed, this reference is a brief, but excellent introduction to the formulation of PDE's as abstract Cauchy problems in Banach space. The reference list in [5] covers most of the subject until 1975. It does not deal though with the asymptotic behavior of semigroups. Accretive operators are, in general, multivalued. As a simple example of such, let $f: \mathbb{R} \to \mathbb{R}$ be an increasing function. Define $F: \mathbb{R} \to 2^{\mathbb{R}}$ by $F(x) = [f(x^-),f(x^+)] \cap \mathbb{R}$. Then F is accretive. For a comprehensive study of nonlinear semigroups, see Barbu [1].

Section 2. The major result, Theorem 2.4 is due to Crandall and Liggett [1]. The proof presented here also holds for multivalued accretive operators. An important question not treated here is the converse question: determine whether a given $T \in Q_\omega$ on a closed convex set $\mathscr{C} \subset X$ has a generator. The problem was first treated by Kōmura [1] in Hilbert space. Recently Baillon [1] showed there is a densely defined accretive operator A on $\mathscr{C}$ which generates T provided the dual X^* is uniformly convex. The demonstration that $-\Delta\phi(u)$ of Example 2.10 is m-accretive is based on Crandall [2,5]. For a sampling of further applications of accretive operators to PDE's of the form $\frac{du}{dt} + Au = 0$ see Crandall [3] and Webb [3]. Theorem 2.13 was first proved by Brezis and Pazy [1] for the case $\omega = 0$. The approximation Theorem 2.16 is established in Kato [2]. The argument used here is based on Crandall and Liggett [1].

Section 3. The generalized domain $\hat{\mathscr{D}}(A)$ is due to Crandall [4]. Lemma 3.6 is from Crandall and Liggett [1]. Theorem 3.10 and the preceeding lemmas are also due to Crandall [4].

Section 4. Theorem 4.3 is due to Walker and Infante [1]. Theorem 4.4 was first stated and proved by Dafermos and Slemrod [1]. The proof here, as well as Theorem 4.5 is from Walker and Infante [1]. Example 4.6 is based on Dafermos [8]. See Hale and Infante [1] for another approach to establish precompactness of positive orbits.

Section 5. There are many notions for solutions of the Cauchy problem in a Banach space. A recent definition of Bénilan [2] appears to be the most appropriate for the case of accretive operators. Indeed, the problem arose from the fact that $T(\cdot)x$, whether generated by A via Theorem 2.4 or as $\lim_{\lambda \downarrow 0} T_\lambda(t)x$ via Theorem 2.16, need not be differentiable in t anywhere. But it can be shown that if u is a strong solution of Equation (5.2), then

$$\|u(t)-x\| - \|u(s)-x\| \leq \int_s^t \langle u(\tau)-x, f(\tau)-Ax\rangle_s d\tau$$

whenever $x \in \mathcal{D}(A)$ and $s \leq t$. Bénilan [2] takes this estimate as defining an integral solution of Equation (5.2). He shows that $T(\cdot)x$ is the unique integral solution of Equation (5.1). The "only if" part of Theorem 5.2 is from Crandall [2]. The "if" part is from Miyadera [1]. Example 5.12 is based on Dafermos [10].

In the event A is linear other modes of solutions are appropriate. Ball [2] develops a variation of constants formula for solutions of Equation (5.2) in the weak topology of X. Pazy [2] considers a similar problem. Other matters not touched on here which are appropriate for a discussion of solutions of differential equations are continuous dependence on A, f, and $u_0 \in \mathcal{D}(A)$. Again, see Crandall [5] for references. The case for time dependent accretive operators $A(t)$ has also been studied; c.f. Crandall and Liggett [1], Crandall and Pazy [2], Evans and Massey [1], and Kato [1].

The question of existence of solutions in Banach space is important. Examples by Cellina [1], Dieudonné [1], and Yorke [2] show that solutions of $\dot{x} + A(t)x = 0$ need not exist in non-reflexive spaces, even when $A(t)x$ is continuous. For some positive results, see Fitzgibbon [1], Li [1], and Martin [1]. For applications to integral equations, see Miller [1]; for delay equations, see Fitzgibbon [2] for example.

Section 6. Theorems 6.1, 6.2, and 6.3 are due to Dafermos and Slemrod [1]. Note that the set $\mathscr{L}$ in the proof of Theorem 6.3 that $T|_{\overline{\text{co}}\,\Omega}$ is an affine group of isometries and consists of the set of Poission stable points of X. (See Bhatia and Szegö [1] for definitions and characterization.) The argument here is based on Edelstein [1]. Theorem 6.5 as well as Example 6.7 are from Dafermos [8]. Additional asymptotic properties of contracting semigroups may be found in Baillon, Bruck and Reich [1], Ball [3], Bruck [1], Pao [1], Pazy [1,3], Reich [1,2], Walker [1], Yen [1], and Zaidman [1]. For applications to the asymptotic behavior of some special PDE's see Chafee [1,2], Chafee and Infante [1,2], Pao [2], and Pao and Vogt [1].

Section 8. The results here are due to Crandall and Liggett [1].

CHAPTER VI

FUNCTIONAL DIFFERENTIAL EQUATIONS

1. Why Hereditary Dependence; Some Examples From Biology, Mechanics, and Electronics

A distinguishing feature of ordinary differential equations is that the future behavior of solutions depends only upon the present (initial) values of the solution. Numerous physical, economic, biological, and social systems, though, exhibit hereditary dependence. That is, the future state of the system depends not only upon the present state, but also upon past states. Models of such systems must take into account this hereditary effect. We illustrate this with some examples.

Example 1.1. Elementary models of population growth or epidemiology do not take into account the effect of the lifetime of members of the population, the gestation period, or the delay before a member is able to reproduce. The simple models studied in the calculus do not adequately characterize the observed properties of many species. A model studied by Cooke and Yorke [1] assumes population growth according to

$$\dot{x}(t) = g(x(t)) - g(x(t-L)), \tag{1.1}$$

where $x(t)$ is the size of the population at time t, and $g(x)$ is the growth rate of the population when at size x. It is assumed here that each member has constant lifetime of L units of time. If the lifespan L is not constant but is a random variable with density $p(s)$, then Equation (1.1) becomes

$$\dot{x}(t) = g(x(t)) - g\left(\int_0^{L_m} x(t-s)p(s)ds\right),$$

where L_m is the maximum lifespan possible. Thus the growth rate at time t depends upon the population during $[t-L_m,t]$.

Example 1.2. The equation

$$\dot{x}(t) = -\alpha x(t-1)[1 + x(t)] \tag{1.2}$$

occurs in a number of areas. In Cunningham [1] it appears as a nonlinear model of population growth. It also appears in Wright [1] dealing with probability methods for the density of prime numbers. Finally ecological applications of (1.2) are presented in May [1].

Example 1.3. The theory of the mechanics and thermodynamics of materials with gradually fading memories (e.g., linear visco-elasticity) provides an example where the hereditary dependence is on the infinite interval $(-\infty,t]$. If the strain at time t is denoted by $x(t)$, and $h(t)$ is an external force,

$$\ddot{x}(t) = -G(0)x(t) - \int_{-\infty}^{0} \dot{G}(-\theta)x(t+\theta)d\theta + h(t). \tag{1.3}$$

This equation is treated later in Section 5. (See Example 5.2.)

Example 1.4. The "flip-flop" circuit is a basic element of a digital computer. Its importance lies in the fact that it possesses multiple equilibria, and thus is ideally suited for its role as a memory storage device. A distributed parameter model gives rise to the following equation

$$C \frac{d}{dt} [x(t) + Kx(t-r)] = -\frac{1}{z}x(t) + \frac{K}{z}x(t-r) - f(x(t)) - Kf(x(t-r)), \tag{1.4}$$

where $x(t)$ is the voltage output of the circuit at time t, f gives the current through a tunnel diode across the output, and C, K, and z are circuit constants. This example is studied in Section 9.

Example 1.5. A model of atomic reactor dynamics leads to the equation for neutron density x, namely

$$\dot{x}(t) = -\int_{-\infty}^{0} a(-\theta)g(x(t+\theta))d\theta. \tag{1.5}$$

This example is treated in Section 5. (See Example 5.4.)

2. Definitions and Notation: Functional Differential Equations with Finite or Infinite Delay. The Initial Function Space.

The equations in Examples 1.1 through 1.5 exhibited a common characteristic: the time derivative of $x(t)$ depended on an "earlier" value of x, namely $x(t-r)$ for some r, $0 < r \leq \infty$. Upon consideration of equation (1.2) in particular, it is not sufficient to specify an initial value $x(0)$ in order to determine $x(t)$ for $t \in \mathbb{R}^+$. Indeed, we observe upon integration of (1.2) over an interval $[0,t] \subset [0,1]$ that $x(\theta)$ must be specified over $[-1,0]$ in order to define $x(t)$. Thus an initial function $\phi\colon [-1,0] \to \mathbb{R}$ is required to insure that the problem is well posed. These ideas are made more precise below.

Denote by $\mathscr{B}$ the collection of all functions mapping $[-r,0]$ into $\mathbb{R}^d$, where $0 < r \leq \infty$. If $r = \infty$, $[-r,0]$ means $(-\infty,0]$, otherwise denoted by $\mathbb{R}^-$. Let $X \subset \mathscr{B}$ be a complete metric space with metric d. If x maps $[-r,a)$ into $\mathbb{R}^d$ with $a > 0$ for $r < \infty$ (respectively a is arbitrary for $r = \infty$), then x_t, $t \in [0,a)$ for $r < \infty$ (respectively $t \in (-\infty,a)$ for $r = \infty$) denotes a function in $\mathscr{B}$ defined by

$$x_t(\theta) = x(t+\theta), \quad \theta \in [-r,0].$$

Thus x_t is the "segment of x" on $[t-r,t]$ translated to $[-r,0]$. In particular, x_0 is the restriction of x to $[-r,0]$. We make the following assumptions concerning the space X.

(2.1i) $r < \infty$ (finite delay): If $\{\phi_n\}$ is a d-bounded sequence in X, then $\phi_n(\theta) \to \phi(\theta)$ uniformly in $\theta \in [-r,0]$ implies $\phi \in X$ and $d(\phi_n,\phi) \to 0$.

(2.1ii) $r = \infty$ (infinite delay): Suppose $y: \mathbb{R} \to \mathbb{R}^d$ so that

a) $|y(t)| \leq M$ for all $t \in \mathbb{R}^+$ and some $M > 0$,

b) $y_t \in X$ for every $t \in \mathbb{R}^+$,

c) there exists $\psi \in X$ so that $d(y_t,\psi)$ is bounded on $\mathbb{R}^+$.

Note that y_t represents the restriction of y to $(-\infty,t]$. If $\{t_n\} \subset \mathbb{R}^+$ with $t_n \to \infty$, then $y_{t_n}(\theta) \to \phi(\theta)$ uniformly for θ in compact subsets of $\mathbb{R}^-$ implies $\phi \in X$ and $d(y_{t_n},\phi) \to 0$.

(2.2) For every $\alpha > 0$ there exists $\beta > 0$ so that $d(\phi,\psi) < \alpha$ implies $|\phi(0) - \psi(0)| < \beta$ whenever $\phi,\psi \in X$ are continuous at $\theta = 0$.

Let $f: X \to \mathbb{R}^d$ and henceforth assume that

(2.3) f is continuous on X

(2.4) f is bounded on d-bounded subsets of X.

Definition 2.1. An (autonomous) *retarded functional differential equation* (RFDE) is a relationship

$$\dot{x}(t) = f(x_t). \tag{2.5}$$

A function $x: [-r,a) \to \mathbb{R}^d$ is a *solution* on $[0,a)$ of (2.5) with initial value $\phi \in X$ if

(i) $x_t \in X$ for every $t \in [0,a)$,

(ii) $x(t)$ satisfies equation (2.5) for every $t \in [0,a)$,

(iii) $x_0 = \phi$.

We write $x(\cdot;\phi)$ for a solution x such that $x_0 = \phi$. $x_t(\phi)$ denotes the corresponding element of X with $x_0(\phi) = \phi$. Note that $x(t;\phi) \in \mathbb{R}^d$, whereas $x_t(\phi) \in X$. Also observe that $x_t(\phi)(\cdot) = x(t+\cdot;\phi)$ for $t \in [0,a)$, $\theta \in [-r,0]$.

It is easy to see that some of the equations in Section 1 are RFDE's. Indeed Equation (1.1) can be expressed in the form (2.5) with $f(\phi) = g(\phi(0)) - g(\phi(-L))$. Likewise for Equation (1.2) we have $f(\phi) = -\alpha\phi(-1)[1+\phi(0)]$. Then $f(x_t) = -\alpha x_t(-1)[1+x_t(0)] = -\alpha x(t-1)[1+x(t)]$. We can write Equation (1.3) in system form: $\dot{x}(t) = f(x_t,y_t)$, $\dot{y}(t) = g(x_t,y_t)$, where $f(\phi,\psi) = \psi(0)$ and $g(\phi,\psi) = -G(0)\phi(0) - \int_{-\infty}^{0} \dot{G}(-\theta)\phi(\theta)d\theta$. Note that Examples 1.1 and 1.2 have finite delays, whereas Example 1.3 has infinite delay. Example 1.5 also is an RFDE with infinite delay.

Example 1.4 is of a different nature than the rest. Equation (1.4) indicates a hereditary dependence on the derivative of x. Such an equation is called a neutral functional differential equation (NFDE). This class will be discussed in Section 8, and Example 1.4 will be explored in Section 9.

We now present some candidates for spaces X which satisfy the conditions (2.1) and (2.2).

Example 2.2. For $0 < r < \infty$, consider the Banach space $C([-r,0];\mathbb{R}^d)$ with norm $\|x\| = \sup_{-r\leq\theta\leq 0} |x(\theta)|$. Let X be an open ball centered at the origin. Then X is a space of initial functions with finite delay and trivially satisfies (2.1) and (2.2).

Example 2.3. Endow the set $C_b(\mathbb{R}^-;\mathbb{R}^d)$, the collection of all bounded continuous functions from $\mathbb{R}^-$ to $\mathbb{R}^d$, with the topology of uniform convergence on compact subsets of $\mathbb{R}^-$. This topology is metrizable with metric d given by

$$d(\phi,\psi) = |\phi(0) - \psi(0)| + \sum_{m=0}^{\infty} d_m(\phi,\psi),$$

where

$$d_m(\phi,\psi) = \min\{2^{-m}, \sup_{-m-1\leq\theta\leq -m} |\phi(\theta) - \psi(\theta)|\}.$$

The space is complete under the metric d. Take for X an open ball centered at the origin.

We verify conditions (2.1) and (2.2). Let $\{y_{t_n}\}$ be a sequence to which the hypothesis in (2.1ii) applies so that $y_{t_n} \to \phi$ uniformly on compact subsets of $\mathbb{R}^-$. Then ϕ is continuous. As $y(t)$ is bounded on $\mathbb{R}^+$, then so is $\{y_{t_n}\}$ bounded. Thus $\phi \in X$. The definition of d ensures

$d(y_{t_n}, \phi) \to 0$. It is also evident that $|\phi(0) - \psi(0)| \le d(\phi,\psi)$; hence condition (2.2) is fulfilled.

Example 2.4. Consider the collection X of all measurable functions from $\mathbb{R}^-$ to $\mathbb{R}^d$ for which

$$\int_{-\infty}^{0} |\psi(\theta)|\ell(\theta)d\theta < \infty, \quad \psi \in X$$

where $\ell(\theta) > 0$ and $\dot{\ell}(\theta) \ge 0$ for all $\theta \in \mathbb{R}^-$, and $\int_{-\infty}^{0} \ell(\theta)d\theta < \infty$. Then X is a complete metric space with

$$d(\phi,\psi) = |\phi(0) - \psi(0)| + \int_{-\infty}^{0} |\phi(\theta) - \psi(\theta)|\ell(\theta)d\theta$$

provided we identify in the usual manner those functions $\phi,\psi \in X$ for which $\phi = \psi$ a.e. in $\mathbb{R}^-$ and $\phi(0) = \psi(0)$. (X is indeed a Banach space, but we shall not need this fact here.) This is an example of the spaces considered by Coleman and Mizel [4] called history spaces. We now verify that X satisfies conditions (2.1) and (2.2).

Suppose $y\colon \mathbb{R} \to \mathbb{R}^d$ and $\{y_{t_n}\}$ is a sequence in X for which the hypotheses of (2.1ii) apply. It will be sufficient to show that

$$\lim_{n\to\infty} \int_{-\infty}^{0} |y_{t_n}(\theta) - \phi(\theta)|\ell(\theta)d\theta = 0.$$

Let $\varepsilon > 0$, and choose $T > 0$ so that

$$(2.6) \quad \int_{-\infty}^{-T} |y_0(\theta)|\ell(\theta)d\theta < \varepsilon/5, \quad \int_{-\infty}^{-T} \ell(\theta)d\theta < \varepsilon/5M.$$

As $\{y_{t_n}\}$ converges pointwise to ϕ, then ϕ is also bounded by M on $\mathbb{R}^-$. Therefore

$$\int_{-\infty}^{0} |y_{t_n}(\theta) - \phi(\theta)|\ell(\theta)d\theta$$

$$\leq \int_{-\infty}^{-T} |y_{t_n}(\theta)|\ell(\theta)d\theta + \int_{-\infty}^{-T} |\phi(\theta)|\ell(\theta)d\theta$$

$$+ \int_{-T}^{0} |y_{t_n}(\theta) - \phi(\theta)|\ell(\theta)d\theta$$

$$\leq \int_{-\infty}^{-T} |y_{t_n}(\theta)|\ell(\theta)d\theta + \int_{-T}^{0} |y_{t_n}(\theta) - \phi(\theta)|\ell(\theta)d\theta + \varepsilon/5.$$

Choose an integer N such that $n \geq N$ implies

$$\int_{-T}^{0} |y_{t_n}(\theta) - \phi(\theta)|\ell(\theta)d\theta < \varepsilon/5, \tag{2.7}$$

$$\frac{\ell(-t_n)}{\ell(-T)} d(0,y_0) < \varepsilon/5, \quad \text{and} \quad t_n > T. \tag{2.8}$$

Then

$$\int_{-\infty}^{-T} |y_{t_n}(\theta)|\ell(\theta)d\theta = \int_{-\infty}^{-T} |y(\theta)|\ell(\theta - t_n)d\theta$$

$$+ \int_{-T-t_n}^{-T} |y(t_n + \theta)|\ell(\theta)d\theta.$$

The first integral on the right side is bounded by $\varepsilon/5$ in view of (2.6). We can transform the second integral into

$$\int_{-T}^{0} |y(\theta)|\frac{\ell(\theta - t_n)}{\ell(\theta)} \ell(\theta)d\theta + \int_{-t_n}^{T} |y(t_n + \theta)|\ell(\theta)d\theta.$$

In view of (2.8) this first integral is less than $\varepsilon/5$, and from (2.6) we see that the second integral is less than $\varepsilon/5$. Consequently,

$$\int_{-\infty}^{0} |y_{t_n}(\theta) - \phi(\theta)|\ell(\theta)d\theta < \varepsilon \quad \text{for} \quad n \geq N.$$

Thus we have demonstrated condition (2.1) holds for the history space X.

Condition (2.2) is trivially satisfied in view of the fact that the metric d satisfies for $\phi,\psi \in X$,

$$|\phi(0) - \psi(0)| \le d(\phi,\psi).$$

Example 2.5. As a final example of a space of initial functions, take X to be the collection of all measurable functions from $\mathbb{R}^-$ to $\mathbb{R}^d$ which are continuous on $[-\alpha,0]$, $0 \le \alpha \le r$. The norm is defined by $(p \ge 1)$

$$\|\psi\|^p = [\sup_{-\alpha \le \theta \le 0} |\psi(\theta)|]^p + \int_{-\infty}^{0} |\psi(\theta)|^p \ell(\theta)d\theta < \infty,$$

where $\ell(\theta) \ge 0$, $\int_{-\infty}^{0} \ell(\theta)d\theta < \infty$, and $\dot{\ell}(\theta) \ge 0$. For $r = -\infty$, $\alpha = 0$, and $p = 1$, X is the space of Example 2.4. For $r = \alpha < \infty$ and $\ell = 0$, then X is the space $C([-r,0];\mathbb{R}^d)$ of Example 2.2.

We turn to the question of existence of solutions to the RFDE (2.5).

Theorem 2.6. Let the space X satisfy condition (2.1) and (2.2). Suppose $f: X \to \mathbb{R}^d$ where for each positive integer n there is an $L_n > 0$ so that if $d(\phi,0) \le n$ and $d(\psi,0) \le n$, we have

$$|f(\phi) - f(\psi)| \le L_n d(\phi,\psi). \tag{2.9}$$

Then for each $\phi \in X$ there is a unique solution $x(\cdot;\phi)$ of (2.5) on some interval $[0,a)$. Furthermore $x(t;\phi)$ is continuous on $[-r,a) \times X$.

The proof of the theorem is deferred to the next section. We present only the case for $r < \infty$. Furthermore, we suppose f satisfies a global Lipschitz condition on X instead of the local condition of relation (2.9). But this allows us to apply the semigroup theory of Chapter V, Section 2. Consequently we obtain a solution continued to all of $\mathbb{R}^+$. Continuous dependence of the solution follows from the continuity of the semigroup constructed in the course of the proof.

3. Existence of Solutions of Retarded Functional Equations

We apply the theory of nonlinear semigroups of Chapter V to obtain existence, uniqueness, and continuous dependence of solutions to the RFDE

$$(3.1) \qquad \begin{aligned} \dot{x}(t) &= f(x_t) \\ x_0 &= \phi \in X, \end{aligned}$$

where f is Lipschitz on $X = C([-r,0]; \mathbb{R}^d)$, $\theta < r < \infty$. To this end we will exhibit a quasi-contraction semigroup T so that $x_t(\phi) = T(t)\phi$. Though the proof to follow is highly technical and is based upon numerous complicated estimates, it has the virtue of providing global existence, as well as the other desired properties of the solution announced in Theorem 2.6.

Theorem 3.1. Let $f: X \to \mathbb{R}^d$ be ω-Lipschitz. Then the operator A defined in X by

$$A\phi = -\dot{\phi}$$

with domain

$$\mathscr{D}(A) = \{\phi \in X: \dot{\phi} \in X, \dot{\phi}(0) = f(\phi)\}$$

generates a semigroup $T \in Q_\omega$ on X so that for every $t \in \mathbb{R}^+$,

$$x(t) \overset{\text{def}}{=} T(t)\phi(0)$$

is the unique solution to (3.1).

Before turning to the proof of Theorem 3.1 we outline its main points and provide the reader with a guide to the proof.

(1) Define linear m-accretive operators A_0, A_1 in $X_0 = \{\phi \in X: \phi(0) = 0\}$ by

$$A_0\phi = -\dot{\phi},\ \mathscr{D}(A_0) = \{\phi \in X_0: \dot{\phi} \in X_0\}$$

$$A_1\phi = -\dot{\phi},\ \mathscr{D}(A_1) = \{\phi \in X_0: \dot{\phi} \in X\}.$$

Observe that the operators are the same but their domains differ. For $i = 0,1$, the resolvent operator $J_{i,\lambda}$ $(\lambda > 0)$ given by

$$J_{i,\lambda} = (I + \lambda A_i)^{-1}$$

exists everywhere in X_i, where we have set $X_1 = X$ (Lemma 3.2).

(2) The linear semigroup T_0 defined by

$$T_0(t)\phi = \lim_{n\to\infty} J^n_{0,t/n}\phi$$

has domain X_0 and behaves as a translation operator: for $t \in (0,r]$, $T_0(t)\phi$ is the function which is zero on $[-t,0]$, and is the segment of ϕ on $[t-r,0]$ translated to $[-r,-t)$. See Figure 3.1 (Lemma 3.3).

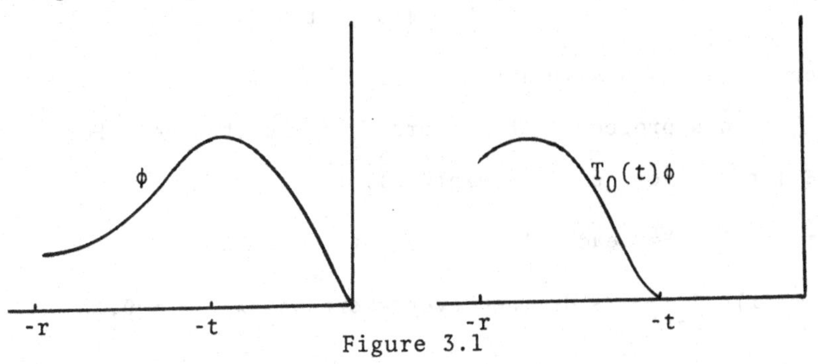

Figure 3.1

(3) The domain of $T_0(t)$, for $t \in (0,r]$, extends to X and for each $\phi \in X$,

$$\lim_{n\to\infty} J^n_{1,t/n}\phi(\theta),$$

(Lemma 3.6).

(4) $A + \omega I$ is m-accretive, hence $J_\lambda = (I + \lambda A)^{-1}$ is defined everywhere in X. For each $\phi \in X$ and $t > 0$, the limit

$$T(t)\phi = \lim_{n\to\infty} J^n_{t/n}\phi$$

exists (Lemmas 3.7, 3.8).

(5) For each $\phi \in X$, $t \in (0,r]$,

$$\lim_{n\to\infty} |J^n_{1,t/n}\phi(\theta) - J^n_{t/n}\phi(\theta)| = 0$$

uniformly for $\theta \in [-r,-t]$. Thus $T(t)$ operates on ϕ by translating it to the left by an amount t and then adds something new in the interval $[-t,0]$. So if $t \in (0,r]$ and $\phi \in X$, then

$$T(t)\phi(\theta) = \phi(\theta+t), \quad \theta \in [-r,-t],$$

(Lemma 3.11)

(6) For $\phi \in X$ define $x \in C([-r,\infty);\mathbb{R}^d)$ by

$$x(t) = \begin{cases} \phi(t), & t \in [-r,0] \\ T(t)\phi(0), & t > 0. \end{cases}$$

Then $x(t)$ is a solution of (3.1).

We now proceed with the proof of the theorem. For $\theta \in [-r,0]$ set $e_\lambda(\theta) = \exp(\theta/\lambda)$.

<u>Lemma 3.2</u>. For each $\lambda > 0$ we have

(i) $J_{i,\lambda}$ is defined everywhere on X_i, $i = 0,1$.

(ii) $\|J_{i,\lambda}\| = 1$, $i = 0,1$.

(iii) For $\phi \in X_i$, $\theta \in [-r,0]$,

$$J_{i,\lambda}\phi(\theta) = \lambda^{-1}e_\lambda(\theta)\int_\theta^0 e_{-\lambda}(s)\phi(s)ds, \quad i = 0,1.$$

(iv) A_0 is densely defined in X_0.

(v) For each $\phi \in X_0$ $\lim_{\lambda \downarrow 0} J_{0,\lambda}\phi = \phi$.

(vi) For each $\phi \in X$ we have

$$J_{i,\lambda}\phi = J_{0,\lambda}(\phi - \phi(0)) + \phi(0)(1-e_\lambda).$$

<u>Proof</u>: (i) Let $(I + \lambda A_i)\phi = 0$ for some $\lambda > 0$ and $\phi \in X_0$. Then $\phi - \lambda\dot{\phi} = 0$, so $\phi(\theta) = \phi(0)e_\lambda(\theta)$ for $\theta \in [-r,0]$. As $\phi(0) = 0$, we have $\phi \equiv 0$. Thus $J_{i,\lambda}$ is well defined on X_i, $i = 0,1$.

(ii) This is left as an exercise for the reader.

(iii) Apply $I + \lambda A_i$ to the function on the right hand side of the expression in (iii) of the lemma to obtain $\phi(\theta)$.

(iv) In as much as every continuous function on a finite interval can be uniformly approximated by a continuously differentiable one, we can conclude that if $\phi \in X_0$, then there exists $\psi \in X$, $\dot{\psi} \in X$ such that $\|\phi-\psi\| < \varepsilon/2$. But $\psi(0)$ need not be zero. So replace ψ by $\psi-\psi(0)$. Then this function is within ε of ϕ on $[-r,0]$.

(v) If $\phi \in \mathscr{D}(A_0)$, then $A_0J_{0,\lambda}\phi = J_{0,\lambda}A_0\phi = \lambda^{-1}(I - J_{0,\lambda})\phi$. So $\|\phi - J_{0,\lambda}\phi\| = \|\lambda J_{0,\lambda}A_0\phi\| = \lambda\|A_0\phi\| \to 0$ as $\lambda \downarrow 0$.

(vi) Apply Lemma 3.2(iii) for $J_{0,\lambda}$ to $\phi-\phi(0) \in X_0$. □

We turn to the semigroup T_0. As A_0 is a densely defined linear operator in X_0 with $\mathscr{R}(I + \lambda A_0) = X_0$, then

A_0 is m-accretive and by Theorem 2.4 of Chapter V generates a strongly continuous linear semigroup T_0 on X_0 given by

$$T_0(t)\phi = \lim_{n\to\infty} J^n_{0,t/n}\phi,$$

where the limit is taken in X_0.

<u>Lemma 3.3</u>. The linear semigroup T_0 on X_0 satisfies for every $t \in [0,r]$

$$T_0(t)\phi(\theta) = \begin{cases} \phi(\theta+t), & \theta \in [-r,-t) \\ 0, & \theta \in [-t,0]. \end{cases} \tag{3.2}$$

<u>Proof</u>: In view of the uniqueness property, Theorem 2.9 of Chapter V, we need only show that A_0 is the infinitesimal generator of the semigroup defined by expression (3.2). In fact,

$$\frac{T_0(t)\phi(\theta)-\phi(\theta)}{t} = \begin{cases} \dfrac{\phi(\theta+t)-\phi(\theta)}{t}, & \theta \in [-r,-t) \\ \dfrac{\phi(\theta)}{t}, & \theta \in [-t,0] \end{cases}$$

$$\to \begin{cases} \phi(\theta), & \theta \in [-r,0) \\ 0, & \theta = 0 \end{cases} = A_0\phi(\theta), \ \theta \in [-r,0],$$

as $t \downarrow 0$, since each $\theta < 0$ is eventually less than $-t$. □

The next lemma is a technical one. It admits a probabilistic interpretation and proof, which for example, may be found in Feller [1], p. 229.

<u>Lemma 3.4</u>. For each $\alpha \geq 0$ and positive integer n, define $M_n(\alpha) = \sum_{j=0}^{n-1} \frac{1}{j!}(\alpha n)^j e^{-\alpha n}$. Then

$$\lim_{n\to\infty} M_n(\alpha) = \begin{cases} 1, & 0 \leq \alpha < 1 \\ 0, & \alpha > 1 \end{cases}.$$

Lemma 3.5. For each $t > 0$, positive integer n, and $\phi \in X$ we have

$$J^n_{1,t/n}\phi(\theta) = J^n_{0,t/n}(\phi-\phi(0))(\theta) + \phi(0)[1-e_{t/n}(\theta)M_n(-\tfrac{\theta}{t})].$$

Proof: For the case $n = 1$ just use (vi) of Lemma 3.2(vi). For $n > 1$ use an inductive argument on the iterates $J^n_{0,\lambda}$ and replace λ by t/n. □

We next show that the iterates $J^n_{1,t/n}\phi$, $t \in (0,r]$, converge to the function defined in expression (3.2) for $\theta \in [-r,0]$, except perhaps $\theta = -t$.

Lemma 3.6. For any $t \in (0,r]$ and $\phi \in X$,

$$\lim_{n\to\infty} J^n_{1,t/n}\phi(\theta) = \begin{cases} \phi(\theta+t), & \theta \in [-r,-t) \\ 0, & \theta \in (-t,0]. \end{cases}$$

Proof: For $\theta \in [-r,-t)$, $\alpha = -\theta/t > 1$, so using Lemmas 3.3, 3.4, and 3.5 we obtain

$$\lim_{n\to\infty} J^n_{1,t/n}\phi(\theta) = T_0(t)(\phi-\phi(0))(\theta) + \phi(0) = \phi(\theta+t).$$

When $\theta \in (-t,0]$,

$$\lim_{n\to\infty} J^n_{1,t/n}\phi(\theta) = T_0(t)(\phi-\phi(0))(\theta) = 0.$$ □

The next step is to show that $A + \omega I$ is m-accretive and to prove that A generates the semigroup T which gives rise to the solution of (3.1).

Lemma 3.7. The operator A defined in the statement of Theorem 3.1 is densely defined in X. Moreover, $\mathscr{R}(I + \lambda A) = X$ for every $\lambda > 0$.

Proof: Let $\psi \in X$, $\varepsilon > 0$. Then f is bounded on $B_1 = B_\varepsilon(\psi)$ by some $Q > 0$. Choose λ_1 so that $0 < \lambda < \lambda_1$

implies (according to Lemma 3.2(v)) $||(J_{0,\lambda} - I)(\psi-\psi(0)|| < \varepsilon/2$. Set $B_2 = B_{\varepsilon/2}(\psi(0))$ in $\mathbb{R}^d$. If $b \in B_2$ and $0 < \lambda < \lambda_1$ then using Lemma 3.2(vi) we get

$$||\psi - (e_\lambda b+J_{1,\lambda}\psi)|| \leq ||\psi-\psi(0) - J_{0,\lambda}(\psi-\psi(0))|| + ||b-\psi(0)|| < \varepsilon.$$

Thus $e_\lambda b + J_{1,\lambda}\psi \in B_1$. Let $0 < \lambda < \min\{\lambda_1, \varepsilon/2Q\}$, and define $F: B_2 \to \mathbb{R}^d$ by

$$F(b) = \psi(0) + \lambda f(e_\lambda b + J_{1,\lambda}\psi). \tag{3.3}$$

Then $F: B_2 \to B_2$ and is continuous on B_2. The Brouwer fixed point theorem (see Appendix A) implies there exists $b_0 \in B_2$ so that

$$b_0 = \psi(0) + \lambda f(e_\lambda b_0 + J_{1,\lambda}\psi). \tag{3.4}$$

Set $\phi_0 = e_\lambda b_0 + J_{1,\lambda}\psi$. We claim $\phi_0 \in \mathcal{D}(A)$. Indeed, upon using Lemma 3.2(iii) we obtain

$$\phi_0(\theta) = \lambda^{-1}e_\lambda(\theta)b_0 + \lambda^{-2}e_\lambda(\theta)\int_\theta^0 e_{-\lambda}(s)\psi(s)ds - \lambda^{-1}\psi(\theta).$$

Moreover, $\phi_0(0) = \lambda^{-1}(b_0-\psi(0)) = f(e_\lambda b_0+J_{1,\lambda}\psi) = f(\phi_0)$. Thus $\phi_0 \in \mathcal{D}(A)$. Furthermore we have that $(I+\lambda A)\phi_0 = \psi$. Indeed, $(I+\lambda A)\phi_0 - \lambda\phi_0 = \phi_0 - [e_\lambda b_0+J_{1,\lambda}\psi-\psi] = \psi$. In particular, $\mathcal{R}(I + \lambda A) = X$.

Use the definition of ϕ_0, (3.4) and Lemma 3.2(vi) to obtain

$$\phi_0 = \psi(0) + \lambda e_\lambda f(\phi_0) + J_{0,\lambda}(\psi-\psi(0)). \tag{3.5}$$

Then, $||\psi-\psi_0|| \leq ||\psi-\psi(0) - J_{0,\lambda}(\psi-\psi(0))|| + \lambda|f(\phi_0)| < \varepsilon/2 + \lambda Q < \varepsilon$. This establishes $\overline{\mathcal{D}(A)} = X$. □

Lemma 3.8.

(i) If $0 < \lambda < \omega^{-1}$, then $J_\lambda \overset{\text{def}}{=} (I+\lambda A)^{-1}$ is defined on all of X.

(ii) J_λ is $(1-\lambda\omega)^{-1}$ Lipschitz

(iii) $\lim\limits_{\lambda \downarrow 0} J_\lambda \phi = \phi$ for each $\phi \in X$.

Proof: (i) Let $0 < \lambda < \omega^{-1}$, $\psi \in X$. From the definition of F (Equation (3.3)) and the fact that f is ω-Lipschitz, we have

$$|F(b_1) - F(b_2)| \le \lambda\omega\|e_\lambda b_1 - e_\lambda b_2\| \le \lambda\omega|b_1 - b_2|.$$

Thus F is a strict contraction from $\mathbb{R}^d$ to $\mathbb{R}^d$, so F has a fixed point $b_0 = b_0(\psi)$. Moreover,

$$\phi_0 = e_\lambda b_0 + J_{1,\lambda}\psi \tag{3.6}$$

is the unique solution to $\phi - \lambda\dot{\phi} = \psi$ with initial value $\dot{\phi}(0) = f(\phi)$. Thus $(I+\lambda A)\phi = \psi$ admits a unique solution, so $J_\lambda\psi$ is well defined. Consequently $\mathscr{D}(J_\lambda) = X$.

(ii) Let $\psi_1, \psi_2 \in X$. Then $J_\lambda\psi_i = {}_i$, $i = 1,2$, where

$$\phi_i = \psi_i(0) + \lambda e_\lambda f(\phi_i) + J_{0,\lambda}(\psi_i - \psi_i(0))$$

from equation (3.5). Using Lemma 3.2(iii) for $J_{0,\lambda}$ and the ω-Lipschitz of f we obtain

$$\|J_\lambda\psi_1 - J_\lambda\psi_2\| \le \|\psi_1 - \psi_2\| + \lambda\omega\|J_\lambda\psi_1 - J_\lambda\psi_2\|.$$

It follows that J_λ is $(1-\lambda\omega)^{-1}$-Lipschitz.

(iii) It was established in Chapter V, Lemma 2.15(i) that $\lim\limits_{\lambda \downarrow 0} J_\lambda\phi = \phi$ for every $\phi \in \mathscr{D}(A)$. We show the limit holds for every $\phi \in X$. If $0 < \lambda < \omega^{-1}$, then

$$|f(J_\lambda\phi)| \le |f(J_\lambda\phi) - f(\phi)| + |f(\phi)| \le \omega\|J_\lambda\phi - \phi\| + |f(\phi)|$$

and (3.5) implies

$$\|J_\lambda\phi-\phi\| = \|\phi(0) + \lambda e_\lambda f(J_\lambda\phi) + J_{0,\lambda}(\phi-\phi(0)) - \phi\|$$

$$\leq \|J_{0,\lambda}(\phi-\phi(0)) - (\phi-\phi(0))\| + \lambda|f(J_\lambda\phi)|.$$

So

$$\|J_\lambda\phi-\phi\| \leq \|J_{0,\lambda}(\phi-\phi(0)) - (\phi-\phi(0))\| + \lambda\omega\|J_\lambda\phi-\phi\| + \lambda|f(\phi)|.$$

Thus

$$\|J_\lambda\phi-\phi\| \leq (1-\lambda\omega)^{-1}\|J_{0,\lambda}(\phi-\phi(0)) - (\phi-\phi(0))\| + \lambda(1-\lambda\omega)^{-1}|f(\phi)|.$$

The result now follows from Lemma 3.2(v). □

The last two lemmas imply that A is densely defined and $A + \omega I$ is m-accretive. Theorem 2.4 of Chapter V shows that A generates a quasi-contracting semigroup T on X given by

$$T(t)\phi = \lim_{n\to\infty} J^n_{t/n}\phi, \quad \phi \in X.$$

Lemma 3.9. For $t \in (0,r]$ and any positive integer n we have for each $\phi \in X$,

(i) $J^n_{t/n}\phi = \sum_{k=0}^{n-1} J^{n-k}_{t/n}\phi(0)J^k_{1,t/n}e_{t/n} + J^n_{1,t/n}\phi$,

(ii) $J^k_{1,t/n}e_{t/n}(\theta) = M_n(-\frac{\theta}{t})$.

Proof: (i) Iterate $J_\lambda\phi = e_\lambda b_0 + J_{1,\lambda}\psi$ from expression (3.6), and set $\lambda = t/n$. For (ii) just apply the iterates $J^n_{1,\lambda}$ to e_λ via the formulas in Lemma 3.2. □

Lemma 3.10. Fix $t \in (0,r]$. For each $\phi \in \mathcal{D}(A)$, the set $\{J^k_{t/n}\phi(0)\}^n_{k=1}$ is bounded, uniformly in $n \in \mathbb{N}$.

Proof: Apply Lemma 2.15 of Chapter V, letting $\lambda = t/n$. Then

$$|J^k_{t/n}\phi(0) - \phi(0)| \le \|J^k_{t/n}\phi - \phi\| \le \frac{kt}{n}(1-\frac{\omega t}{n})^{-k} \|\dot{\phi}\|$$
$$\le t(1-\frac{\omega t}{n})^{-n} \|\dot{\phi}\| < te^{2\omega t} \|\dot{\phi}\| . \quad \square$$

Lemma 3.11. For each $t \in [0,r]$ and $\phi \in X$,

$$(3.7) \qquad T(t)\phi(\theta) = \phi(\theta+t), \ \theta \in [-r,-t].$$

Proof: Let $\phi \in \mathcal{D}(A)$. Then from Lemma 3.9(i),

$$|J^n_{t/n}\phi(\theta) - J^n_{1,t/n}\phi(\theta)| \le \sum_{k=0}^{n-1} Q|J^k_{1,t/n}e_{t/n}(\theta)| = QM_n(-\frac{\theta}{t})$$

which tends to zero as $n \to \infty$, uniformly for $\theta \in [-r,-t)$, $t \in (0,r)$. Combine this with Lemma 3.6 to obtain the expression (3.7) for the case $\phi \in \mathcal{D}(A)$ when $\theta \in [-r,-t)$. As $\mathcal{D}(A)$ is dense in X and $T(t)$ is continuous on X, we conclude that expression (3.7) holds for every $\phi \in X$. As both the right and left hand sides of expression (3.7) are continuous functions of $\theta \in [-r,-t]$, (3.7) holds for $\theta \in [-r,-t]$, $t \in [0,r]$. $\square$

We complete the proof of Thoerem 3.1. Let $\phi \in X$ and define $x \in C([-r,\infty);\mathbb{R}^d)$ by

$$(3.8) \qquad x(t) = \begin{cases} \phi(t) &, \ t \in [-r,0] \\ T(t)\phi(0), & t \in \mathbb{R}^+. \end{cases}$$

We show that x is a solution of (3.1). First, it is clear that $x_t \in X$ for every $t \in \mathbb{R}^+$ and $x_0 = \phi$. Now we must prove for each $t \ge 0$ that

$$(3.9) \qquad x_t(\theta) = T(t)\phi(\theta), \ \ \theta \in [-r,0].$$

We will only verify (3.9) for $t \in [0,r]$. The extension to

$\mathbb{R}^+$ is immediate.

If $\theta \in [-r,-t]$, then $x_t(\theta) = x(t+\theta) = \phi(t+\theta) = T(t)\phi(\theta)$, where the second equality derives from (3.8) and the third equality derives from (3.7). On the other hand if $\theta \in [-t,0]$, then $x_t(\theta) = x(t+\theta) = T(t+\theta)\phi(0) = T(-\theta)[T(t+\theta)\phi](\theta) = T(t)\phi(\theta)$. The third equality drives from (3.7) and the fourth equality comes from the semigroup property of T. Clearly $x_0 = \phi$.

It remains to prove that $x(t)$ satisfies (3.1) for every $t \in \mathbb{R}^+$. This is accomplished by constructing a differentiable semigroup which uniformly approximates $T(t)\phi$, and which yields

$$x(t) = \phi(0) + \int_0^t f(x_s)ds.$$

According to Theorem 2.16 of Chapter V, there exists a strongly continuous semigroup $\{T_\lambda(t): t \in \mathbb{R}^+\}$ for small $\lambda > 0$ so that for every $t \in \mathbb{R}^+$

$$T_\lambda(t)\phi = \phi - \int_0^t AJ_\lambda T_\lambda(s)\phi \, ds.$$

Evalute both sides at $\theta = 0$. As the definition of A implies $AJ_\lambda T_\lambda(s)\phi(0) = -f(J_\lambda T_\lambda(s)\phi)$, we have

$$T_\lambda(t)\phi(0) = \phi(0) + \int_0^t f(J_\lambda T_\lambda(s)\phi)ds. \tag{3.10}$$

Fix $s \in \mathbb{R}^+$. Then

$$\begin{aligned}\|J_\lambda T_\lambda(s)\phi - T(s)\phi\| &\le \|J_\lambda T_\lambda(s)\phi - J_\lambda T(s)\phi\| \\ &\quad + \|J_\lambda T(s)\phi - T(s)\phi\| \\ &\le (1-\lambda\omega)^{-1}\|T_\lambda(s)\phi - T(s)\phi\| + \|J_\lambda T(s)\phi - T(s)\phi\|.\end{aligned}$$

In view of Theorem 2.16(iii) of Chapter V and Lemma 3.8(iii) we have

$$\lim_{\lambda \downarrow 0} f(J_\lambda T_\lambda(s)\phi) = f(T(s)\phi), \quad \lim_{\lambda \downarrow 0} T_\lambda(s)\phi(0) = T(s)\phi(0),$$

An application of the Lebesgue dominated convergence theorem to Equation (3.10) yields (as $\lambda \downarrow 0$)

$$T(t)\phi(0) = \phi(0) + \int_0^t f(T(s)\phi)ds.$$

This completes the proof of the theorem. □

4. Some Remarks on the Semidynamical System Defined by the Solution to an Autonomous Retarded Functional Differential Equation: The Invariance Principle and Stability

It is readily verified that the pair (X,T), where T is the semigroup constructed in Section 3, is a semidynamical system on X. Indeed, the mapping $T: X \times \mathbb{R}^+ \to X$ is given by $T(\phi,t) = T(t)\phi$. The notation $x_t(\phi) = T(t)\phi$ is employed. T enjoys the following property in the event r is finite.

Theorem 4.1. Suppose $0 < r < \infty$. For every $t \geq r$, the map $T(t)$ is locally compact. That is, if $\phi \in X$, there exists neighborhood W of ϕ such that $T(t)W$ is precompact.

Proof: The continuity of T implies there exists a neighborhood W of ϕ such that $T(t)W$ is a bounded set uniformly in compact t-intervals. As f is bounded on bounded subsets of X, then for any $\tau \geq r$, the set $\{\dot{x}(t;\psi): \psi \in W\} = f(T(t)W)$ is bounded for $t \in [0,\tau]$. Thus $\{x(\cdot;\psi): \psi \in W\}$ is an equicontinuous family on $[0,\tau]$. As $T(t)\psi(\theta) = x(t+\theta;\psi)$, then for every $t \geq r$, $\{T(t)\psi: \psi \in W\}$ is an equicontinuous and uniformly bounded family of X. Accordingly, the Ascoli theorem implies that $T(t)W$ is precompact. □

Corollary 4.2. Suppose $0 < r < \infty$. If f is linear then $T(t)$ is a compact operator on X for every $t \geq r$.

Proof: If f is linear then so is $T(t)$. Then $T(t)W$ is bounded whenever W is bounded. Now the proof of the last theorem shows that $T(t)W$ is precompact. □

The autonomous RFDE provides us with examples of start points and the nonuniqueness of negative solutions of semidynamical systems. The first example demonstrates the existence of start points.

Example 4.3. Consider the RFDE (2.5) with initial function $\phi \in X = C([-r,0]; \mathbb{R}^d)$, $0 < r < \infty$. If a solution through ϕ can be defined in the past, there must exist $\psi \in X$ and $t \in \mathbb{R}^+$ so that $x_t(\psi) = \phi$. Consequently ϕ must be differentiable on $(-\min\{r,t\},0]$. Otherwise, ϕ is a start point. Note that the collection of non-start points is dense in X.

The next example provides a semidynamical system without negative uniqueness. Thus $T(t)$ need not be one-to-one on X.

Example 4.4. Consider the equation

$$\dot{x}(t) = [x(t-1) + 1]x(t).$$

For every $\phi \in X = C([-1,0]; \mathbb{R})$ with $\phi(0) = 0$, the solution $x(t;\phi)$ is zero for all $t \in \mathbb{R}^+$. Consequently $T(t)\phi$ is the zero function for all $t \geq 1$. Thus for all initial values $\phi \in X$ with $\phi(0) = 0$, $T(t)\phi \equiv 0$ on $[1,\infty)$.

We turn to some properties of the motion $x_t(\phi)$. As we wish to study the limit set $L^+(\phi)$ of this motion, it would be nice if the positive orbit $\gamma^+(\phi)$ were precompact. If the phase space X were finite dimensional, one could infer the precompactness of $\gamma^+(\phi)$ provided $\gamma^+(\phi)$ were bounded. The fact is, that even though X is infinite dimensional, a bounded $\gamma^+(\phi)$ is sufficient for precompactness. This is a consequence of the smoothing property of solutions of (2.5). We accomplish these results for both finite and infinite delays.

Theorem 4.5. Suppose $x(\cdot;\phi)$ is a solution of (2.5) with d-bounded positive orbit $\gamma^+(\phi)$ in X. Then $\gamma^+(\phi)$ is precompact.

Proof: There exists a constant $k > 0$ and a function $\eta \in X$ so that $d(x_t,\eta) < k$ for all $t \in \mathbb{R}^+$. Then in view of the conditions (2.3) and (2.4), $|\dot{x}(t;\phi)| < m$ for some constant $m > 0$ and each $t \in \mathbb{R}^+$. From condition (2.2) there exists $\bar{m} > 0$ so that $\|x_t(0) - \eta(0)\| < \bar{m}$ for every $t \in \mathbb{R}^+$. Set $M = \bar{m} + |\eta(0)|$. Then $|x(t;\phi)| < M$ for each $t \in \mathbb{R}^+$. Denote by Δ_N the interval $[-r,0] \cap [-N,0]$, $N = 1,2,\ldots$. Let $\{t_n\} \to \infty$. Then we must have $\{x_{t_n}\}$ is an equicontinuous family on each Δ_N. For each N there exists a subsequence of $\{t_n\}$, which we denote by $\{t_n^N\}$, so that $\{x_{t_n^N}\}$ converges uniformly on Δ_N to some (norm) bounded continuous function ψ on Δ_N. The usual diagonalization procedure yields a subsequence $\{t_n'\} \subset \{t_n\}$ with $x_{t_n'}(\theta) \to \psi(\theta)$, uniformly on compact intervals in $[-r,0]$. Condition (2.1) implies $\psi \in X$ and $d(x_{t_n'},\psi) \to 0$. Consequently the motion $x_t(\phi)$ is precompact. □

The following property of limit sets is an immediate consequence of condition (2.1).

Corollary 4.6. Suppose $x = x(\cdot;\phi)$ is a solution of (2.5) with $\gamma^+(\phi)$ d-bounded. Then $\psi \in L(\phi)$ if and only if ψ is a bounded continuous function on $[-r,0]$, and there exists

$\{t_n\} \subset \mathbb{R}^+$ with $t_n \to \infty$ such that $x_{t_n}(\theta) \to \psi(\theta)$, uniformly on compact intervals of $[-r,0]$.

Accordingly we can state that if $x(\cdot;\phi)$ is a solution of (2.5) with d-bounded positive trajectory $\gamma^+(\phi)$, then $L(\phi)$ is nonempty, connected, compact, and weakly invariant. The notion of weak invariance allows us to conclude that the motion $T(t)\psi$ through each $\psi \in L^+(\phi)$ extends backwards in time (perhaps nonuniquely). But it is not obvious that this extension of the semigroup generates a solution of Equation (2.5) which is defined for all $t \in \mathbb{R}$. Indeed, the result on weak invariance, Theorem 3.5 Chapter II, does not depend on the underlying structure given by Equation (2.5). Once the semidynamical system (X,T) has been obtained, then the results of the general theory can be applied without reference back to the defining Equation (2.5). Thus we must take a different approach. It will be necessary to produce a negative continuation of $x(t;\phi)$ which satisfies Equation (2.5). These ideas are now made more precise.

Definition 4.7. A subset $\Gamma \subset X$ is called *weakly invariant relative to solutions of Equation (2.5)* if for each $\psi \in \Gamma$, there exists function $y: \mathbb{R} \to \mathbb{R}^d$ so that

(i) $y_t \in \Gamma$ for all $t \in \mathbb{R}$,

(ii) $y_0 = \psi$,

(iii) $y(t)$ is continuously differentiable on $\mathbb{R}$, and

(iv) $\dot{y}(t) = f(y_t)$ for all $t \in \mathbb{R}$.

Theorem 4.8. Let $x = x(\cdot;\phi)$ be a solution of Equation (2.5) whose positive orbit $\gamma^+(\phi)$ is d-bounded. Then $L^+(\phi)$ is non-empty, compact, connected, and weakly invariant relative to solutions of (Equation (2.5). Moreover

$$\lim_{t\to\infty} d(x_t(\phi), L(\phi)) = 0.$$

Proof: We need only establish weak invariance relative to solutions of Equation (2.5). So let $\psi \in L^+(\phi)$. There exists a sequence $\{t_n\} \subset \mathbb{R}^+$ with $d(x_{t_n}, \psi) \to 0$. For each positive integer N the functions defined by

$$x^*_{t_n+N}(\theta) = \begin{cases} 0 & , \quad \theta \in (-\infty, -t_n-N) \\ x(t_n+N+\theta), & \theta \in [-t_n-N, 0] \end{cases}$$

form a sequence which is uniformly bounded and equicontinuous on compact intervals of $\mathbb{R}^-$. An application of the diagonal procedure yields a subsequence $\{\tau_n\} \subset \{t_n\}$ so that for each N, $\{x^*_{\tau_n+N}\}$ converges to a function $y^{(N)}$, uniformly on compact intervals of $\mathbb{R}^-$. Observe that $y^{(N)}$ is bounded and continuous on $\mathbb{R}^-$, and $y^{(N)}(t) = y^{(N+1)}(t-1)$ for each $t \in \mathbb{R}^-$. In view of this we may define a continuous function $y: \mathbb{R} \to \mathbb{R}^d$ by

$$y(t) = y^{(N)}(t-N), \quad t \leq N.$$

Then $\lim_{n\to\infty} x^*_{\tau_n+t}(\theta) = \lim_{n\to\infty} x_{\tau_n+t}(\theta) = y_t(\theta)$, uniformly on compact intervals of $\mathbb{R}^-$. Condition (2.1) implies $y_t \in X$ and $d(x_{\tau_n+t}, y_t) \to 0$ as $n \to \infty$. Thus $y_t \in L^+(\phi)$ for every $t \in \mathbb{R}$. Moreover $y_0 = \psi$. This establishes (i) and (ii) of Definition 4.7.

It remains to show that $y(t)$ is continuously differentiable and that $\dot{y}(t) = f(y_t)$. We begin by proving the mapping $t \to f(y_t)$ is continuous on $\mathbb{R}$. It will be sufficient to prove that the map $t \to y_t$ is continuous on $\mathbb{R}$. So let the sequence $\{s_j\}$ converge to $s_0 \in \mathbb{R}$. Then

$$(4.1) \qquad d(y_{s_j}, y_{s_0}) \le d(y_{s_j}, x_{\tau_n+s_j}) + d(x_{\tau_n+s_j}, y_{s_0}).$$

$\{\tau_n\}$ is the sequence chosen earlier. We estimate this expression.

Suppose $\varepsilon_j \downarrow 0$. For each ε_j and $\Delta_N = [-r,0] \cap [-N,0]$, there exists $n(j,N) \ge j + N$ so that $n \ge n(j,N)$ implies

$$(4.2) \qquad d(y_{s_j}, x_{\tau_n+s_j}) < \varepsilon_j$$

and

$$(4.3) \qquad |y_{s_j}(\theta) - x_{\tau_n+s_j}(\theta)| < \varepsilon_j$$

for every $\theta \in \Delta_N$. We turn to an estimate of $d(x_{\tau_n+s_j}, y_{s_0})$. Consider the sequence $\{x_{\tau_{n(j,j)}}\}$. We have

$$(4.4) \qquad |x_{\tau_{n(j,j)}+s_j}(\theta) - y_{s_0}(\theta)|$$

$$\le |x_{\tau_{n(j,j)}+s_j}(\theta) - y_{s_j}(\theta)| + |y_{s_j}(\theta) - y_{s_0}(\theta)|.$$

As $y_{s_j}(\theta) - y_{s_0}(\theta) = y(s_j+\theta) - y(s_0+\theta)$, and $y(t)$ is uniformly continuous on compact intervals of $\mathbb{R}$, then

$$\lim_{j\to\infty} |y_{s_j}(\theta) - y_{s_0}(\theta)| = 0$$

uniformly on every compact interval of $\mathbb{R}^-$. Combining expressions (4.3) and (4.4) we find that

$$\lim_{j\to\infty} |x_{\tau_{n(j,j)}+s_j}(\theta) - y_{s_0}(\theta)| = 0$$

uniformly on compact intervals of $\mathbb{R}^-$. From condition (2.1) and expressions (4.1) and (4.2) we finally obtain $\lim_{j\to\infty} d(y_{s_j}, y_{s_0}) = 0$. Thus the mapping $s \to f(y_s)$ is continuous.

We conclude by showing $\dot{y}(t) = f(y_t)$. This will also guarantee that $\dot{y}(t)$ is continuous. So fix $t_0 < t$ in $\mathbb{R}$. As $x(\cdot) = x(\cdot;\phi)$ is a solution of Equation (2.5) with bounded positive motion $x_s(\phi)$, $s \in \mathbb{R}^+$, then $x(s)$ is bounded on every interval $[t_k+t_0, t_k+t]$, uniformly with respect to sufficiently large k. Thus

$$x(t_k+t)-x(t_k+t_0) = \int_{t_k+t_0}^{t_k+t} \dot{x}(s)ds = \int_{t_0}^{t} f(x_{t_k+s})ds.$$

Since f is uniformly bounded on the motion $x_s(\phi)$, then letting $k \to \infty$ we obtain

$$y(t) - y(t_0) = \int_{t_0}^{t} f(y_s)ds.$$

This completes the proof of the theorem. □

Remark 4.9. The requirement that X be complete under a metric d (see Sec. 2) cannot be dropped. For instance, take X to be the set of all bounded continuous functions $\phi: \mathbb{R}^- \to \mathbb{R}$ with norm $\|\phi\| = \sup_{\theta \in \mathbb{R}^+} |\phi(\theta)|$. Denote by $\mathscr{C}$ the subset of X which consists of all constant functions. Set $\rho(\phi,\mathscr{C}) = \inf_{\alpha \in \mathscr{C}} \|\phi-\alpha\|$ for each $\phi \in X$. Then consider the scalar equation

$$\dot{x}(t) = -\min\{\rho(x_t,\mathscr{C}),1\}x(t) + 1.$$

If $\phi \in X$ is an initial function with $\rho(\phi,\mathscr{C}) \geq 1$, then $x(t;\phi) \to 1$ as $t \to \infty$. (Observe for any solution x, that $\rho(x_t,\mathscr{C})$ is nondecreasing in t.) Indeed, for every solution x we have $x(t;\phi) \to 1$ as $t \to \infty$. But $x(t) \equiv 1$ is not a solution. Thus $L^+(\phi) = \emptyset$ for each $\phi \in X$. However, every positive orbit $\gamma^+(\phi)$ is bounded.

Lemma 4.10. The union of a collection of sets which are weakly invariant relative to solutions of Equation (2.5) is weakly invariant relative to solutions of Equation (2.5). The closure of a set which is weakly invariant relative to solutions of Equation (2.5) is weakly invariant relative to solutions of Equation (2.5).

Proof: Left as an exercise. □

We can now state the invariance principle for autonomous RFDE's. The proof is obvious.

Theorem 4.11. (Invariance Principle) Suppose there exist subsets $E, H \subset X$ with the following property: for each $\phi \in H$, $\lim_{t\to\infty} d(x_t(\phi), E) = 0$. If $\gamma^+(\phi)$ is bounded, then $\lim_{t\to\infty} d(x_t(\phi), M) = 0$ where M is the largest subset of E which is weakly invariant relative to solutions of Equation (2.5).

The set E is usually obtained from a Lyapunov function. Let $W \subset X$ be open and $G \subset W$ be positively invariant with $\overline{G} \subset W$. Recall that $V\colon W \to \mathbb{R}$ is a Lyapunov function provided V is continuous on $\overline{G}$ and $V(x_t(\phi)) \le V(\phi)$ for all $\phi \in G$, $t \in \mathbb{R}^+$.

Define

$$\dot{V}(\phi) = \limsup_{t \downarrow 0} \frac{1}{t}\,[V(x_t(\phi)) - V(\phi)],$$

and let M be the largest subset of

$$E = \{\phi \in \overline{G}\colon \dot{V}(\phi) = 0\}$$

which is weakly invariant relative to solutions of Equation (2.5).

Theorem 4.12. Let V be a Lyapunov function on the positively invariant set G. If $\phi \in G$ has a bounded positive orbit $\gamma^+(\phi)$, then $\lim_{t\to\infty} d(x_t(\phi),M) = 0$. If, in addition, there is a continuous nonnegative function w on $\mathbb{R}^+$ with $w(s) \to \infty$ as $s \to \infty$ so that $w(|\phi(0)|) \leq V(\phi)$ for every $\phi \in G$, then each solution $x(\cdot;\phi)$ is bounded on $\mathbb{R}^+$. Hence $\gamma^+(\phi)$ is also bounded.

Proof: $\gamma^+(\phi)$ is precompact, hence $V(x_t(\phi))$ is bounded from below for $t \in \mathbb{R}^+$. As $V(x_t(\phi))$ is nonincreasing in t, then $\lim_{t\to\infty} V(x_t(\phi)) = v_0$ exists. Thus $V(\psi) = v_0$ for all $\psi \in L^+(\phi)$. The weak invariance of $L^+(\phi)$ relative to solutions of Equation (2.5) implies $\dot{V}(\psi) = 0$. Therefore $L^+(\phi) \subset M$, so $x_t(\phi) \to M$ as $t \to \infty$.

Suppose w satisfies the conditions of the theorem, yet there is some $\phi \in G$ for which $x(t) = x(t;\phi)$ is not bounded on $\mathbb{R}^+$. Let $t_n \to \infty$ so that $|x(t_n)| \to \infty$. As

$$w(|x(t)|) = w(|x_t(0)|) \leq V(x_t)$$

Then $V(x_{t_n}) \to \infty$, which contradicts the fact that $V(x_t)$ is nonincreasing. □

Definition 4.13. A solution $y(\cdot;\phi)$ of Equation (2.5) is *stable (asymptotically stable)* if the positive orbit $\gamma^+(\phi)$ is a positively stable (asymptotically stable) set of the semi-dynamical system (X,T).

Corollary 4.14. If $f(0) = 0$, and V is a Lyapunov function on X for which $V(0) = 0$, $\dot{V}(\phi) < 0$ for $\phi \neq 0$, then all solutions of Equation (2.5) approach zero as $t \to \infty$, and the zero solution is globally asymptotically stable.

5. Some Examples of Stability of RFDE's

The first example derives from a model of Volterra [1] involving the interaction of two species. The system presented here has been modified to account for finite hereditary dependence. For a modern treatment of this example without delay see Hirsh and Smale [1].

Example 5.1. Let there be two species, A and B. Denote by $N_A(t)$, $N_B(t)$ the numbers of species A and B respectively, at time t. Suppose species A has an unlimited food supply and species B depends upon A for its development. The populations of species A and B are assumed to evolve according to the relations

$$
\begin{aligned}
\dot{N}_A(t) &= [\varepsilon_A - \gamma_A N_B(t) - \int_{-r}^{0} F_A(\theta) N_B(t+\theta)\,d\theta] N_A(t) \\
\dot{N}_B(t) &= [-\varepsilon_B + \gamma_B N_A(t) + \int_{-r}^{0} F_B(\theta) N_A(t+\theta)\,d\theta] N_B(t),
\end{aligned}
\tag{5.1}
$$

where r, ε_A, ε_B, γ_A, γ_B are positive constants and F_A, F_B are nonnegative continuously differentiable functions on $[-r,0]$. We see that the growth rate of species A depends linearly on the past population of species B. We will show that the System (5.1) has an asymptotically stable solution under fairly general assumptions on F_A and F_B.

The equilibrium points of (5.1) are $(0,0)$ and (K_A,K_B), where $K_A = \varepsilon_B/(\gamma_B+\Gamma_B)$, $K_B = \varepsilon_A/(\gamma_A+\Gamma_A)$, $\Gamma_\alpha = \int_{-r}^{0} F_\alpha(\theta)\,d\theta$, $\alpha = A,B$. The point $(0,0)$ is unstable. So consider solutions of System (5.1) in a neighborhood of (K_A,K_B). Define x and y by $N_A = K_A(1+x)$, $N_B = K_B(1+y)$. We then obtain the linear variational system

$$\text{(5.2)}\qquad \begin{aligned} \dot{x}(t) &= -py(t) - \int_{-r}^{0} G(\theta)[y(t+\theta)-y(t)]d\theta \\ \dot{y}(t) &= qx(t) + \int_{-r}^{0} F(\theta)[x(t+\theta)-x(t)]d\theta \end{aligned}$$

where $p = \gamma_A K_B + \int_{-r}^{0} G(\theta)d\theta$, $q = \gamma_B K_A + \int_{-r}^{0} F(\theta)d\theta$, and

$$G(\theta) = K_B F_A(\theta),\ F(\theta) = K_A F_B(\theta),\ \text{for}\ \ -r \le \theta \le 0.$$

Claim: If $p > 0$, $q > 0$, $F(\theta) \ge 0$, $G(\theta) \ge 0$, $\dot{F}(\theta) \le 0$, $\dot{G}(\theta) \le 0$ for $-r \le \theta \le 0$, and there exists $\theta_0 \in [-r,0]$ such that either $\dot{F}(\theta_0) < 0$ or $\dot{G}(\theta_0) < 0$, then every solution of (5.2) approaches zero as $t \to \infty$. Thus the equilibrium (K_A,K_B) of (5.1) is globally asymptotically stable.

To establish this result we need only produce a Lyapunov function V which satisfies the conditions of Corollary 4.14. To this end, define $V\colon X \to \mathbb{R}$, where $X = C([-r,0];\mathbb{R}^2)$ by

$$\begin{aligned} V(\phi,\psi) = \tfrac{1}{2}p\psi^2(0) + \tfrac{1}{2}q\phi^2(0) &= \tfrac{1}{2}\int_{-r}^{0} G(\theta)[\psi(\theta)-\psi(0)]^2 d\theta \\ &+ \tfrac{1}{2}\int_{-r}^{0} F(\theta)[\phi(\theta)-\phi(0)]^2 d\theta \end{aligned}$$

A calculation shows that

$$\begin{aligned} \dot{V}(\phi,\psi) = {}'-\tfrac{1}{2}G(-r)[\psi(-r)-\psi(0)] &+ \tfrac{1}{2}\int_{-r}^{0} \dot{G}(\theta)[\psi(\theta)-\psi(0)]^2 d\theta \\ -\tfrac{1}{2}F(-r)[\phi(-r)-\phi(0)] &+ \tfrac{1}{2}\int_{-r}^{0} \dot{F}(\theta)[\phi(\theta)-\phi(0)]^2 d\theta \le 0. \end{aligned}$$

Thus all solutions of (5.2) are bounded. If $\dot{F}(\theta_0) < 0$ for some $\theta_0 \in [-r,0]$, then for $x_t(\phi)$ to belong to the largest weakly invariant subset (relative to solutions of (5.2)) of $\{(\phi,\psi) \in X\colon \dot{V}(\phi,\psi) = 0\}$, we must have $x(t+\theta;\phi) - x(t;\phi) = 0$ for θ in some interval about θ_0. Thus $x(t;\phi)$ must be a constant for all $t \in \mathbb{R}^+$. In view of the Equations (5.2) we must have $x(t;\phi) = y(t;\psi) = 0$ for all $t \in \mathbb{R}^+$. Thus all solutions of Equation (5.2) approach $(0,0)$ as $t \to \infty$. Therefore the equilibrium (K_A,K_B) of (5.1) is globally asymptotically stable.

The next example comes from continium mechanics. It was introduced in Example 1.3 as a case of infinite delay.

Example 5.2. Consider the motion of a unit mass attached to an elastic filament and acted on by a prescribed force h. If $x \in \mathbb{R}$ denotes the strain (i.e., stretch) of the filament, then the equation of motion of the mass is

$$\ddot{x}(t) = -\sigma(t) + h(t) \tag{5.3}$$

where σ is the stress (i.e., force). The classical theory of elasticity states that σ is a function $F(x)$ of the strain, where F is assumed to be an increasing function which vanishes at zero. One disadvantage of this model is that x need not tend to a constant when h tends to a constant.

The modern theory of elastic materials with memory assumes that $\sigma(t)$ is a functional

$$\sigma(t) = F(x_t) \tag{5.4}$$

of the history of x, namely $x_t(\theta) = x(t+\theta)$, $\theta \leq 0$. The histories belong to a space X as described in Example 2.4. That is, take X to be the space of all real valued measurable functions on $\mathbb{R}^-$ with norm

$$||\psi|| = |\psi(0)| + \int_{-\infty}^{0} |\psi(\theta)|\ell(\theta)d\theta < \infty,$$

where $\ell(\theta) = ae^{\theta/N}$ for some positive constants a and N

(to be specified later). We identify in the usual manner functions $\psi,\phi \in X$ with $\|\psi-\phi\| = 0$.

We will require the functional in (5.4) to be linear and have the form

$$F(\psi) = G(0)\psi(0) + \int_{-\infty}^{0} \dot{G}(-\theta)\psi(\theta)d\theta. \tag{5.5}$$

Thus we obtain the second order RFDE

$$\ddot{x}(t) = -G(0)x(t) - \int_{-\infty}^{0} \dot{G}(-\theta)x(t+\theta)d\theta + h(t). \tag{5.6}$$

For simplicity assume the external force h is zero.

The term G is called the relaxation function; $G(0)$ is called the instantaneous modulus; and the limit

$$G(\infty) \overset{\text{def}}{=} \lim_{s\to\infty} G(s)$$

which we will assume exists (and is finite) is called the equilibrium modulus. In a special case of linear viscoelasticity G has the form

$$G(s) = \int_0^\infty k(\tau)e^{-s/\tau}d\tau + G(\infty),$$

with k a nonnegative, measurable function of bounded support which satisfies

$$0 < \int_0^\infty k(\tau)d\tau < \infty, \quad 0 < \int_0^\infty \frac{1}{\tau}\, k(\tau)d\tau < \infty.$$

The function k is referred to as the relaxation spectrum. Consequently if the support of k lies in $[0,N]$, $N > 0$, then

$$-\dot{G}(s) = \int_0^\infty \frac{1}{\tau}k(\tau)e^{-s/\tau}d\tau \le ae^{-s/N},$$

where $a = \int_0^N \frac{1}{\tau}k(\tau)d\tau$. Thus $-\dot{G}(s)$ is a positive, bounded, increasing function on $\mathbb{R}^-$ and dominated by $\ell(\theta) = ae^{\theta/N}$,

$\theta \in \mathbb{R}^-$.

In view of the properties of $-\dot{G}$ it is easily verified that the linear functional F given in Equation (5.5) is continuous in the topology of X. The linearity of Equation (5.6) insures the existence of solutions for all $t \in \mathbb{R}^+$.

We now show that the zero solution of Equation (5.6) is asymptotically stable. First we need to re-evaluate Equation (5.5) using the expression for $\dot{G}$. We obtain

$$F(\psi) = G(0)\psi(0) + \int_{-\infty}^{0}\left[-\int_{0}^{\infty}\frac{1}{\tau}\,k(\tau)e^{\theta/\tau}d\tau\right]d\theta$$

$$(5.7)\qquad = \left[\int_0^\infty k(\tau)d\tau + G(\infty)\right]\psi(0) - \int_0^\infty k(\tau)\left[\int_{-\infty}^{0}\frac{1}{\tau}e^{\theta/\tau}\psi(\theta)d\theta\right]d\tau$$

$$= G(\infty)\psi(0) + \int_0^\infty k(\tau)\left[\psi(0) - \frac{1}{\tau}\int_{-\infty}^{0}e^{\theta/\tau}\psi(\theta)d\theta\right]d\tau.$$

Let B be any bounded neighborhood of the origin in X. Define the (Lyapunov) function V on B by

$$(5.8)\qquad \begin{aligned} V(\psi) &= \frac{1}{2}G(\infty)\psi^2(0) + \frac{1}{2}\left[\int_{-\infty}^{0}G(-\theta)\psi(\theta)d\theta\right]^2 \\ &\quad + \frac{1}{2}\int_0^\infty k(\tau)\left[\psi(0) - \frac{1}{\tau}\int_{-\infty}^{0}e^{\theta/\tau}\psi(\theta)d\theta\right]^2 d\tau. \end{aligned}$$

It is readily verified that V is continuous on B. From

$$\begin{aligned} V(x_t) &= \frac{1}{2}G(\infty)x^2(t) - \frac{1}{2}\left[\int_{-\infty}^{t}G(t-\theta)x(\theta)d\theta\right]^2 \\ &\quad + \frac{1}{2}\int_0^\infty k(\tau)\left[x(t) - \frac{1}{\tau}e^{-t/\tau}\int_{-\infty}^{t}e^{\theta/\tau}x(\theta)d\theta\right]^2 d\tau, \end{aligned}$$

we obtain

$$\dot{V}(x_t) = -\int_0^\infty \frac{1}{\tau}\,k(\tau)\left[x(t) - \frac{1}{\tau}\int_{-\infty}^{0}e^{\theta/\tau}x(t+\theta)d\theta\right]^2 d\tau.$$

So

$$(5.9)\qquad \dot{V}(\psi) = -\int_0^\infty \frac{1}{\tau}k(\tau)\left[\psi(0) - \frac{1}{\tau}\int_{-\infty}^{0}e^{\theta/\tau}\psi(\theta)d\theta\right]^2 d\tau \le 0.$$

Observe that $\frac{1}{\tau}\int_{-\infty}^{0} e^{\theta/\tau}\psi(\theta)d\theta = \psi(0)$ for almost all $\tau > 0$ if and only if $\psi(\theta) = \psi(0)$ for almost all $\theta \leq 0$. Indeed this follows from the relationship

$$\frac{1}{\tau}\int_{-\infty}^{0} e^{\theta/\tau}\psi(\theta)d\theta - \psi(0) = \frac{1}{\tau}\int_{-\infty}^{0} e^{\theta/\tau}[\psi(\theta) - \psi(0)]d\theta.$$

Thus we see from Equation (5.7) that if $G(\infty) > 0$, then $\psi = 0$ is the unique critical point of the System (5.6). (Note that we identify functions which are almost everywhere the same.) It also follows from Equation (5.8) that if $G(\infty) > 0$, the Lyapunov function V satisfies

$$V(\psi) > 0 \text{ if and only if } \psi \neq 0,$$

and from the relation (5.9) that

$$\dot{V}(\psi) = 0 \text{ if and only if } \psi \text{ is a constant function.}$$

Consequently $\psi = 0$ is asymptotically stable provided $G(\infty) > 0$. This concludes Example 5.2.

The next example comes from nuclear reactor kinetics (DiPasquantonio and Kappel [1]). A problem though arises in the choice of Lyapunov function V. It turns out that $V(\phi)$ is not continuous in ϕ, although $V(x_t)$ is continuous in t. Consequently we may not use Theorem 4.12. This difficulty can be removed if $\ddot{V}(x_t)$ exists, as is demonstrated in the next lemma.

<u>Lemma 5.3</u>. Suppose $G \subset X$, where X is a space of the kind defined by conditions (2.1) and (2.2) and G is positively invariant. Let $V: G \to \mathbb{R}$ have the following properties:

(i) $g(t) \overset{\text{def}}{=} V(x_t)$ is continuous in t and is bounded below for every solution $x(\cdot;\phi)$ of Equation (2.5) with $\phi \in G$,

(ii) $\dot{V}(x_t) = \frac{dg(t)}{dt}$ exists for every solution $x(\cdot;\phi)$ of Equation (2.5) with $\phi \in G$,

(iii) There is a continuous and nonpositive function $W: \overline{G} \to \mathbb{R}$ with $W(\phi) = \dot{V}(\phi) = \frac{dg(0^+)}{dt}$ for each $\phi \in G$,

(iv) $\ddot{V}(x_t) = \frac{d^2g(t)}{dt^2}$ exists for every solution $x(\cdot;\phi)$ of Equation (2.5) with $\phi \in G$ and is uniformly bounded on $\mathbb{R}^+$.

Let M denote the largest subset of

$$E = \{\phi \in \overline{G}: W(\phi) = 0\}$$

which is weakly invariant relative to solutons of Equation (2.5). If $x(\cdot;\phi)$ is a solution of Equation (2.5) for which $\gamma^+(\phi)$ is bounded, $\phi \in G$, then

$$\lim_{t\to\infty} d(x_t(\phi),M) = 0.$$

Proof: Condition (iv) implies $\frac{dg}{dt}$ is uniformly continuous on $\mathbb{R}^+$. So let $\varepsilon > 0$ and choose $\delta > 0$ according to the uniform continuity of $\frac{dg}{dt}$. Now as $g(t)$ is nonincreasing according to (iii) and bounded from below on $\mathbb{R}^+$ we have $\lim_{t\to\infty} g(t)$ exists. Set

$$\ell = \lim_{n\to\infty} g(n\delta) - g(0).$$

We can rewrite this as

$$\ell = \sum_{n=0}^{\infty} [g((n+1)\delta) - g(n\delta)].$$

Use the intermediate value theorem to obtain a sequence $\{t_n\} \subset \mathbb{R}^+$ so that

$$\ell = \sum_{n=0}^{\infty} \frac{dg(t_n)}{dt} \delta, \; n\delta \le t_n \le (n+1)\delta.$$

Thus

$$\lim_{n\to\infty} \frac{dg(t_n)}{dt} = 0.$$

Let N be a positive integer so that $n \ge N$ implies $\left|\frac{dg(t_n)}{dt}\right| < \frac{1}{2}\varepsilon$. For each $t \ge N$ there exists some t_n so that $|t-t_n| < \delta$. Consequently $\left|\frac{dg(t)}{dt} - \frac{dg(t_n)}{dt}\right| < \frac{1}{2}\varepsilon$. Finally for each $t \ge N$

$$\left|\frac{dg(t)}{dt}\right| \le \left|\frac{dg(t)}{dt} - \frac{dg(t_n)}{dt}\right| + \left|\frac{dg(t_n)}{dt}\right| < \frac{1}{2}\varepsilon + \frac{1}{2}\varepsilon = \varepsilon.$$

Thus $\dot{V}(x_t) \to 0$ along every solution $x(\cdot;\phi)$, $\phi \in G$. As W is continuous on $\overline{G}$ we must have $W(\psi) = 0$ for each $\psi \in L^+(\phi)$. The proof now concludes as in the case for Theorem 4.12. □

<u>Example 5.4</u>. We consider the point kinetic reactor equations for m groups of delayed neutrons

$$\begin{aligned} \dot{p}(t) &= -\sum_{i=1}^{m} \frac{\beta_i}{\Lambda}[p(t) - c_i(t)] - \frac{P_1}{\Lambda}[1 + p(t)]F(p_t), \\ \dot{c}_i(t) &= \lambda_i[p(t) - c_i(t)], \quad i = 1,\dots,m, \end{aligned} \tag{5.10}$$

where p and c_i, $i = 1,\dots,m$ denote the normalized, dimensionless and incremental reactor-power and delayed neutron precursors densities respectively. F is the linear feedback functional given by

$$F(p_t) = \int_{-\infty}^{0} k(-\theta)p(t+\theta)d\theta.$$

P_1, Λ, β_i, λ_i are positive constants. For the variables $(p,c) = (p,c_i,\ldots,c_m)$ only the following domain Q is physically significant:

$$Q = \{(p,c) \in \mathbb{R}^{m+1}: p \geq -1,\ c_i \geq -1 \text{ for } i = 1,\ldots,m\}.$$

We suppose that $\int_{-\infty}^{0} k(-\theta)d\theta \neq 0$. Thus the point $(p,y) = 0$ is an isolated equilibrium and is called the power equilibrium state. The kernel k is assumed to have the following properties: there exists a continuous function ℓ on $\mathbb{R}^-$ so that

(i) $|k(-\theta)| \leq \ell(\theta)$ for all $\theta \in \mathbb{R}^-$,

(ii) $\ell(\theta) > 0$ for all $\theta \in \mathbb{R}^-$,

(iii) $\dot{\ell}(\theta) \geq 0$ for all $\theta \in \mathbb{R}^-$,

(iv) $\int_{-\infty}^{0} \ell(\theta)d\theta < \infty$.

Let $x = (x_0,x_1,\ldots,x_m) \in \mathbb{R}^{m+1}$ have the norm $|x| = \sum_{i=0}^{m} |x_i|$. The set X will consist of all measurable functions $\phi\colon \mathbb{R}^- \to \mathbb{R}^{m+1}$ which satisfy

(a) $\phi_i(\theta) \geq -1$ for all $\theta \in \mathbb{R}^-$, $i = 0,1,\ldots,m$,

(b) $\int_{-\infty}^{0} \ell(\theta)|\phi(\theta)|d\theta < \infty$.

Then X is a complete metric space with metric

$$d(\phi,\psi) = |\phi(0) - \psi(0)| + \int_{-\infty}^{0} \ell(\theta)|\phi(\theta) - \psi(\theta)|d\theta$$

and satisfies conditions (2.1) and (2.2) for a space of histories (see Example 2.4).

We turn to the question of existence of solutions of the System (5.10). Write $f = (f_0,f_1,\ldots,f_m)$, $\phi = (\phi_0,\phi_1,\ldots,\phi_m)$, where

$$f_0(\phi) = -\sum_{i=1}^{m} \frac{\beta_i}{\Lambda}[\phi_0(0) - \phi_i(0)] - \frac{P_1}{\Lambda}[1 + \phi_0(0)]F(\phi_0),$$

$$f_i(\phi) = \lambda_i[\phi_0(0) - \phi_i(0)], \quad i = 1,\ldots,m.$$

Suppose $\alpha > 0$ and let $\phi,\psi \in X$, with $d(0,\phi) < \alpha$ and $d(0,\psi) < \alpha$. We have

$$\begin{aligned}|f(\phi) - f(\psi)| &= \sum_{i=0}^{m} |f_i(\phi) - f_i(\psi)| \\ &\le \sum_{i=1}^{m} \left(\frac{\beta_i}{\Lambda} + \lambda_i\right)|\phi_0(0) - \phi_i(0) - \psi_0(0) + \psi_i(0)| \\ &\quad + \frac{P_1}{\Lambda}|[1 + \phi_0(0)]F(\phi_0) - [1 + \psi_0(0)]F(\psi_0)|.\end{aligned}$$

Set $B = \sum_{i=1}^{m} \left(\frac{\beta_i}{\Lambda} + \lambda_i\right)$. Then

$$\begin{aligned}|f(\phi) - f(\psi)| &\le B\sum_{i=0}^{m} |\phi_i(0) - \psi_i(0)| + \frac{P_1}{\Lambda}|F(\phi_0) - F(\psi_0)| \\ &\quad + \frac{P_1}{\Lambda}|\phi_0(0)F(\phi_0) - \phi_0(0)F(\psi_0)| + \frac{P_1}{\Lambda}|\phi_0(0)F(\psi_0) - \psi_0(0)F(\psi_0)| \\ &\le Bd(\phi,\psi) + \frac{P_1}{\Lambda}d(\phi,\psi) + \frac{P_1}{\Lambda}\phi_0(0)d(\phi,\psi) \\ &\quad + \frac{P_1}{\Lambda}\psi_0(0)d(\phi,\psi) \\ &\le \left[B + \frac{P_1}{\Lambda}(1 + 2\alpha)\right]d(\phi,\psi).\end{aligned}$$

Thus f is Lipschitz on bounded subsets of X.

Denote by $A \subset X$ the class of all "physically admissible" initial values; namely $\phi \in A$ provided

(i) $\phi(\theta) = 0$ for $\theta \le -T \le 0$ for some $T = T(\phi)$,

(ii) $|\phi(\theta)| \le K$ for all $\theta \in \mathbb{R}^-$ for some $K = K(\phi)$,

(iii) ϕ is piecewise continuous on $\mathbb{R}^-$ with left and right hand limits at every point of $\mathbb{R}^-$.

The justification for the choice A reflects the behavior of the reactor due to external reactivity changes and to external sources ($t \in \mathbb{R}^-$). For every $\phi \in A$ there exists a unique (local) solution $x(\cdot;\phi)$ of the System (5.10). Then $x_t(\phi) \in A$ whenever the solution exists.

Our candidate for a Lyapunov function V is

$$V(\phi) = \phi_0(0) - \ln[1 + \phi_0(0)] - \frac{1}{2a}\phi_0^2(0) + \sum_{i=1}^{m} \frac{\beta_i}{\lambda_i \Lambda}\{\phi_i(0) - \ln[1 + \phi_i(0)] - \frac{1}{2a}\phi_i^2(0)\} + \frac{P_1}{a\Lambda}\int_{-\infty}^{0} [a-1 - \phi_0(s)]\phi_0(s)\{\int_{-\infty}^{0} k(-\theta)\phi_0(\theta+s)d\theta\}ds.$$

If $x(\cdot;\phi) = (p(\cdot),c_1(\cdot),\ldots,c_m(\cdot))$ is the solution of the System (5.10) with $\phi \in A$, we have

$$g(t) = V(x_t(\phi)) = p(t) - \ln[1 + p(t)] - \frac{1}{2a}p^2(t) + \sum_{i=1}^{m} \frac{\beta_i}{\lambda_i \Lambda}\{c_i(t) - \ln[1 + c_i(t)] - \frac{1}{2a}c_i^2(t)\} + \frac{P_1}{a\Lambda}\int_{-\infty}^{0} [a-1 - p(s)]p(s)F(p_s)ds.$$

As $x(t;\phi)$ is continuous in t then so is $g(t)$. Now compute, using System (5.10),

$$\frac{dy(t)}{dt} = -\sum_{i=1}^{m} \frac{\beta_i}{\Lambda}[p(t)-c_i(t)]^2\{\frac{1}{[1+p(t)][1+c_i(t)]} - \frac{1}{a}\}. \tag{5.11}$$

Then $g(t)$ is differentiable provided $p(t) > -1$ and $c_i(t) > -1$, $i = 1,\ldots,m$. This suggests that the functional $W(\phi)$ be defined as $\frac{dg(0^+)}{dt}$; namely,

$$W(\phi) = -\sum_{i=1}^{m} \frac{\beta_i}{\Lambda}[\phi_0(0)-\phi_i(0)]^2\{\frac{1}{[1+\phi_0(0)][1+\phi_i(0)]} - \frac{1}{a}\}. \tag{5.12}$$

Finally we see that $\frac{d^2g(t)}{dt^2}$ exists; namely,

$$\frac{d^2g(t)}{dt^2} = \sum_{i=1}^{m} \frac{\beta_i}{\Lambda}[p(t)-c_i(t)]^2 \frac{\dot{p}(t)[1+c_i(t)]+\dot{c}_i(t)[1+p(t)]}{[1+p(t)]^2[1+c_i(t)]^2}$$

(5.13)

$$-2 \sum_{i=1}^{m} \frac{\beta_i}{\Lambda}[p(t)-c_i(t)][\dot{p}(t)-\dot{c}_i(t)]\{\frac{1}{[1+p(t)][1+c_i(t)]} - \frac{1}{a}\}.$$

We must find a subset $G \subset A$ which satisfies the requirements of Lemma 5.3. In order that $W(\phi) \leq 0$ we require that

$$-1 < \phi_i(0) < \sqrt{a} - 1, \; i = 0,1,\ldots,m.$$

This also ensures W is continuous and that $g(t)$ is twice differentiable whenever $x(t;\phi)$ exists.

Set

$$h(z) = z - \ell n(1+z) - z^2/2a, \quad z \in \mathbb{R},$$

and let

$$\alpha = h(\sqrt{a} - 1)$$

$$b = \min\{\alpha, \frac{\beta_1}{\lambda_1\Lambda}\alpha,\ldots, \frac{\beta_m}{\lambda_m\Lambda}\alpha\}.$$

Denote by q the negative root of $h(z) = b$. Figure 5.1 illustrates the behavior of $h(z)$. If we define

$$G = \{\phi \in A\colon V(\phi) < b, \; q < \phi_i(0) < \sqrt{a} - 1 \;\text{ for }\; i = 0,1,\ldots,m\}$$

and require for each $\phi \in G$ that

$$(5.14) \quad L_a \overset{\text{def}}{=} \int_{-\infty}^{0} [a-1-\phi_0(s)]\, \phi_0(s)\{\int_{-\infty}^{0} k(-\theta)\phi_0(\theta+s)d\theta\}ds \geq 0,$$

then clearly $V(\phi) \geq 0$ and $W(\phi) \geq 0$. Now suppose $x(\cdot;\phi) = (p(\cdot),c_1(\cdot),\ldots,c_m(\cdot))$ is a solution of the System (5.10) with $\phi \in G$. We claim $x_t = x_t(\phi) \in G$ for all $t \in \mathbb{R}^+$.

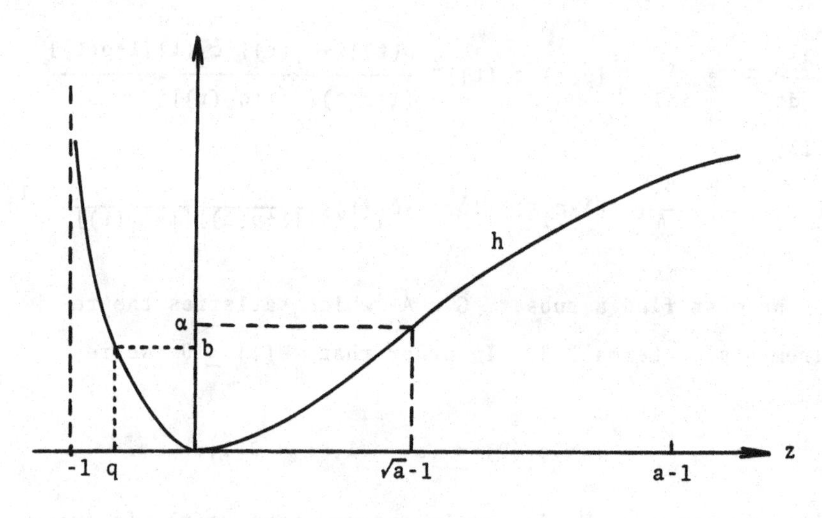

Figure 5.1

Indeed we must show that $V(x_t) < b$ and

$$q < p(t) < \sqrt{a} - 1, \quad q < c_i(t) < \sqrt{a} - 1, \quad i = 1,\dots,m \tag{5.15}$$

$$\text{for every } t \in \mathbb{R}^+.$$

As V is nonincreasing along x_t, then $V(x_t) < b$ for each $t \in \mathbb{R}^+$. In this event since we can write

$$V(x_t) = h(p(t)) + \sum_{i=1}^{m} \frac{\beta_i}{\lambda_i \Lambda} h(c_i(t)) + L_a < b,$$

then

$$h(p(t)) < b, \quad \frac{\beta_i}{\lambda_i \Lambda} h(y_i(t)) < b \quad \text{for } i = 1,\dots,m.$$

These inequalities imply the relations (5.15) hold. This means that $x_t \in G$ for every $t \in \mathbb{R}^+$. In other words, G is positively invariant.

There remains only a few more conditions of Lemma 5.3 to verify. Again suppose $\phi \in G$ with $x(\cdot;\phi)$ a solution of System (5.10). As $h(z) \geq 0$ for $z > -1$, then $g(t)$ is bounded from below by zero. Next as G is bounded, the coordinates of $x(t;\phi)$ are uniformly bounded in $\mathbb{R}^+$. This means $\dot{p}(t)$ and $\dot{c}_i(t)$, $i = 1,\ldots,m$ are uniformly bounded in $\mathbb{R}^+$. In view of expression (5.13) $\frac{d^2g(t)}{dt^2}$ is also uniformly bounded in $\mathbb{R}^+$. As $\gamma^+(\phi)$ is a d-bounded trajectory, then using Lemma 5.3 we obtain that $x(t;\phi)$ tends to the largest subset M of $E = \{\phi \in \overline{G}: W(\phi) = 0\}$ which is weakly invariant relative to solutions of System (5.10). But if $y(t)$ is a solution of (5.10) with $y_t \in M$ for each $t \in \mathbb{R}$, then $W(y_t) = 0$. Thus from (5.11) we have $p(t) = c_i(t)$ for $i = 1,\ldots,m$ and all $t \in \mathbb{R}^+$. Use the System (5.10) to show $\dot{c}_i(t) \equiv 0$ for $i = 1,\ldots,m$. This means $p(t) \equiv c_i(t) \equiv c$, $i = 1,2,\ldots,m$ for some $c > -1$. A final calculation shows

$$0 = -\frac{P_1}{\Lambda}\ 1+c \int_{-\infty}^{0} ck(-\theta)d\theta,$$

which implies $c = 0$ in view of the requirement that $\int_{-\infty}^{0} k(-\theta)d\theta \neq 0$.

Summarizing these results we have

<u>Proposition 5.5</u>. Suppose for some $a > 1$ we have $L_a \geq 0$, where L_a is given by (5.14). Then every solution $x(\cdot;\phi)$ of the System (5.10) with initial value $\phi \in G$ tends to 0 as $t \to \infty$.

<u>Remark 5.6</u>. Though for $\phi \in A$, $V(x_t(\phi))$ is continuous in t, it is not the case that V is continuous on A. Indeed define

$$\psi_0(\theta) = \begin{cases} c \neq 0, & \text{for } \theta \in [-n-1,-n] \\ 0\ , & \text{otherwise.} \end{cases}$$

Then

$$\int_{-\infty}^{0} [a-1-\psi_0(\theta)]\,\psi_0(\theta)\{\int_{-\infty}^{0} k(-\theta)\psi_0(\theta+s)d\theta\}ds$$

$$= (a-1-c)c^2 \int_{-1}^{0}\int_{-1-\tau}^{0} k(-\theta)d\theta d\tau.$$

Now define the sequence $\{\psi^{(n)}\}$ by

$$\psi_0^{(n)}(\theta) = \begin{cases} c, & \theta \in [-n-1,-n] \\ 0, & \text{otherwise}, \end{cases}$$

$$\psi_i^{(n)}(\theta) = 0, \quad \theta \in \mathbb{R}^-, \; i = 1,2,\ldots,m.$$

Now $\psi^{(n)} \to 0$ in the d-metric, but

$$V(\psi^{(n)}) = \frac{P_1}{a\Lambda}(a-1-c)c^2 \int_{-1}^{0}\int_{-1-\tau}^{0} k(\theta)d\theta d\tau.$$

As $V(0) = 0$ we see that V is not continuous at $\psi = 0$.

6. Remarks on the Asymptotic Behavior of Nonautonomous Retarded Functional Differential Equations

We propose to indicate how semidynamical systems may be used to study the behavior of solutions to nonautonomous systems of the initial value problem

$$(6.1) \qquad \dot{x}(t) = f(x_t,t), \quad t \in \mathbb{R}^+, \quad x_0 = \phi$$

where $f\colon X \times \mathbb{R}^+ \to \mathbb{R}^d$, $\phi \in X = C([-r,0]; \mathbb{R}^d)$. The notation is the same as that introduced in Section 2. Proofs of most results are omitted.

Chapter IV was devoted to a similar problem for ordinary differential equations in Euclidean space. The reader should refer to the opening sections of Chapter IV to obtain the necessary motivation for the semidynamical system we are about to define. Let $\mathscr{F}$ denote the set of all continuous functions $f\colon X \times \mathbb{R}^+ \to \mathbb{R}^d$ which are bounded on sets of the form $B \times \mathbb{R}^+$, B a bounded subset of X. Endow $\mathscr{F}$ with the topology given by the metric

$$d(f,g) = \sum_{n=1}^{\infty} 2^{-n} \min\{1, \sup_{(\phi,t)\in B_n \times \mathbb{R}^+} \|f(\phi,t) - g(\phi,t)\|\}$$

where B_n is the ball of radius n centered at zero in X. Before proceeding any further we need to establish existence and continuous dependence for solutions of (6.1).

Proposition 6.1. (Existence) Let $f \in \mathcal{J}$ and $\phi \in X$. Then there exists $\alpha > 0$ and a function $x \in C([-r,\alpha); \mathbb{R}^d)$ so that $x(t)$ satisfies Equation (6.1) on $[0,\alpha)$ with $x_0 = \phi$.

Proposition 6.2. (Continuous dependence) Suppose all solutions of Equation (6.1) exist on $\mathbb{R}^+$. Consider sequences $f^{(k)} \to f$ in $\mathcal{J}$ and $\phi^{(k)} \to \phi$ in X so that $x^{(k)}$ is a solution of $\dot{x}(t) = f^{(k)}(x_t,t)$ with $x_0^{(k)} = \phi^{(k)}$, $k = 1,2,\ldots$ Then a subsequence of $x^{(k)}$ converge uniformly on compact subsets of $\mathbb{R}^+$ to a solution of $\dot{x}(t) = f(x_t,t)$, $x_0 = \phi$.

For each $f \in \mathcal{F}$ let f_t be the function in $\mathcal{J}$ given by $f_t(\phi,s) = f(\phi,s+t)$, $t \in \mathbb{R}^+$. f_t is called the *t-translate* of f. Fix $f \in \mathcal{J}$ and define the hull of f, $\mathcal{H}^+(f)$, by

$$\mathcal{H}^+(f) = \overline{\{f_t : t \in \mathbb{R}^+\}}.$$

Assume the following conditions from now on.

Global Existence: solutions of (6.1) exist on all of $\mathbb{R}^+$.

Uniqueness: for every $g \in \mathcal{H}^+(f)$, there exists a unique solution of the initial value problem

$$\dot{x}(t) = g(x_t,t), \quad x_0 = \phi \in X. \tag{6.2}$$

A semidynamical system with phase space $\mathcal{H}^+(f) \times X$ can now be constructed as in Chapter IV. Denote by $F(g,\phi;t)$

the solution of Equation (6.2), $t \in \mathbb{R}^+$, with $F_0(g,\phi;\cdot) = \phi(\cdot)$. We reserve the notation $F_t(g,\phi)$ for the usual time translation of $F(g,\phi;\cdot)$ to $[-r,0]$. Then define $\pi: \mathscr{A}^+(f) \times X \times \mathbb{R}^+ \to \mathscr{A}^+(f) \times X$ by

$$\pi(g,\phi,t) = (g_t, F_t(g,\phi)).$$

<u>Theorem 6.3</u>. $(\mathscr{A}^+(f) \times X, \pi)$ is a semidynamical system.

<u>Proof</u>: The identity axiom as well as the semigroup axiom hold as can be seen from Lemma 2.10 of Chapter IV. The continuity axiom is easy to verify. In fact, the mapping $(g,t) \to g_t$ from $\mathscr{A}^+(f) \times \mathbb{R}^+$ into $\mathscr{A}^+(f)$ is continuous. Also the mapping of $\mathscr{A}(f) \times X \times \mathbb{R}^+$ into X by $(g,\phi,t) \to F_t(g,\phi)$ is continuous in $(g,\phi) \in \mathscr{A}^+(f) \times X$, uniformly with respect to t in compact subsets of $\mathbb{R}^+$. This observation comes from Proposition 6.2. The reader may fill in the remaining details of the proof.

Unlike the case in Chapter IV, $\mathscr{A}^+(f)$ generally is not compact. This is because the topology used here is stronger than that given by the convergence Lemma 2.9 of Chapter IV. The following proposition (due to J. Kato [1]) characterizes those $f \in \mathscr{J}$ for which $\mathscr{A}^+(f)$ is compact. The proof is omitted.

<u>Proposition 6.4</u>. Let $f \in \mathscr{J}$. Then $\mathscr{A}^+(f)$ is compact if and only if f is uniformly continuous on subsets of $X \times \mathbb{R}^+$ of the form $K \times \mathbb{R}^+$, K compact. In particular, if f is almost periodic in t, uniformly for ϕ in compact subsets of X, then $\mathscr{A}^+(f)$ is compact.

We now turn to a discussion of the asymptotic properties of solutions $F(g,\phi;\cdot)$, $g \in \mathscr{A}^+(f)$, under the hypothesis that $\mathscr{A}^+(f)$ is compact. Fix $(f,\phi) \in \mathscr{J} \times X$ and write

$$\mathscr{L}^+(f) = \{g \in \mathscr{A}^+(f): g = \lim_{k\to\infty} f_{t_k} \text{ for some } \{t_k\} \subset \mathbb{R}^+ \text{ with } t_n \to \infty\}.$$

The following positive limit sets also are required.

$$\mathscr{L}_f^+(\phi) = \{\psi \in X: \psi = \lim_{k\to\infty} F_{t_k}(f,\phi) \text{ for some } t_k \to \infty\}$$

$$\Omega_f^+(\phi) = \{x \in \mathbb{R}^d: x = \lim_{k\to\infty} F(f,\phi;t_k) \text{ for some } t_k \to \infty\}$$

A subset $M \subset X$ is called *quasi-invariant* with respect to Equation (6.1) if for each $\psi \in M$ there exists $f^* \in \mathscr{L}^+(f)$ so that $F_t(f^*,\psi) \in M$ for all $t \in \mathbb{R}$. We have the following result which is similar to Theorem 4.11 of Chapter IV.

Theorem 6.5. If $F(f,\phi;\cdot)$ is a bounded solution of Equation (6.1), then $\mathscr{L}_f^+(\phi)$ is a nonempty quasi-invariant set.

An invariance principle for solutions of Equation (6.1) can now be formulated. It is an extension of Theorem 4.17 of Chapter IV for autonomous RFDE's.

Theorem 6.6. Suppose there exist subsets $H, E \subset X$ with the following property: for each $\phi \in H$, $F_t(f,\phi) \to E$ as $t \to \infty$. If $F(f,\phi;t)$ is bounded on $\mathbb{R}^+$, then $F_t(f,\phi) \to M$ as $t \to \infty$, where M is the largest quasi-invariant subset of E.

Again, the set E is usually determined by a Lyapunov function. Most of the remaining results of Section 4 and 5 of Chapter IV extend to Equation (6.1).

7. Critical Points and Periodic Solutions of Autonomous Retarded Functional Differential Equations

Consider the linear autonomous RFDE

$$\dot{x}(t) = L(x_t) \tag{7.1}$$

and the perturbed RFDE

$$\dot{x}(t) = L(x_t) + f(t), \tag{7.2}$$

where $L: X \to \mathbb{R}^d$ is a continuous linear function, $f \in L^1_{loc}(\mathbb{R}^+; \mathbb{R}^d)$ and X is the space $C([-r,0]; \mathbb{R}^d)$, $0 < r < \infty$. We have the following representation for solutions of Equation (7.2). Note that we are dealing with a special case of the nonautonomous RFDE, $\dot{x}(t) = f(x_t,t)$.

Lemma 7.1. For every $\phi \in X$, (7.2) has a unique solution $x(t;\phi)$, $t \in \mathbb{R}^+$, which satisfies $x_0 = \phi$. Moreover if we set $x_t(\phi)(\theta) = x(t+\theta;\phi)$, then

$$x_t(\phi) = T(t)\phi + \int_0^t T(t-s)\Lambda_0 f(s)ds \tag{7.3}$$

where T is the linear semigroup generated by Equation (7.1) and Λ_0 is the $d \times d$ matrix function on $[-r,0]$ given by

$$\Lambda_0(\theta) = \begin{cases} 0, & -r \leq \theta < 0 \\ I, & \theta = 0 \end{cases}.$$

Proof: Define the operator A by $A\phi = -\dot{\phi}$, with domain $\mathcal{D}(A) = \{\phi \in X: \dot{\phi} \in X, \dot{\phi}(0) = L(\phi)\}$. A is linear, and as shown in Chapter V, $\mathcal{D}(A)$ is dense in X. A also is the infinitesimal generator of a uniquely defined strongly continuous linear semigroup T (Theorem 2.9 and Corollary 2.10 of Chapter V. Moreover the mapping $t \to T(t)\phi$ is differentiable. It is a straightforward calculation to verify that

$$(7.4) \qquad T(t)\phi(0) + \int_0^t T(t-s)\Lambda_0(0)f(s)ds$$

satisfies Equation (7.2) for the initial function $\phi \in X$.

The linearity of L and an application of Gronwall's inequality ensures that for each $\phi \in X$, there is a unique solution $x(\cdot;\phi)$ of Equation (7.2) for which $x_0 = \phi$. Thus we must have that $x(t;\phi)$ is given by (7.4). Consequently the representation (7.3) is valid. (Note that we do not require the existence theorem of Section 6.) □

<u>Definition 7.2</u>. A solution x of the RFDE $\dot{x}(t) = f(x_t,t)$ is called *ω-periodic* if $x(t+\omega) = x(t)$ for all $t \in \mathbb{R}^+$.

Before proceeding to establish results on periodic solutions, we need an additional lemma.

<u>Lemma 7.3</u>. Suppose $f \in C(\mathbb{R}^+;\mathbb{R}^d)$ is ω-periodic, $\omega \geq r$. Then a solution x of Equation (7.2) is ω-periodic if and only if $x_\omega = x_0$.

<u>Proof</u>: Inasmuch as (7.2) is a nonautonomous system, the solutions $x_t(\phi)$ (in X) do not give rise to a semidynamical system except as in the context of Section 6. Instead, we will establish the lemma by appeal to the integral representation (7.3). We omit the details, but observe that as $T(t)$ is a linear operator, it commutes with the Riemann integral operator. □

We now state and prove a theorem on the existence of periodic solutions.

<u>Theorem 7.4</u>. Consider the RFDE

$$(7.5) \qquad \dot{x}(t) = L(x_t) + f(t)$$

where L is a linear continuous map from $X = C([-r,0]; \mathbb{R}^d)$ to $\mathbb{R}^d$, and $f \in C(\mathbb{R}^+; \mathbb{R}^d)$ is ω-periodic, $\omega \geq r$. If Equation (7.5) possesses a bounded solution, then Equation (7.5) possesses a (nontrivial) ω-periodic solution.

<u>Proof</u>: An ω-periodic solution exists if and only if there exists $\phi \in X$ so that $\phi = T(\omega)\phi + \psi$, where $\psi = \int_0^\omega T(\omega-s)\Lambda_0 f(s)ds$. As L is linear, $T(\omega)$ is a compact operator (Corollary 4.2). Now suppose no such ϕ exists. Thus 1 is an eigenvalue of $T(\omega)$ and $I - T(\omega)$ is singular. So there is some $\phi^* \in X^*$ with $((I-T(\omega))\phi,\phi^*) = 0$ for all $\phi \in X$. Let ϕ_0 be the initial function for a bounded solution to Equation (7.5), and define

$$\phi_n = T^n(\omega)\phi_0 + [T^{n-1}(\omega) + \dots + T(\omega) + I]\psi$$

for each $n = 1,2,\dots$. By a direct calculation we have that

$$\begin{aligned}\phi_2 &= T^2(\omega)\phi_0 + [T(\omega) + I]\psi \\ &= T(2\omega)\phi_0 + \int_0^\omega T(2\omega-s)\Lambda_0 f(s)ds + \int_\omega^{2\omega} T(2\omega-s)\Lambda_0 f(s)ds \\ &= x_{2\omega}(\phi_0).\end{aligned}$$

Proceed by induction to establish that

$$x_{n\omega}(\phi_0) = \phi_n, \quad n = 1,2,\dots,$$

which, by hypothesis, is a bounded sequence in X.

Now

$$(\phi_n,\phi^*) = (\phi_0, T^{*n}(\omega)\phi^*) + (\psi, \sum_{k=0}^{n-1} T^{*k}(\omega)\phi^*),$$

where $T^*(\omega)$ is the adjoint of $T(\omega)$. As $T(\omega)$ is compact, so is each $T^{*k}(\omega)$, $k = 1,2,\dots$. Since 1 is also an eigenvalue of $T^*(\omega)$, then $T^{*k}(\omega)\phi^* = \phi^*$. Therefore

$$(\phi_n,\phi^*) = (\phi_0,\phi^*) + n(\psi,\phi^*).$$

This implies $\lim_{n\to\infty} (\phi_n,\phi^*) = \infty$, which is impossible. □

The last theorem may be applied to establish the existence of critical points of motions corresponding to one-dimensional linear autonomous RFDEs. We show by an example that the same does not hold for nonlinear RFDEs. Note that $\psi \in X = C([-r,0];\mathbb{R}^d)$ is a critical point of the semidynamical system determined by solutions of the general RFDE $\dot{x}(t) = f(x_t)$ if and only if $x_t(\psi) = \psi$ for all $t \in \mathbb{R}^+$. Equivalently, the solution $x(\cdot;\psi)$ is constant on $[-r,\infty)$.

Theorem 7.5. Consider the scalar RFDE

$$\dot{x}(t) = L(x_t) + b, \tag{7.6}$$

where $L\colon X = C([-r,0];\mathbb{R}) \to \mathbb{R}$ is a continuous linear function and $b \neq 0$ is a constant. If Equation (7.6) possesses a bounded solution, then it possesses a critical point.

Proof: Suppose (7.6) possesses no critical points. Then $L(\alpha) = 0$ for every constant function α. For otherwise the constant solution $x(t) \overset{\text{def}}{=} -bL(\alpha)^{-1}\alpha$ is a critical point of Equation (7.6). In view of Theorem 7.4, for each positive integer n there exists a periodic solution $x^{(n)}(t)$ of Equation (7.6) with period 2^{-n}. Set $\alpha_n = \inf\{x^{(n)}(t)\colon t \in \mathbb{R}^+\}$ and define $y^{(n)}(t) = x^{(n)}(t) - \alpha_n$ for each n. Then $y^{(n)}(t)$ is a non-negative 2^{-n}-periodic solution of Equation (7.6). We claim there exists a subsequence of $\{y^{(n)}\}$ which is uniformly bounded on $\mathbb{R}^+$. Otherwise as

$$\frac{\dot{y}^{(n)}(t)}{\|y^{(n)}\|} = L\left(\frac{y_t^{(n)}}{\|y^{(n)}\|}\right) + \frac{b}{\|y^{(n)}\|},$$

$$\|y_t^{(n)}\| \leq \|y^{(n)}\|,$$

then the linearity and boundedness of L implies $\dot{y}^{(n)}(t)\|y^{(n)}\|^{-1}$ is uniformly bounded in $t \in \mathbb{R}^+$. But for each n it is easy to see that there exists $t_n \in \mathbb{R}^+$ so that $|\dot{y}^{(n)}(t_n)| \geq 2^n \|y^{(n)}\|$. This is a contradiction. Hence $\{y^{(n)}\}$ is uniformly bounded. As $|\dot{y}^{(n)}(t)| \leq \|L\| \|y^{(n)}\| + b$, then $\{\dot{y}^{(n)}\}$ is also uniformly bounded. Thus $\{y^{(n)}\}$ is equicontinuous. The Ascoli theorem implies that $\{y^{(n)}\}$ has a convergent subsequence whose limit must be a constant function, since each $y^{(n)}$ is 2^{-n}-periodic. Taking such limits in Equation (7.6) yields $b = 0$, a contradiction. □

Example 7.6. We show that for a nonlinear RFDE, even the existence of infinitely many bounded solutions need not imply the existence of a critical point. For $\alpha \in \mathbb{R}$ define

$$u^\alpha(t) = \begin{cases} \alpha, & t \in [-r,0] \\ \alpha + \sin t, & t \in \mathbb{R}^+, \end{cases} \tag{7.7}$$

$$U = \{u_t^\alpha : \alpha \in \mathbb{R},\ t \in \mathbb{R}^+\} \subset X = C([-r,0];\mathbb{R}).$$

Then U is a closed subset of X. Define $g: U \to [-1,1]$ by

$$g(u_t^\alpha) = \dot{u}^\alpha(t^+),$$

the right hand derivative of u^α at t. Then g is continuous on U. According to the Tietze extension theorem there exists a continuous function $\hat{g}: X \to [-1,1]$ such that the restriction of $\hat{g}$ to U is given by g. Then the RFDE

$$\dot{x}(t) = \hat{g}(x_t) \tag{7.8}$$

has infinitely many bounded solutions, as the functions $u^\alpha(t)$ are solutions. On the other hand, Equation (7.7) has no critical points as $\hat{g}(\alpha) = 1$ for all constant functions α.

Theorem 7.4 admits the following generalization.

Theorem 7.7. Consider the RFDE

$$\dot{x}(t) = L(x_t) + h(x_t,t), \tag{7.9}$$

where L is a continuous linear map from $X = C([-r,0]; \mathbb{R}^d)$ to $\mathbb{R}^d$, and $h: X \times \mathbb{R}^+ \to \mathbb{R}^d$ is continuous and ω-periodic in t for each fixed $\phi \in X$, $\omega \geq r$. Assume h is bounded on bounded subsets of $X \times \mathbb{R}^+$ and that $|h(\phi,t)|/\|\phi\| \to 0$ as $\|\phi\| \to \infty$, uniformly in $t \in \mathbb{R}^+$. If the only ω-periodic solution of Equation (7.9) when $h \equiv 0$ is $x \equiv 0$, then Equation (7.9) possesses a (nontrivial) ω-periodic solution.

Proof: Let P_ω denote the Banach space of ω-periodic functions from $[-r,\infty)$ into $\mathbb{R}^d$ with the supremum norm. Then $\dot{x}(t) = L(x_t)$ has an ω-periodic solution if and only if $\phi = T(\omega)\phi$ for some $\phi \in X$. T is the linear semigroup which was established in Lemma 7.1. If $\dot{x}(t) = L(x_t)$ admits only the trivial ω-periodic solution, then $I-T(\omega)$ is invertible.

For any $\psi \in P_\omega$, consider the system

$$\dot{x}(t) = L(x_t) + h(\psi_t,t). \tag{7.10}$$

We can express the unique solution of Equation (7.10) satisfying $x_0 = \phi$ (unique in view of Lemma 7.1) as

$$x_t = T(t)\phi + \beta_t, \quad \beta_t = \int_0^t T(t-s)\Lambda_0 h(\psi_s,s)ds. \tag{7.11}$$

Then Equation (7.10) admits a unique periodic solution $x(\cdot;\phi)$, where $x_0 = \phi = [I - T(\omega)]^{-1}\beta_\omega \overset{\text{def}}{=} \phi_\omega$.

Define an operator $Q: P_\omega \to P_\omega$ by

$$Q\psi(\cdot) = x(\cdot;\phi_\omega),$$

the solution of Equation (7.9) through $\phi_\omega \in X$, where now

$$\phi_\omega = [I - T(\omega)]^{-1}\int_0^\omega T(\omega-s)\Lambda_0 h(\psi_s,s)ds.$$

It is easily established that Q is continuous, so that a fixed point of Q will yield an ω-periodic solution of Equation (7.9).

Set $f(t) = h(\psi_t,t)$. An application of Gronwall's inequality to Equation (7.4) yields

$$|x(t;\phi)| \leq [\|\phi\| + \int_0^t |f(s)|ds]e^{\|L\|t}, \quad t \geq 0.$$

Thus,

$$|Q\phi(t)| \leq k \sup_{0\leq s\leq\omega} |h(\psi_s,s)|, \quad t \geq 0,$$

where

$$k = \omega e^{\|L\|\omega}[\|(I - T(\omega))^{-1}\| e^{\|L\|\omega}+1].$$

Now we construct a closed convex subset $H \subset P_\omega$ with the property that $\overline{Q(H)}$ is compact and lies in H. The Schauder fixed point theorem then asserts that Q has a fixed point in H.

Let $0 < \varepsilon < k^{-1}$. There exists $b > 0$ such that if $\phi \in X$ with $\|\phi\| \geq b$, then $|h(\phi,t)| \leq \varepsilon\|\phi\|$ for all $t \in \mathbb{R}^+$. Set $m = \sup\{|h(\phi,t)|: \|\phi\| \leq b,\ t \in \mathbb{R}^+\}$, and $c = \max\{mk,b\}$. Set

$$H = \{\psi \in P_\omega: \|\phi\| \leq c\}.$$

Let $\psi \in H$, and fix $s \in [0,\omega]$. Either $\|\psi_s\| \leq b$ or

$b < \|\psi_s\| \leq c$. In the first case $|Q\psi(t)| \leq km \leq c$, and in the second, $|Q\psi(t)| \leq k\varepsilon\|\psi_s\| \leq \|\psi_s\| \leq c$ for all $t \geq -r$. Therefore $Q(H) \subset H$. Also

$$\left|\frac{d}{dt} Q\psi(t)\right| = |\dot{x}(t;\phi_\omega)| \leq |L(x_t)| + |h(x_t,t)|$$

$$\leq \|L\| \|x_t\| + ck^{-1} \leq \|L\| \, |Q\psi(t)| + ck^{-1} \leq c(\|L\| + k^{-1}).$$

Thus $\frac{d}{dt} Q\psi(t)$ is uniformly bounded on H, so $Q(H)$ is an equicontinuous family. The Ascoli theorem assures that $Q(H)$ is precompact. □

8. Neutral Functional Differential Equations

We investigate a class of autonomous FDEs of the type considered in Example 1.4; that is, the derivative of x occurs not only at t but at $t-r$. As in the case of RFDEs, the phase space will be given by $X = C([-r,0]; \mathbb{R}^d)$, $r > 0$.

Definition 8.1. Let $f,D: X \to \mathbb{R}^d$ be continuous. An (autonomous) *neutral functional differential equation* (NFDE) is a relation of the form

$$\frac{d}{dt} D(x_t) = f(x_t). \tag{8.1}$$

For every $\phi \in X$ a *solution of Equation (8.1) through* ϕ is a continuous function $x: [-r,a) \to \mathbb{R}^d$ for some $a > 0$ so that

(i) $x_0 = \phi$,

(ii) $D(x_t)$ is continuously differentiable on $[0,a)$,

(iii) x_t satisfies (8.1) on $[0,a)$.

As in the case of RFDE's, we call $x_t(\phi)$ a motion of Equation (8.1) through ϕ.

Remark 8.2. $x = x(\cdot;\phi)$ is a solution of Equation (8.1) through $\phi \in X$ if and only if x satisfies the integral equation

$$(8.2) \qquad D(x_t) = D(\phi) + \int_0^t f(x_s)ds.$$

Definition 8.3. Suppose the continuous function $D: X \to \mathbb{R}^d$ can be expressed as

$$D(\phi) = \phi(0) - g(\phi),$$

where

(i) $g: X \to \mathbb{R}^d$ is linear and continuous; thus there exists an $d \times d$ matrix function μ of bounded variation in $\theta \in [-r,0]$ so that

$$g(\phi) = \int_{-r}^{0} [d\mu(\theta)]\phi(\theta),$$

(ii) there exists a continuous, nonnegative scalar function β on $\mathbb{R}^+$ with $\beta(0) = 0$ so that

$$\left|\int_{-s}^{0} [d\mu(\theta)]\phi(\theta)\right| \leq \beta(s) \sup_{-s\leq\theta\leq 0} |\phi(\theta)|$$

$$\text{for } s \in [0,r], \ \phi \in X.$$

Then the linear functional differential operator D is called *atomic at 0*.

Property (ii) implies that the point $s = 0$ has zero measure (with respect to the measure generated by μ). Thus g only depends upon values of $\phi(\theta)$ for $-r \leq \theta < -\varepsilon < 0$ (ε is arbitrarily small). Consequently the measure determined by the continuous linear function g is nonatomic at 0.

The following theorem guarantees that the solution of the initial value problem

$$(8.3) \qquad \frac{d}{dt} D(x_t) = f(x_t), \quad x_0 = \phi$$

generates a semi-dynamical system on X. The assumptions of the theorem are assumed to hold throughout the remainder of this section. Its proof is omitted (see Hale and Cruz [2]).

Theorem 8.4. Suppose $f,D\colon X \to \mathbb{R}^d$ are continuous where

(i) f is Lipschitz on closed bounded subsets of X,

(ii) D is linear and atomic at 0.

Given $\phi \in X$ there exists $a \in \mathbb{R}^+$ and a unique solution $x(\cdot;\phi)$ of Equation (8.3) on $[-r,a)$. If we can extend the solution to $[-r,\infty)$ then there exists a strongly continuous semigroup T on X so that

$$(8.4) \qquad x(t;\phi) = T(t)\phi(0), \quad \phi \in X.$$

Remark 8.5. (i) We will henceforth assume throughout this section that each function $x(t;\phi)$ is defined for all $t \in \mathbb{R}^+$.

(ii) If $g \equiv 0$, so that $D(\phi) = \phi(0)$, the relation Equation (8.3) reduces to the RFDE $\dot{x}(t) = f(x_t)$.

(iii) If $f \equiv 0$, the relation Equation (8.3) reduces to the *functional difference equation*

$$(8.5) \qquad D(x_t) = D(\phi)$$

Theorem 8.4 guarantees a unique solution to the functional difference Equation (8.5).

We now proceed to derive a suitable representation for solutions of the NFDE (8.3). Denote by T_D the linear semigroup which is determined by Theorem 8.4 in the event $f \equiv 0$. Thus $T_D(\cdot)\phi$ is the unique solution of Equation (8.5). Let

$$X_D = \{\phi \in X\colon D(\phi) = 0\}.$$

Then X_D is a Banach space with the relative topology of X

and remains invariant under the semigroup T_D.

Proposition 8.6. The (unique) solution $x(\cdot;\phi)$ of Equation (8.3) through $\phi \in X$ admits the representation

$$x_t = T_D(t)\phi + \int_0^t T_D(t-s)\Lambda_0 f(x_s)ds, \tag{8.6}$$

where Λ_0 is the $d \times d$ matrix function on $[-r,0]$ given by

$$\Lambda_0(\theta) = \begin{cases} 0, & -r \leq \theta < 0 \\ I, & \theta = 0. \end{cases}$$

Proof: Let $x(\cdot;\phi)$ be the unique solution of Equation (8.3). Define $y_t = T_D(t)\phi + \int_0^t T_D(t-s)\Lambda_0 f(x_s)ds$. As $T_D(\cdot)\phi$ is the solution of Equation (8.5), then by the linearity of D we have

$$D(y_t) = D(\phi) + \int_0^t D(\Lambda_0)f(x_s)ds.$$

But

$$D(\Lambda_0) = \Lambda_0(0) - g(\Lambda_0) = I - \int_{-s}^0 d\mu(\theta)\,\Lambda_0(\theta)$$

for every $s \in [-r,0)$. Thus

$$|g(\Lambda_0)| \leq \beta(s)\|\Lambda_0\| = \beta(s), \quad s \in [-r,0).$$

In view of the special properties of $\beta(s)$ from Definition 8.3 we have $g(\Lambda_0) = 0$. Consequently $D(\Lambda_0) = I$. In view of the fact that x is a solution of the integral equation (8.2) we must have

$$D(y_t) = D(x_t).$$

Set $z_t = y_t - x_t$. Then $z_0 = 0$. So we must have $z_t = 0$, $t \in \mathbb{R}^+$. Thus $y_t = x_t$ and so the representation (8.6) is valid. □

Corollary 8.7. The motion through $\phi \in X_D$ of

$$D(x_t) = h(t)$$

where $h \in C(R^+; \mathbb{R}^d)$ is denoted by $x_t(\phi,h)$ and satisfies

$$x_t = T_D(t)\phi + T_D(t)\Lambda_0 h(t).$$

Though for every solution x of Equation (8.3) we have that $D(x_t)$ is continuously differentiable, it does not follow that $x(t)$ is differentiable. Thus unlike RFDE's, the solution x of Equation (8.3) gets no smoother than its initial function ϕ (c.f. Theorems 4.5, 4.7). So we cannot expect to use the Ascoli theorem to obtain precompact positive orbits $\gamma^+(\phi) = \bigcup_{t \in \mathbb{R}^+} x_t(\phi)$ whenever $x(t;\phi)$ is bounded on $\mathbb{R}^+$.

This dilemma can be resolved by one of two ways. Firstly we might embed the semidynamical system in a space wherein the positive orbits are precompact. This was done in Chapter V, Section 4. Alternatively, we might restrict our class of NFDE's so that some smoothing takes place. It is this latter approach which we take. This leads us now to develop the concept of a stable linear functional differential operator D.

There exist numbers $\alpha \in \mathbb{R}$ and $K = K(\alpha) > 0$ so that the solution $T_D(t)\phi$ of

$$D(x_t) = 0, \quad x_0 = \phi \in X_D$$

satisfies

$$\|T_D(t)\phi\| \leq Ke^{\alpha t}\|\phi\|, \; t \in \mathbb{R}^+, \; \phi \in X_D. \tag{8.7}$$

(c.f. Yosida, [1], p. 232.) Set

$$\alpha_D = \inf\{\alpha \in \mathbb{R}: \exists K = K(\alpha) \text{ satisfying (8.7)}\}.$$

Definition 8.8. We say that D is *stable* provided $\alpha_D < 0$.

Example 8.9. Let $A_1, A_2, \ldots, A_m$ be $d \times d$ matrices, $0 \le \tau_k \le r$ with τ_j/τ_k rational. Suppose $D(\phi)$ is given by

$$D(\phi) = \phi(0) - \sum_{k=1}^{m} A_k \phi(-\tau_k).$$

If every root λ of the equation

$$\det\left[I - \sum_{k=1}^{m} A_k \lambda^{-\tau_k}\right] = 0 \tag{8.8}$$

satisfies $|\lambda| < 1$, then D is stable.

Proof: Without loss of generality we can assume the τ_k are integers, say $\tau_k = k$ for $k = 1,2,\ldots,m$. Our hypothesis ensures the matrix $I - \sum_{k=1}^{m} A_k$ (the case for $\lambda = 1$) is non-singular.

Now consider the functional difference equation $D(y_t) = 0$, $y_0 = \phi \in X_D$. Then $y(t) = \sum_{k=1}^{m} A_k y(t-k)$. We make the following transformation. Let $z = (z^{(1)}, z^{(2)}, \ldots, z^{(m)})$, where $z^{(k+1)}(t) = y(t-k)$ for $k = 0,1,\ldots,m-1$. Then the difference equation can be written as

$$z(t) = Az(t-1),\ z_0 = \psi \in C([-1,0]; \mathbb{R}^{dm})$$

where

$$A = \begin{bmatrix} A_1 & A_2 & & A_{m-1} & A_m \\ I & 0 & & 0 & 0 \\ 0 & I & & 0 & 0 \\ & & & & \\ 0 & 0 & & I & 0 \end{bmatrix}, \quad \|\psi\| \le M\|\phi\|$$

for some $M > 0$.

The eigenvalues of A have modulii less than 1, so $|A| < 1$. For $t \in \mathbb{R}^+$ let $j = [t]$, the greatest integer in t. If $\theta \in [-1,0]$, then $z(t+\theta) = A^j z(t+\theta-j) = A^j \psi(t+\theta-j)$. Thus

$$\|z_t\| \leq |A|^j\|\psi\| \leq e^{-\alpha t}\|\psi\|,$$

where $\alpha = -\ell n|A| > 0$. In particular

$$\|y_t(\phi)\| = \|z_t^{(1)}\| \leq Me^{-\alpha t}\|\phi\|,$$

so D is stable. □

Remark 8.10. The significance of relation (8.8) is that for any linear functional difference operator D, the spectrum of the infinitesimal generator A of the semigroup generated by solutions of $D(x_t) = D(\phi)$, $x_0 = \phi$, consists of those complex numbers λ for which $\det \Delta(\lambda) = 0$ where

$$(8.9) \qquad \Delta(\lambda) = I - \int_{-r}^{0} [d\mu(\theta)]\lambda e^{\lambda\theta}.$$

Lemma 8.11. There exists functions $\phi_j \in X$, $1 \leq j \leq d$, so that the matrix function $\Phi = (\phi_1,\phi_2,\ldots,\phi_d)$ satisfies $D(\Phi) = I$, the identity matrix. Thus $D\colon X \to \mathbb{R}^d$ is onto.

Proof: The function $\det A$ is a continuous map of the $d \times d$ matrices $A \in \mathbb{R}^{d^2}$ into $\mathbb{R}$. As $\det I = 1$, there exists $\varepsilon > 0$ so that $\det A \neq 0$ whenever $|A-I| \leq \varepsilon$. For $s \in [0,r]$ define $\psi^{(s)} \in C([-r,0];\mathbb{R})$ by

$$\psi^{(s)}(\theta) = \begin{cases} 0, & -r \leq \theta \leq -s \\ 1 + \frac{\theta}{s}, & -s \leq \theta \leq 0. \end{cases}$$

Then

$$D(\psi^{(s)}I) = I - \int_{-s}^{0} [d\mu(\theta)](1 + \tfrac{\theta}{s}).$$

As D is atomic at zero, there exists $s > 0$ so that $|D(\psi^{(s)}I) - I| \leq \varepsilon$. Thus $D(\psi^{(s)}I)$ is nonsingular and the columns of $D(\psi^{(s)}I)$ form a basis in $\mathbb{R}^d$. A change of basis completes the proof of the lemma. □

Lemma 8.12. Suppose D is stable. There exist constants $\alpha > 0$, $K > 0$ and $B > 0$ such that for each $h \in C(\mathbb{R}^+; \mathbb{R}^d)$, the motion $x_t(\psi,h)$ of

$$(8.10) \qquad D(x_t) = h(t), \quad x_0 = \psi \in X$$

satisfies

$$(8.11) \qquad \|x_t(\psi,h)\| \leq Ke^{-\alpha t}\|\psi\| + B \sup_{0\leq s\leq t} |h(s)|.$$

Proof: We first make a transformation which assures we can take $h(0) = 0$. Indeed by Lemma 8.11 there exists a continuous matrix valued function Φ on $[-r,0]$ which satisfies $D(\Phi) = I$. Define

$$y(t) = \begin{cases} \Phi(t)h(0), & -r \leq t \leq 0 \\ \Phi(0)h(t), & t \geq 0. \end{cases}$$

Then $y: [-r,\infty) \to \mathbb{R}^d$ is continuous. If x is a solution of Equation (8.10), then we have

$$D(x_t - y_t) = D(x_t) - D(y_t) = h(t) - D[\Phi(0)h_t]$$

for $t \in \mathbb{R}^+$. Set for $t \in \mathbb{R}^+$

$$z(t) = x(t) - y(t),$$

$$h^*(t) = h(t) - D[\Phi(0)h_t].$$

Then $h^*(0) = h(0) - D[\Phi(\cdot)h(0)] = h(0) - D(\Phi)h(0) = 0$. So if x is a solution of Equation (8.10), $z = x - y$ is a solution of

$$D(z_t) = h^*(t), \quad h^*(0) = 0.$$

Note that $h^* \in C(\mathbb{R}^+; \mathbb{R}^d)$. We estimate $h^*(t)$.

$$|h^*(t)| \le |h(t)| + |D[\Phi(0)h_t]|$$

$$\le |h(t)| + |\Phi(0)h(t)| + |g[\Phi(0)h_t]|$$

$$\le |h(t)| + |\Phi(0)||h(t)| + |\Phi(0)|G \sup_{-r\le\theta\le 0} |h_t(\theta)|$$

for some $G > 0$ in view of the linearity of g. (Observe that h is not defined on $\mathbb{R}^-\setminus\{0\}$. Thus we can find some constant $M > 0$ so that

$$|h^*(t)| \le M \sup_{0\le s\le t} |h(s)|.$$

Now set $\phi = x_0 - y_0 = \psi - y_0 \in X_D$. According to Corollary 8.7 the (unique) motion through ϕ of $D(z_t) = h^*(t)$ can be written

$$z_t = z_t(\phi,h^*) = T_D(t)\phi + T_D(t)\Lambda_0 h^*(t).$$

As we can write

$$\|x_t\| \le \|x_t - y_t\| + \|y_t\| = \|z_t\| + \|y_t\|,$$

there exists $\alpha > 0$ and $K > 0$ so that

$$\|x_t\| \le Ke^{-\alpha t}\|\phi\| + Ke^{-\alpha t} \sup_{0\le u\le t} |h^*(u)| + \|\Phi(0)h_t\|$$

$$\le Ke^{-\alpha t}\|\psi\| + K|\Phi(0)h(0)|$$

$$+ (KM + |\Phi(0)|) \sup_{0\le s\le t} |h(t)|.$$

Choose $B = K|\Phi(0)| + KM + |\Phi(0)|$ so that the relation (8.11) is satisfied. □

Lemma 8.13. Suppose D is stable. If the positive orbit $\gamma^+(\phi)$ of the solution of Equation (8.3) through $\phi \in X$ is bounded, there exist constants $\alpha > 0$ and $M > 0$ so that

(8.12) $$\|x_{t+\tau} - x_t\| \le Me^{-\alpha t}\|x_\tau - \phi\| + M\tau,$$

for every $t,\tau \in \mathbb{R}^+$.

Proof: As f is bounded on bounded subsets of X, there exists a constant $L > 0$ so that $|f(x_s)| \le L$ for all $s \in \mathbb{R}^+$. Rewrite Equation (8.3) as

$$D(x_{t+\tau} - x_t) = \int_t^{t+\tau} f(x_s)ds,$$

for all $t,\tau \in \mathbb{R}^+$. Set $z_t = x_{t+\tau} - x_t$, and $h(t) = \int_t^{t+\tau} f(x_s)ds$. Now apply Lemma 8.12 to obtain the relation (8.12). □

We now present a stability result analogous to Theorem 4.5.

Theorem 8.14. Suppose D is stable. If either the positive orbit $\gamma^+(\phi)$ of the solution of Equation (8.3) through $\phi \in X$ is bounded or $D(x_t(\phi))$ is bounded on $\mathbb{R}^+$, then $\gamma^+(\phi)$ is precompact.

Proof: Write $x_t = x_t(\phi)$. If $D(x_t)$ is bounded on $\mathbb{R}^+$, then so is $\int_0^t f(x_s)ds$, say by $B > 0$. There exist constants $\alpha > 0$, $K > 0$ so that for every $t \in \mathbb{R}^+$

$$\|x_t\| \le Ke^{-\alpha t}\|\phi\| + \int_0^t Ke^{-\alpha(t-s)}|f(x_s)|ds$$

according to Proposition 8.6. Thus

$$\|x_t\| \le K\|\phi\| + KB, \quad t \in \mathbb{R}^+.$$

Consequently $\gamma^+(\phi)$ is bounded. Now apply Lemma 8.13 to obtain for some $M > 0$

$$\|x_{t+\tau} - x_t\| \le M\|x_\tau - \phi\| + M\tau$$

for $\phi \in X$ and every $t,\tau \in \mathbb{R}^+$. Let $\varepsilon > 0$. Choose $\delta \in (0,\varepsilon/2M)$ so that $\|x_\tau - \phi\| < \varepsilon/2M$ whenever $0 \le \tau \le \delta$. Therefore the bounded positive orbit $\gamma^+(\phi)$ is precompact, as $\{x_t\}$ is an equicontinuous family. □

We have not used the full force of Lemma 8.13 in establishing the last theorem. We are able, though, to show that some smoothing takes place in the limit sets of a bounded positive trajectory. In particular, we obtain the following theorem, the proof of which is left as an exercise.

<u>Theorem 8.15.</u> Let D be stable. If for some $\phi \in X$, $\gamma^+(\phi)$ is bounded or $D(x_t(\phi))$ is bounded on $\mathbb{R}^+$, then $L^+(\phi)$ consists of equi-Lipschitz functions; that is, there is a constant $k = k(\phi)$ so that $|\psi(\theta_1) - \psi(\theta_2)| \le k|\theta_1 - \theta_2|$ for all $\psi \in L(\phi)$, $\theta_1,\theta_2 \in [-r,0]$.

We establish a fundamental result on the structure of the semigroup T which generates the solution of the initial value problem (8.3) according to the formula (8.4).

<u>Theorem 8.16</u>. Suppose for each $t \in \mathbb{R}^+$ that $T(t)$ takes bounded subsets of X into bounded subsets of X. If D is stable, there is a continuous linear map $H\colon X \to X_D$ and a compact operator $W(t)\colon X \to X$ so that

$$T(t) = T_D(t)H + W(t), \quad t \in \mathbb{R}^+. \tag{8.13}$$

Consequently $T(t)$ is the sum of a contraction operator and a compact operator for sufficiently large t.

<u>Proof</u>: Let Φ be the matrix function established in the last lemma. Define $H\colon X \to X_D$ by

$$H\phi = \phi - \Phi D(\phi).$$

In view of the linearity of H and the stability of D there exist constants $\alpha > 0$, $k > 0$, so that

$$\|T_D(t)\phi\| \le Ke^{-\alpha t}\|\phi\|, \quad t \in \mathbb{R}^+, \ \phi \in X.$$

Choose $q > 0$, so that $Ke^{-\alpha q} < 1$. Thus $T_D(t)H$ is a contraction on X for $t \ge q$.

Now write for $\phi \in X$,

$$T(t)\phi = T_D(t)H\phi + W(t)\phi$$

where

$$W(t)\phi = T_D(t)\Phi D(\phi) + \int_0^t T_D(t-s)\Lambda_0 f(T(s)\phi)ds.$$

The range of the map $T_D(t)\Phi D(\cdot)$ is finite dimensional (lies in $\mathbb{R}^d$), hence this map is compact. Finally, for any bounded set $U \subset X$, $f(T(s)U)$ is a bounded subset of $\mathbb{R}^d$, uniformly in $s \in [0,t]$ (see proof of Theorem 4.1). Therefore the set $\int_0^t T_D(t-s)\Lambda_0 f(T(s)U)ds$ is a bounded equicontinuous family of functions in X, hence precompact by the Ascoli theorem. Thus the mapping $W(t)$ is compact. □

It follows by a fixed point theorem of Krasnoselskii (see Appendix A), that under the hypotheses of Theorem 8.14, the operators $T(t)$ have the fixed point property for t sufficiently large. This may be used to obtain the existence of periodic solutions and critical points of Equation (8.3). The reader is referred to the paper of Hale and Lopes [1], and to the references contained therein for results on periodic solutions.

Invariant sets for solutions of a NFDE are defined in a similar manner to that for an RFDE.

Definition 8.17. A subset $\Gamma \subset X$ is called *weakly invariant relative to solutions of Equation* (8.3) if for each $\psi \in \Gamma$, there exists a function $y: \mathbb{R} \to \mathbb{R}^d$ so that

(i) $y_t \in \Gamma$ for all $t \in \mathbb{R}$

(ii) $y_0 = \psi$

(iii) $D(y_t)$ is continuously differentiable on $\mathbb{R}$

(iv) $\frac{d}{dt} D(y_t) = f(y_t)$ for all $t \in \mathbb{R}$.

We now obtain the usual characterization of positive limit sets. The proof is omitted as it is similar to the case for RFDE's (Theorem 4.8). The conclusion of Theorem 8.14 is also included.

Theorem 8.18. Suppose D is stable. If for some $\phi \in X$, $\gamma^+(\phi)$ is bounded, then $L^+(\phi)$ is nonempty, compact, connected, and weakly invariant with respect to solutions of Equation (8.3). Moreover, $L^+(\phi)$ consists of equi-Lipschitz functions.

As in the case of RFDE's the Invariance Principle (Theorem 4.11) still holds for autonomous NFDE's when the functional difference operator D is stable. Thus we establish the counterpart to Theorem 4.12 as well. The proofs are obvious. Recall the definition of a Lyapunov function V on $G \subset X$ from Section 4.

Theorem 8.19. Suppose D is stable and V is a Lyapunov function for Equation (8.3) on $G \subset X$. If $x(\cdot;\phi)$ is a solution of (8.3) for which $\gamma^+(\phi)$ is a bounded subset of G, then $d(x_t(\phi),M) \to 0$ as $t \to \infty$, where M is the largest subset of $E = \{\phi \in \overline{G}: \dot{V}(\phi) = 0\}$ which is weakly invariant relative to solutions of Equation (8.3).

The next two corollaries are useful for applications. The set M is defined in Theorem 8.19.

Corollary 8.20. Suppose D is stable and V is a Lyapunov function for Equation (8.3) on $G_\beta = \{\phi \in X: V(\phi) \leq \beta\}$ for some $\beta > 0$. If there exists a number $k = k(\beta)$ so that $|D(\phi)| \leq k$ whenever $\phi \in G_\beta$, then any solution $x(\cdot;\phi)$ of Equation (8.3) with $\phi \in G_\beta$ has the property that $x_t(\phi)$ approaches M as $t \to \infty$.

Corollary 8.21. Assume the same hypotheses as in Corollary 8.20. If in addition

$$\dot{V}(\phi) \leq -w(|D(\phi)|) \leq 0$$

for some continuous function $w: \mathbb{R}^+ \to \mathbb{R}$, then every solution $x(\cdot;\phi)$ of (8.3) with $\phi \in G_\beta$ has the property that $x_t(\phi)$ approaches $\{\phi \in \overline{G}_\beta: w(|D(\phi)|) = 0\}$ as $t \to \infty$. In particular, if $w(s) > 0$, for every $s > 0$, then $x \equiv 0$ is an asymptotically stable solution of Equation (8.3) with G_β contained within its region of attraction.

Proof: As $M \subset \{\phi \in \overline{G}_\beta: w(|D(\phi)|) = 0\}$, the first conclusion of the corollary is clear. If $w(s) > 0$ when $s > 0$, then $M \subset \{\phi \in \overline{G}_\beta: D(\phi) = 0\}$. Now suppose $x(\cdot;\phi)$ is a solution of Equation (8.3) with $x_t = x_t(\phi) \in M$ for all $t \in \mathbb{R}$. We claim $x_t = 0$ for all $t \in \mathbb{R}$. First note that x_t is bounded, say by $L > 0$, uniformly in $t \in \mathbb{R}$. Suppose x $\|x_\tau\| \neq 0$ for some $\tau \in \mathbb{R}$. We can choose $t_0 \in \mathbb{R}$ so that $Ke^{-\alpha(\tau - t_0)}L < \|x_\tau\|/2$, where $\alpha > 0$, $K > 0$ are constants arising from the stability of D. As $D(x_t) = 0$, $x_0 = \phi \in X_D$, we use the stability of D to conclude

$$\|x_\tau\| = \|T_D(\tau)\phi\| = \|T_D(\tau-t_0)T_D(t_0)\phi\|$$
$$\le Ke^{-\alpha(\tau-t_0)}\|x_{t_0}(\phi)\| \le Ke^{-\alpha(\tau-t_0)}L < \|x_\tau\|/2,$$

a contradiction. Thus the corollary is established. □

Example 8.22. Consider the scalar equation

$$\dot{x}(t) - c\dot{x}(t-r) + ax(t) = 0,$$

where $a > 0$, $|c| < 1$. The operator $D(\phi) = \phi(0) - c\phi(-r)$ is stable. Define

$$V(\phi) = |D(\phi)|^2 + ac^2\int_{-r}^{0}\phi^2(\theta)d\theta.$$

Then

$$\dot{V}(\phi) = -a[|D(\phi)|^2 + (1-c)^2\phi^2(0)] \le -a|D(\phi)|^2.$$

In view of Corollary 8.21, the solution $x(t) \equiv 0$ is asymptotically stable.

9. A Flip-Flop Circuit Characterized by a NFDE - The Stability of Solutions

We consider below a nonlinear distributed network studied by Slemrod [3]. Such networks are used to model flip-flop circuits in digital computers. It is important in applications to establish criteria for the nonexistence of oscillations in these circuits. To this end the network is first characterized by a NFDE. Then the equilibrium solution is shown to be asymptotically stable.

The network under consideration is shown in Figure 9.1. The section between 0 and 1 is a lossless transmission line with specific inductance L_s and capacitance C_s. The parameter ξ is the position measure along the transmission line. The voltage v across this line and the current

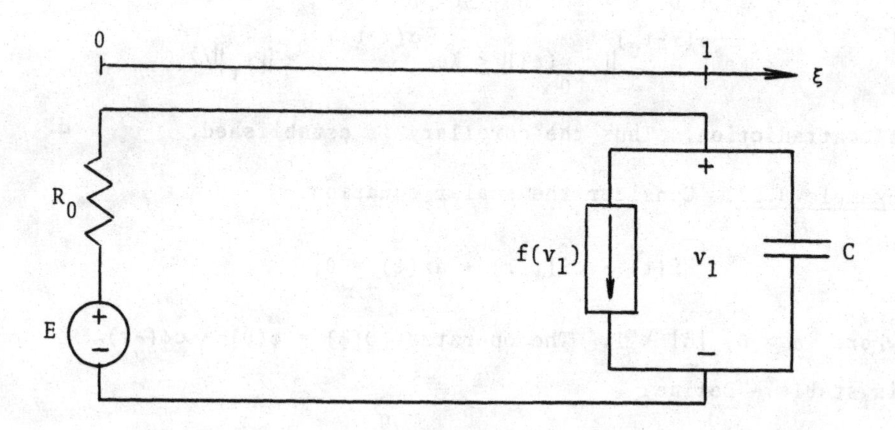

Figure 9.1

flowing through it are functions of ξ and t (time) and satisfy the (wave) equations

$$L_s \frac{\partial i}{\partial t} = -\frac{\partial v}{\partial \xi}$$

(9.1) $$0 < \xi < 1,\ t > 0.$$

$$-C_s \frac{\partial v}{\partial t} = \frac{\partial i}{\partial \xi}$$

The circuits at the ends of the line give rise to the boundary conditions

$$0 = E - v_0 - i_0 R_0,$$

(9.2)

$$-C\frac{dv_1}{dt} = -i_1 + f(v_1),\ t > 0,$$

where we employ the notation

$$v_0(t) = v(0,t),\ v_1(t) = v(1,t),$$
$$i_0(t) = i(0,t),\ i_1(t) = i(1,t).$$

The nonlinear function f which gives the current through the box has a graph of the kind displayed in Figure 9.2. It represents the general characteristic of a tunnel (Esaki)

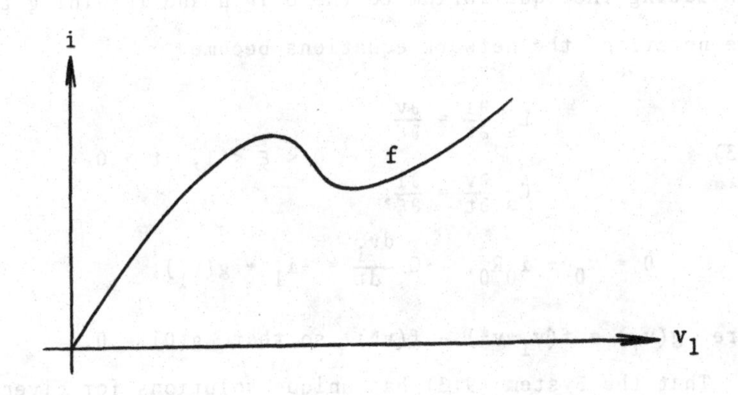

Figure 9.2

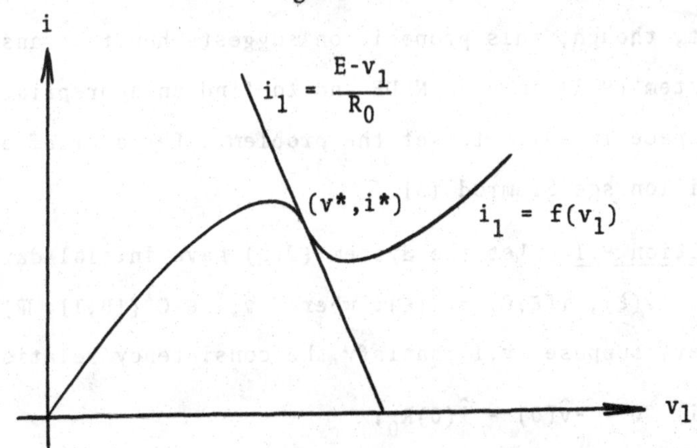

Figure 9.3

diode. We assume f is continuously differentiable and (globally) Lipschitz. The equilibrium states of the System (9.1), (9.2) are solutions (v_1,i_1) of

$$0 = E - v_1 - i_1R_0,$$
$$0 = -i_1 + f(v_1).$$

These obtain when the derivatives in the System (9.1) and (9.2) vanish. We only consider the case of a unique equilibrium point (v^*,i^*) as illustrated in Figure 9.3. By translating the equilibrium to the origin and retaining the same notation, the network equations become

$$(9.3) \qquad \begin{aligned} L_s \frac{\partial i}{\partial t} &= \frac{\partial v}{\partial \xi}, \\ -C_s \frac{\partial v}{\partial t} &= \frac{\partial i}{\partial \xi}, \end{aligned} \qquad 0 < \xi < 1, \quad t > 0$$

$$0 = v_0 + i_0 R_0, \quad -C \frac{dv_1}{dt} = -i_1 + g(v_1),$$

where $g(v_1) = f(v_1 + v^*) - f(v^*)$ so that $g(0) = 0$.

That the System (9.3) has unique solutions for given initial data is a result of the next proposition. More important, though, this proposition suggests how to transform the System (9.3) into an NFDE and to find an appropriate phase space in which to set the problem. For a proof of the proposition see Slemrod [3].

Proposition 9.1. Let the System (9.3) have initial data $v(\xi,0) = \hat{v}(\xi)$, $i(\xi,0) = \hat{i}(\xi)$; where $\hat{v},\hat{i} \in C^1([0,1]; \mathbb{R})$. Moreover, suppose v,i satisfy the consistency relations

(i) $0 = -\hat{v}(0) - \hat{i}(0)R_0$,

(ii) $0 = L_s \dfrac{d\hat{i}(0)}{d\xi} + R_0 C_s \dfrac{d\hat{v}(0)}{d\xi}$,

(iii) $\dfrac{C}{C_s} \dfrac{d\hat{i}(1)}{d\xi} = -\hat{i}(1) + f(\hat{v}(1))$.

Then there exists a unique solution to the System (9.3), $v, i \in C^1([0,1] \times \mathbb{R}^+; \mathbb{R})$ which satisfies the initial data. Furthermore the solution admits the representation

$$(9.4) \qquad \begin{aligned} v(\xi,t) &= \frac{1}{2}[\phi(\xi-\sigma t) + \psi(\xi+\sigma t)], \\ i(\xi,t) &= \frac{1}{2z}[\phi(\xi-\sigma t) - \psi(\xi+\sigma t)], \end{aligned}$$

where $\sigma = (L_s C_s)^{-\frac{1}{2}}$, $z = (L_s/C_s)^{\frac{1}{2}}$, and $\phi \in C^1(-\infty,1]; \mathbb{R})$,

$\psi \in C^1(\mathbb{R}^+; \mathbb{R})$.

This representation suggests a way to reduce the System (9.3) to a NFDE. Indeed, introduce Equations (9.4) into the System (9.3). The wave equations are automatically satisfied and the boundary conditions become

$$v_1(t) + zi_1(t) = K\psi_1(t-r),$$
$$v_1(t) - zi_1(t) = \psi_1(t),$$
$$C\frac{dv_1}{dt} + g(v_1) = i_1,$$

where $K = \frac{R_0 - z}{R_0 + z}$, $r = \frac{2}{\sigma}$, $\psi_1(t) = \psi(1+\sigma t)$ and $\phi_1(t) = \phi(1-\sigma t)$. If we eliminate i_1 and ψ_1 from the above we obtain the (autonomous) NFDE

$$(9.5)\qquad \begin{aligned} &C\frac{d}{dt}[v_1(t) + Kv_1(t-r)] \\ &= -\frac{1}{z}v_1(t) + \frac{K}{z}v_1(t-r) - g(v_1(t)) - Kg(v_1(t-r)). \end{aligned}$$

We now derive the initial values for the NFDE (9.5) from the initial data for the original System (9.3).

<u>Proposition 9.2</u>. The initial data $\hat{v},\hat{i} \in C^1([0,1]; \mathbb{R})$ for the System (9.3) uniquely determines the initial value for the NFDE (9.5) in the phase space $X = C([-r,0]; \mathbb{R})$. Indeed, the initial values are also continuously differentiable on $[-r,0]$.

<u>Proof</u>: The Equations (9.4) yield

$$\phi(\xi) = \hat{v}(\xi) + z\hat{i}(\xi)$$
$$\psi(\xi) = \hat{v}(\xi) - z\hat{i}(\xi).$$

Thus $\phi,\psi \in C^1([0,1]; \mathbb{R})$. ϕ_1 is determined on $[0,r/2]$, and ψ_1 is determined on $[-r/2,0]$. From the boundary conditions (9.2) we have

$$\begin{aligned}(9.6)\qquad & C[\dot{\phi}_1(\theta) - \dot{\psi}_1(\theta)] \\ & = -\frac{r}{2z}[\phi_1(\theta) - \psi_1(\theta)] + f([\phi_1(\theta) + \psi_1(\theta)]/2).\end{aligned}$$

Since ψ_1 is known on $[-r/2,0]$, then Equation (9.6) is an ordinary differential equation for ϕ_1 with initial value specified by $\phi_1(0) = \phi(1) = \hat{v}(1) + z\hat{i}(1)$. As f is Lipschitz, we may solve Equation (9.6) for ϕ_1 on $[-r/2,0]$. A similar argument yields ψ_1 on $[0,r/2]$. Now set

$$v_1 = \frac{1}{2}(\phi_1 + \psi_1).$$

Thus v_1 is determined by the initial data $\hat{v}$ and $\hat{i}$ on $[-r/2,r/2]$. As the NFDE (9.5) is autonomous we can make a time translation so that the initial values of the NFDE (9.5) are given on $[-r,0]$. □

We now turn to an analysis of the stability of the zero solution of the NFDE (9.5). Define the difference operator

$$D(\phi) = \phi(0) + K\phi(-r). \tag{9.7}$$

As $|K| < 1$, D is a stable operator. g is Lipschitz, so we see that the initial value problem for the NFDE (9.5) with initial values in $X = C([-r,0];\mathbb{R})$ admits unique solutions on $[-r,\infty)$. Our main result is

Proposition 9.3. If g satisfies the sector criteria

$$\sup_{x\in\mathbb{R}}\left[\frac{g(x)}{x}\right] \le \frac{1-|K|}{1+|K|}\cdot\frac{1}{z} + \inf_{x\in\mathbb{R}}\left[\frac{g(x)}{x}\right]$$

$$\inf_{x\in\mathbb{R}}\left[\frac{g(x)}{x}\right] \ge -\frac{1-|K|}{1+|K|}\cdot\frac{1}{z},$$

then the equilibrium solution $v_1 \equiv 0$ of the NFDE (9.5) is asymptotically stable.

<u>Proof</u>: Define $V: X \to \mathbb{R}$ by

$$V(\phi) = \frac{1}{2}[D(\phi)]^2 + \alpha\int_{-r}^{0} \phi^2(\theta)d\theta \tag{9.8}$$

where $D(\phi)$ is given by Equation (9.7) and $\alpha \in \mathbb{R}^+$ is to be determined later. Our objective is to find a positive number A so that $\dot{V}(\phi) \leq -A[D(\phi)]^2$; hence we can apply Corollary 8.21 to obtain that $v_1 \equiv 0$ is an asymptotically stable solution of the NFDE (9.5).

For convenience write $x(t) = v_1(t)$ and set

$$h(t) = \frac{g(x(t))}{x(t)},$$

$$a(t) = \frac{1}{Cz} + \frac{h(t)}{C}, \tag{9.9}$$

$$b(t) = -\frac{K}{Cz} + \frac{K}{C}h(t-r).$$

Then the NFDE (9.5) can be written (in nonautonomous form)

$$\frac{d}{dt}D(x_t) + a(t)x(t) + b(t)x(t-r) = 0. \tag{9.10}$$

A computation yields (letting a denote $a(t)$, b denote $b(t)$)

$$\dot{V}(x_t) = -(a-\alpha)x_t^2(0) - (b+aK)x_t(0)x_t(-r) - (bK+\alpha)x_t^2(-r). \tag{9.11}$$

If we can find a constant $A > 0$ and functions $B(t)$, $G(t)$, and $J(t)$ with $B(t) \geq 0$ and $B(t)J(t) \geq G^2(t)$ so that $\dot{V}(x_t)$ can also be written

$$\dot{V}(x_t) = -A[D(x_t)]^2 - Bx_t^2(0) - 2Gx_t(0)x_t(-r) - Jx_t^2(-r), \tag{9.12}$$

then

$$\dot{V}(x_t) \leq -A[D(x_t)]^2.$$

(Note that we have suppressed the dependence of B, G, and J

upon the variable t.) Identification of Equations (9.11) and (9.12) yields

$$a - \alpha = A + B,$$
$$KA + G = (b + ak)/2,$$
$$bK + \alpha = J + AK^2.$$

Now observe since g is Lipschitz and $g(0) = 0$, then $g(x)/x$ is bounded so that $\inf_{t\geq 0} h(t)$ and $\sup_{t\geq 0} h(t)$ are finite. Use the notation

$$(p)_m \overset{\text{def}}{=} \inf_{t\geq 0} p(t), \quad (p)_M \overset{\text{def}}{=} \sup_{t\geq 0} p(t).$$

A sufficient condition for $B \geq 0$ is that

$$(a)_m \geq \alpha + A. \tag{9.13}$$

A sufficient condition for $BJ \geq G^2$ is

$$[K(b)_m + \alpha - AK^2][(a)_m - \alpha - A] \geq \frac{K^2}{4}[-2A + \frac{b(t)}{K} + a(t)]^2, \quad t \in \mathbb{R}^+. \tag{9.14}$$

The left side of the Inequality (9.14) is independent of t. So we get the best estimate if α is chosen in order to maximize the left side of the Inequality (9.14). Set

$$\alpha = \frac{(a)_m - A - K(b)_m + AK^2}{2} \tag{9.15}$$

For this choice of α, $BJ \geq G^2$ provided

$$[(a)_m - A(1+K^2) + K(b)_m]^2 \geq K^2[-2A + \frac{b(t)}{K} + a(t)]^2. \tag{9.16}$$

As $\frac{b(t)}{K} + a(t) = \frac{h(t-r)}{C} + \frac{h(t)}{C}$, then if we require

$$\frac{(h)_M}{C} < A, \tag{9.17}$$

a sufficient condition for the Inequality (9.16) is

$$(9.18)\quad (a)_m - A(1+K^2) + K(b)_m \geq |K|[2A - \frac{h(t-r)}{C} - \frac{h(t)}{C}].$$

Since

$$(9.19)\quad \begin{aligned} (a)_m &= -\frac{1}{Cz} + \frac{(h)_m}{C}, \\ K(b)_m &= -\frac{K^2}{Cz} + \frac{K^2(h)_m}{C}, \end{aligned}$$

then the Inequality (9.18) will be true provided

$$(9.20)\quad \frac{1}{Cz} + \frac{(h)_m}{C} - A(1+K^2) - \frac{K^2}{Cz} + \frac{K^2(h)_m}{C} \geq 2|K|[A - \frac{(h)_m}{C}].$$

Thus if in addition to the Inequality (9.17) we require

$$(9.21)\quad A \leq \frac{1 - K^2}{Cz(1 + 2|K| + K^2)} + \frac{(h)_m}{C}$$

then the Inequality (9.16) is satisfied so $BJ \geq G^2$.

Since we require $\alpha \leq 0$, we obtain from 9.15 that A must satisfy

$$(9.22)\quad 0 < A \leq \frac{(a)_m - K(b)_m}{1 - K^2}.$$

Substitute α from Inequality (9.15) into Inequality (9.13); a sufficient condition for $B \geq 0$ is

$$(9.23)\quad \frac{(a)_m + K(b)_m}{1 + K^2} \geq A > 0.$$

If we replace $(a)_m$ and $(b)_m$ in Inequalities (9.22) and (9.23) by their equivalents in Inequality (9.19), then the following is a sufficient condition for $\alpha \geq 0$, $B(t) \geq 0$, and $B(t)J(t) \geq G^2(t)$:

$$\frac{(h)_m}{C} < A < \min\left\{\frac{1}{Cz}\left(\frac{1+K^2}{1-K^2}\right) + \frac{(h)_m}{C},\right.$$

(9.24)

$$\left.\frac{1}{Cz}\left(\frac{1-K^2}{1+K^2}\right) + \frac{(h)_m}{C},\ \frac{1}{Cz}\,\frac{1-K^2}{(1+2|K|+K^2)} + \frac{(h)_m}{C}\right\}.$$

Since $|K| < 1$, the third inequality in the brackets of the Inequality (9.24) is the minimum, so the Inequality (9.24) is equivalent to

$$\frac{(h)_m}{C} < A < \left(\frac{1-|K|}{1+|K|}\right)\frac{1}{Cz} + \frac{(h)_m}{C}. \tag{9.25}$$

As we also require $A > 0$, it is necessary to have

$$(h)_m \geq -\left(\frac{1-|K|}{1+|K|}\right)\frac{1}{z}. \tag{9.26}$$

In conclusion we see that if the sector criteria of the hypotheses are satisfied, there exists a constant $\alpha > 0$ so that V is a Lyapunov function on X, and there exists a constant $A > 0$ so that $\dot{V}(\phi) \leq -A[D(\phi)]^2$ in X. Now set $G_\beta = \{\phi \in X: V(\phi) < \beta\}$, $\beta > 0$. Then $\phi \in G_\beta$ implies $|D(\phi)| < \sqrt{2\beta}$. Consequently we have from Corollary 8.21 that $v_1 = 0$ is an asymptotically stable solution of the NFDE (9.5). □

10. Exercises

10.1. Show that the infinitesimal generator of the semigroup T established in Theorem 3.1 is indeed the operator A defined therein.

10.2. Let A and T be the operators defined in Theorem 3.1. For each $t \geq r$ show that $AT(t)$ is $\omega e^{\omega t}$-Lipschitz on X.

10.3. ($T(t)$ is not one-to-one) Show for the FDE $\dot{x}(t) = \|x_t\|$, where $X = C([-1,0];\mathbb{R})$, that there are $\phi,\psi \in X$, $\phi \neq \psi$ so that $T(t)\phi = T(t)\psi$ for all $t \geq \frac{1}{2}$.

10.4. (Sufficient conditions for $T(t)$ to be one-to-one) Let $X = C([-r,0];\mathbb{R}^d)$ be the space of initial values for the RFDE $\dot{x}(t) = \int_{-r}^{0} [d\eta(\theta)]x(t+\theta)$, where η is a matrix function of bounded variation on $[-r,0]$. Furthermore, suppose there exists a continuous, non-negative scalar function β on $\mathbb{R}^+$, $\beta(0) = 0$ so that $|\int_{-r}^{-s} [d\eta(\theta)]\phi(\theta) - [\eta(-r)-\eta(-r^+)]\phi(-r)| \leq \beta(s)\|\phi\|$ for $s \in [0,r]$. (This is to say that the continuous linear functional on X, $L(\phi) = \int_{-r}^{0} [d\eta(\theta)]\phi(\theta)$ is atomic at $-r$.) Prove the semigroup T generated by the FDE is one-to-one on X.

10.5. (Negative trajectories do not exist) Show the RFDE $\dot{x}(t) = 2\int_{t-1}^{t} x(s)ds$ with $x_0 = \phi \in X = C([-1,0];\mathbb{R})$ given by $\phi(\theta) = \theta + 1$ does not admit a solution $x(\cdot;\phi)$ on any interval of the form $(-1-\varepsilon,0]$, $\varepsilon > 0$.

10.6. (A different topology on X) Consider the RFDE $\dot{x}(t) = x(t-1)$ with initial function space $X = \{\phi \in C^1([-1,0];\mathbb{R})\colon \dot{\phi}(0) = \phi(-1)\}$ whose norm is $\|\psi\| = \sup_{-1\leq\theta\leq 0} |\psi(\theta)|$. Given an example to show that solutions $x(\cdot;\psi)$ are not continuous with respect to $\psi \in X$. (Hint: Show that the mapping $\psi \to x(-2;\psi)$ is not continuous -- note that $x(-2;\psi)$ is well defined.) How can the norm in X be modified so that $x(t;\psi)$ does depend continuously on $\psi \in X$?

10.7. Prove that the union of a collection of sets, each of which is weakly invariant relative to solutions of Equation (2.5) is weakly invariant relative to solutions of (2.5). Prove that the closure of a set which is weakly invariant relative to solutions of Equation (2.5) is weakly invariant relative to solutions of Equation (2.5).

10.8. (Stability of RFDE with finite delay) Again let $X = C([-r,0];\mathbb{R})$, and consider the system of RFDE's

$$\dot{x}(t) = y(t),\ \dot{y}(t) = -x(t) + \int_{-r}^{0} F(\theta)[x(t+\theta)-x(t)]d\theta,$$

where $F\colon [-r,0] \to \mathbb{R}$ is continuously differentiable, $F(\theta) \geq 0$, $\dot{F}(\theta) \leq 0$ for all $\theta \in [-r,0]$. For $\phi,\psi \in X$ define $V(\phi,\psi) = \frac{1}{2}\phi^2(0) + \frac{1}{2}\psi^2(0) + \int_{-r}^{0} F(\theta)[\phi(\theta)-\phi(0)]d\theta$.

(i) Show that V is a Lyapunov function.

(ii) If there exists $\theta_0 \in [-r,0]$ such that $\dot{F}(\theta_0) < 0$, show that the zero solution $x \equiv 0$, $y \equiv 0$ is asymptotically stable.

(iii) If $\dot{F} \equiv 0$ on $[-r,0]$ and $F(-r) > 0$, show that every solution of the system is bounded, and that for every $\phi,\psi \in X$, $L^+(\phi,\psi)$ consists of periodic solutions of the system of period r.

10.9. (Stability of RFDE with infinite delay) Suppose $a \in C^2(\mathbb{R}^+;\mathbb{R})$ with $a(t) > 0$, $\dot{a}(t) < 0$, $\ddot{a}(t) \geq 0$ for every $t \in \mathbb{R}^+$. Furthermore, assume $\lim_{t\to\infty} t^2\dot{a}(t) = 0$ and $\int_0^\infty t^2 a(t)dt < \infty$. Let $h \in C^1(\mathbb{R}^d;\mathbb{R}^d)$ with $H(x) \overset{\text{def}}{=} \int_{-\infty}^{x} h(s)ds \to \infty$ as $|x| \to \infty$. Consider the

RFDE $\dot{x}(t) = -\int_{-\infty}^{t} a(t-s)h(x(s))ds$, $x_0 = \phi \in X$, where X is the history space of Example 2.4.

(i) Show that Theorem 2.6 is applicable in order to obtain existence, uniqueness, and continuous dependence of solutions of the RFDE for $\phi \in B_\alpha(0) = \{\phi \in X: d(\phi,0) < \alpha\}$.

(ii) Show that $V: B_\alpha(0) \to \mathbb{R}$ is a Lyapunov function for the RFDE, where $V(\phi) = H(\phi(0)) - \frac{1}{2}\int_{-\infty}^{0} a(-\theta)\{\int_{\theta}^{0} h(\phi(s))ds\}^2 d\theta$.

(iii) Prove that if $\phi \in B_\alpha(0)$, then $\gamma^+(\phi)$ is precompact.

(iv) Prove that if $\phi \in B_\alpha(0)$, then $L^+(\phi)$ is the union of trajectories of $\ddot{y} + a(0)h(y) = 0$ which satisfy $\int_{-s}^{0} h(y(t+\theta))d\theta = 0$, $t \in \mathbb{R}$ if $\ddot{a}(s) > 0$.

(v) If h has only a finite number of zeros, show every solution of the RFDE with $\phi \in B_\alpha(0)$ approaches a zero of h; i.e., a critical point of the RFDE.

10.10. (Instability of RFDE) Let $\psi \in X = C([-r,0]; \mathbb{R}^d)$ be a critical point of Equation (2.5), where $\psi \in G \overset{\text{def}}{=} \overline{U} \cap N$, U open in X and N any neighborhood of ψ. Suppose V is a Lyapunov function for (2.5) on G and M is the largest subset of $E = \{\phi \in \overline{G}: \dot{V}(\phi) = 0\}$ which is weakly invariant relative to solutions of Equation (2.5). Assume

(i) $M \cap G = \emptyset$, or $M \cap G = \{\psi\}$,

(ii) There exists $b > 0$ so that $V(\phi) < b$ for $\phi \in G \setminus \{\psi\}$,

(iii) $V(\phi) = b$ for $\phi = \psi$, or $\phi \in G \cap N$.

Show the solution $x(t;\psi)$ of Equation (2.5) is unstable.

10.11. (Instability of RFDE) Let $X = C([-r,0]; \mathbb{R})$ be the space of initial functions for the RFDE $\dot{x}(t) = ax^3(t) + bx^3(t-r)$, where $a > 0$ and $|b| < a$. For $\phi \in X$ set $V(\phi) = -\frac{\phi^4(0)}{4a} + \frac{1}{2}\int_{-r}^{0} \phi^6(\theta)d\theta$. Show that

(i) V is a Lyapunov function on X.

(ii) For E and M defined in Exercise 10.10, show that $M = \{0\}$.

(iii) If $G = \{\phi \in X: V(\phi) < 0\}$, then G is positively invariant.

(iv) $\gamma^+(\phi)$ is unbounded for each $\phi \in G$.

(v) The solution of the RFDE is unstable. (Use Exercise 10.10.)

10.12. (Lemma 7.3) Suppose $L: X \to \mathbb{R}^d$ is continuous, linear where $X = C([-r,0]; \mathbb{R}^d)$, and $h \in C(\mathbb{R}^+; \mathbb{R}^d)$ is ω-periodic with $\omega \geq r > 0$. Then a solution x of $\dot{x}(t) = L(x_t) + f(t)$ is ω-periodic if and only if $x_\omega = x_0$.

10.13. (Periodic solutions) Suppose $L: X \to \mathbb{R}^d$ is continuous, linear where $X = C([-r,0]; \mathbb{R}^d)$ and $h \in C(\mathbb{R}^+; \mathbb{R}^d)$ is continuous and ω-periodic. Show that if the only ω-periodic solution of $\dot{x}(t) = L(x_t)$ is $x \equiv 0$, then there is a unique ω-periodic solution of $\dot{x}(t) = L(x_t) + h(t)$.

10.14. (Theorem 8.15) Suppose D is a stable operator for the NFDE $\frac{d}{dt} D(x_t) = f(x_t)$. If for some $\phi \in X = C([-r,0]; \mathbb{R}^d)$, we have $\gamma^+(\phi)$ is bounded or $D(x_t(\phi))$

is bounded on $\mathbb{R}^+$, prove that $L^+(\phi)$ consists of equi-Lipschitz functions.

10.15. (Stability of difference equations) Suppose D is a stable operator and let $h \in C(\mathbb{R}^+; \mathbb{R}^d)$ with $\lim_{t\to\infty} h(t) = 0$. Show every solution $x(\cdot;\phi)$ of $D(x_t) = h(t)$ approaches zero as $t \to \infty$, uniformly with respect to ϕ in closed bounded subsets of $X = C([-r,0]; \mathbb{R}^d)$.

10.16. (Asymptotic stability of NFDE) Prove that the zero solution of the NFDE $\dot{x}(t) + c\dot{x}(t-r) + ax(t) + bx^3(t-r) = 0$ where $a > 0$, $0 < |c| < 1$, and $b \in \mathbb{R}$ is asymptotically stable. (Use Exercise 10.15.)

10.17. (Stability of NFDE) For $X = C([-r,0]; \mathbb{R}^d)$ let $G_H = \{\phi \in X: \|\phi\| < H\}$ for some $H > 0$. Suppose for every $a,b \in \mathbb{R}$ with $0 < a \le b < H$ there is a Lyapunov function V for the NFDE (8.3) on G_H with the following property: there exists a constant $\delta > 0$ so that $a \le \|\phi\| \le b$ implies $V(\phi) \le -\delta|f(\phi)|$. Show that

(i) the zero solution of (8.3) is stable,

(ii) if there is some $\alpha > 0$, $0 < \alpha < H$ so that $\|\phi\| < \alpha$ implies $\liminf_{t\to\infty} x_t(\phi) = 0$, then the zero solution of Equation (8.3) is asymptotically stable.

11. Notes and Comments

Section 1. Example 1.1 is due to Cooke and Yorke [1]. Also see Yorke [3] for a summary of the problem as well as a good bibliography. Example 1.3 is treated by Coleman and Mizel [4]. Example 1.4 is from Slemrod [3]. Example 1.5 is due to Hale [3]; it is an infinite delay version of the model of Levin and Nohel [1].

Section 2. A basic source is Hale [8]. Stephan [1] offers a brief introduction to FDE's. The characterization of the space X of initial functions by conditions (2.1) and (2.2) is due to Kappel [1]. For the case of infinite delay this characterization generalizes the notion of history spaces first introduced by Coleman and Mizel [1,2,3,4] and later extended by Hale [3], Hino [1,2,3], and MacCamy [1]. Example 2.3 comes from Hale [2]. The proof that the metric space of Example 2.4 satisfies conditions (2.1) and (2.2) is due to Kappel [1]. Example 2.5 is also due to Hale [3,13]. Also see Corduneanu [2,3] for systems with infinite delay. Even in the case for finite delay, various topologies on X have been considered by Delfour and Mitter [1,2], Hale [9], Jones [1], and Melvin [1,2] (these last two for NFDE's also). A proof of Theorem 2.6 for either infinite or finite delay can be found in Hale and Imaz [1]. An especially interesting approach for the case of finite delay is by Costello [1]. Another type of problem not considered in this chapter concerns state dependent delays; c.f. Winston [1,2] and Winston and Yorke [1].

Section 3. Theorem 3.1 is from Flaschka and Leitman [1] and Webb [2]. Also see Webb [1,5] for the semigroup approach to FDE's with other initial function spaces.

Section 4. On the matter of existence and continuation of negative solutions see Hale [4], Hastings [1], and Lillo [1]. On the matter of one-to-oneness of $T(t)$ or uniqueness of solutions, see Hale [6,10], Hale and Oliva [1], Henry [1], and Winston [1]. Halanay and Yorke [1] provide further examples and references. The proofs of Theorems 4.5 and 4.8 are from Kappel [1]. The example in Remark 4.9 is due to Chow [3]. Further results and examples in stability are in Hale [2], Hale and Cruz [1], Jones [2], Kaplan and Yorke [1], and Webb [4].

Section 5. Example 5.1 is based on models of Volterra [1,2]. The analysis here is from Hale [2]. Example 5.2 comes from Coleman and Mizel [4]. Example 5.4 is due to DiPasquantonio and Kappel [1]. Remark 5.6 comes from Kappel [1].

Section 6. For basic existence, uniqueness, and continuous dependence see Costello [1] or Hale [8]. The semidynamical system obtained is like that of Miller [2,3]. Related results are in Seifert [1]. Proposition 6.4 is due to Kato [2]. Also see Kato and Yoshizawa [1] and Yoshizawa [1]. For quasi-invariance of limit sets see Rouché [2].

Section 7. Theorems 7.4, 7.5, and Example 7.6 are from Chow [2]. Theorem 7.7 is due to Fennell [1]. We have limited ourselves to a discussion of results which are based on properties of the semigroup T. There is an extensive literature on the existence of periodic solutions of FDE's. Halanay and Yorke [1] survey the literature prior to 1970. More recent results based upon condensing maps have been obtained by Chow and Hale [1], Jones [3], Lopes [1], and Nussbaum [1]. Nussbaum [1] also has an extensive bibliography of more recent results. In addition to these also see Chow [3] and Chow and Hale [2]. Finally a class of processes called dissipative has been studied by Billotti and LaSalle [1],

Hale and Lopes [1], Hale, LaSalle, and Slemrod [1], and LaSalle [7] for asymptotically stable periodic solutions of FDE's.

Section 8. Theorem 8.4 is established in Hale and Cruz [2]. Other initial function spaces are considered by Hale [9] and Melvin [1,2]. Also see Hale and Meyer [1] and Hale and Cruz [1] for linear systems. The representation given by (8.6) can be found in Hale [7]. Example 8.9, Lemmas 8.11, 8.12 and Theorem 8.14 are due to Cruz and Hale [1]. Lemma 8.13 is from Hale [11]. The representation in Theorem 8.16 is also due to Hale [3].

CHAPTER VII
STOCHASTIC DYNAMICAL SYSTEMS

1. Introduction

We turn to a semidynamical system which is generated by a Markov process. Here again we obtain, in general, a non-differentiable system.

Loosely speaking, a Markov process is a family of random variables $\{\xi(t): t \in \mathbb{R}^+\}$, each of whose values are specified according to some probability. In addition, the probability distribution of $\xi(t+h)$ depends solely on h and $\xi(t)$. The semidynamical system we consider here just reflects the manner in which an initial probability distribution evolves with time. In fact, the random nature of $\xi(t)$ eliminates it as a candidate for a semidynamical system.

We would like to explore the question: where does $\xi(t)$ go as $t \to \infty$? If the process were deterministic (nonrandom), then it would make sense to say (under certain conditions) that $\xi(t)$ approaches some set M (in terms of a metric). Regrettably, we cannot say the same thing when $\xi(t)$ is Markov. The strongest thing we might say is that $\xi(t)$ approaches M with probability one. More likely, though, we can only say that the convergence is in probability or distribution. Our approach will allow us to determine the limiting probability distributions of $\xi(t)$ as $t \to \infty$. Indeed these will be the positive limit sets of the semidynamical system. If we can determine the support of the limit sets, then we can find those values which $\xi(t)$ approaches in probability (or with probability one) as $t \to \infty$.

The reader is referred to Appendix B for the relevant definitions and concepts from probability theory and stochastic processes.

2. The Space of Probability Measures

Denote by $\mathscr{B}^d$ the collection of all Borel subsets of $\mathbb{R}^d$ and $B_b(\mathbb{R}^d;\mathbb{R})$ the set of all real valued, bounded, $\mathscr{B}^d$-measurable functions on $\mathbb{R}^d$. Let $C_b(\mathbb{R}^d;\mathbb{R})$ be the subset of $B_b(\mathbb{R}^d;\mathbb{R})$ consisting of all continuous functions on $\mathbb{R}^d$. Both spaces are Banach spaces under the norm $\|f\| = \sup_{x\in\mathbb{R}^d} |f(x)|$. This defines the topology of uniform convergence on $\mathbb{R}^d$. For convenience, we write C for $C_b(\mathbb{R}^d;\mathbb{R})$ if there is no ambiguity.

Let $\mathscr{M}$ denote the set of all probability measures whose domain is $\mathscr{B}^d$. It is known that $\mathscr{M} \subset C^*$, the dual of C. C^* can be characterized as the collection of all real valued finitely additive set functions on $\mathscr{B}^d$ (cf. Dunford and Schwartz [1]). Then $\mathscr{M} = \{\mu \in C^*: \mu$ is countably additive, non-negative, $\mu(\mathbb{R}^d) = 1\}$. Endow C^* with the w^*-topology, and let $\mathscr{M}$ have the relative topology of C^*. This means that a net $\{\mu_\alpha\} \subset \mathscr{M}$ converges to $\mu \in \mathscr{M}$ if and only if $(f,\mu_\alpha) \to (f,\mu)$ for every $f \in C$. (g,ν) denotes the value of the linear functional $\nu \in C^*$ at $g \in C$; thus

$(g,\nu) = \int_{\mathbb{R}^d} g(x)\nu(dx)$. (Henceforth we will write integrals of this form as $\int g\,d\nu$ when there is no ambiguity.)

We record here some basic facts about $\mathcal{M}$; for a proof see Parthasarathy [1].

<u>Proposition 2.1</u>.

(i) $\mathcal{M}$ is metrizable as a complete metric space with metric ρ given by $\rho(\mu,\nu) = \sum_{k=1}^{\infty} 2^{-k}|(g_k,\mu-\nu)|$, where $\{g_k\}$ is a specially chosen sequence of non-negative functions from C, each with norm one.

(ii) $\mu_k \to \mu$ in $\mathcal{M}$ if and only if $\sup_{f\in F}|(f,\mu_k-\mu)| \to 0$ for every equicontinuous and uniformly bounded family $F \subset C$.

(iii) A subset $\mathcal{S} \subset \mathcal{M}$ is precompact if and only if for each $\varepsilon > 0$ there exists a compact set $K \subset \mathbb{R}^d$ so that $\sup_{\mu\in\mathcal{S}} \mu(K) \geq 1 - \varepsilon$. (A set $\mathcal{S}$ with this property is called *uniformly tight*.)

(iv) Let $Q \subset \mathbb{R}^d$ be compact. The space $\mathcal{M}_Q = \{\mu \in \mathcal{M}: \operatorname{supp}\mu \subset Q\}$ is compact. ($\operatorname{supp}\mu \overset{def}{=} x \in \mathbb{R}^d$: $\mu(N(x)) > 0$ for every neighborhood $N(x)$ of x. This set is called the *support* of μ. If $\mathcal{S}\subset\mathcal{M}$, $\operatorname{supp}\mathcal{S} \overset{def}{=} \bigcup_{\nu\in\mathcal{S}} \operatorname{supp}\nu$.)

3. <u>Markov Transition Operators and the Semidynamical System</u>

The semidynamical system will be based upon the existence of a Markov transition operator. Section 6 discusses ways in which to generate this operator. For the present it is enough to say that if $\{\xi(t): t \in \mathbb{R}^+\}$ is a Markov process, then $\mathcal{P}(t,x,B)$ is the probability that $\xi(t)$ belongs to the Borel set B at time t, given that $\xi(0) = x$. Markov processes are defined in Section 6.

Definition 3.1. A *Markov transition* function is a mapping $\mathscr{P}\colon \mathbb{R}^+ \times \mathbb{R}^d \times \mathscr{B}^d \to [0,1]$ which satisfies

(i) $\mathscr{P}(t,x,\cdot) \in \mathscr{M}$ for each $t \in \mathbb{R}^+$, $x \in \mathbb{R}^d$,

(ii) $\mathscr{P}(t,\cdot,B) \in B_b(\mathbb{R}^d;\mathbb{R})$ for each $t \in \mathbb{R}^+$, $B \in \mathscr{B}^d$,

(iii) $\mathscr{P}(0,x,B) = \chi_B(x)$ for each $x \in \mathbb{R}^d$, $B \in \mathscr{B}^d$,

(iv) $\mathscr{P}(t+s,x,B) = \int \mathscr{P}(t,y,B)\,\mathscr{P}(s,x,dy)$ for each $s,t \in \mathbb{R}^+$, $x \in \mathbb{R}^d$, $B \in \mathscr{B}^d$ (the Chapman-Kolmogorov equations).

For each $t \in \mathbb{R}^+$ define the operator $T(t)\colon B_b(\mathbb{R}^d;\mathbb{R}) \to B_b(\mathbb{R}^d;\mathbb{R})$ by

$$T(t)f(\cdot) = \int f(y)\,\mathscr{P}(t,\cdot,dy), \quad x \in \mathbb{R}^d. \tag{3.1}$$

It is easy to verify that the family $T = \{T(t)\colon t \in \mathbb{R}^+\}$ is a family of linear operators of norm one on $B_b(\mathbb{R}^d;\mathbb{R})$ for which T_0 is the identity, $T(s+t) = T(s)T(t)$, and that for each $x \in \mathbb{R}^d$, $T(t)f(x) \to f(x)$ as $t \to 0$. As this convergence need not be uniform in $x \in \mathbb{R}^d$, we must stop short of claiming T is a linear contraction semigroup. Fortunately in most situations where Markov processes arise, the following conditions are satisfied.

$$T(t)C \subset C \quad \text{for all } t \in \mathbb{R}^+ \tag{3.2}$$

(3.3) $T(\cdot)f(x)\colon \mathbb{R}^+ \to \mathbb{R}^d$, is continuous, uniformly for x in compact subsets of $\mathbb{R}^d$, for every $f \in C$.

We observe that $T(t)f \geq 0$ whenever $f \geq 0$.

The family of operators T defined by Equation (3.1) is called a *Markov semigroup*. If T satisfies condition

(3.2), it is said to possess the *Feller* property. Condition (3.3) is called *stochastic continuity on compacta*. Henceforth we assume conditions (3.2) and (3.3) are satisfied. In Section 6 we provide sufficient conditions for (3.2) and (3.3) to hold.

For each $t \in \mathbb{R}^+$ define the operator $U(t): \mathcal{M} \to \mathcal{M}$ by

$$(3.4) \qquad U(t)\mu(\cdot) = \int \mathcal{P}(t,x,\cdot)\mu(dx).$$

Observe that $U = \{U(t): t \in \mathbb{R}^+\}$ is the dual semigroup to T. In fact, $(f,U(t)\mu) = (T(t)f,\mu)$ for $f \in B_b(\mathbb{R}^d;\mathbb{R})$, $\mu \in \mathcal{M}$.

Theorem 3.2. The mapping from $\mathcal{M} \times \mathbb{R}^+ \to \mathcal{M}$ taking (μ,t) to $U(t)\mu$ defines a semidynamical system on $\mathcal{M}$. It is denoted either by $(\mathcal{M},U)$ or just U.

Proof: It is a straightforward exercise in probability to verify that $U(t)\mu \in \mathcal{M}$ for each $(\mu,t) \in \mathcal{M} \times \mathbb{R}^+$. Moreover, properties (i) and (ii) of a semidynamical system follow easily from Conditions (iii) and (iv) of $\mathcal{P}$ in Definition 3.1. We establish continuity.

Let $\{\mu_i\} \subset \mathcal{M}$ converge to $\mu \in \mathcal{M}$, $\{t_i\} \subset \mathbb{R}^+$ converge to $t \in \mathbb{R}^+$, and let $f \in C$. Then

$$|(f,U(t_i)\mu_i) - (f,U(t)\mu)| = |(T(t_i)f,\mu_i) - (T(t)f,\mu)|$$

$$\leq \int |T(t_i)f - T(t)f|d\mu_i + \left|\int T(t)fd\mu_i - \int T(t)fd\mu\right|.$$

The second term on the right hand side tends to zero by w*-convergence of μ_i to μ. As this sequence is precompact, we may use the criterion of Proposition 2.1 (iii) to show the first term also tends to zero. In particular if $\varepsilon > 0$, choose $K \subset \mathbb{R}^d$ compact so that $\sup_i \mu_i(\mathbb{R}^d \setminus K) < \varepsilon/4\|f\|$, and

$\sup_{x \in K} |T(t_i)f(x) - T(t)f(x)| < \varepsilon/2$ for large enough i. Thus

$$\int_K |T(t_i)f - T(t)f|d\mu_i + \int_{\mathbb{R}^d \setminus K} |T(t_i)f - T(t)f|d\mu_i$$

$$< (\varepsilon/2)\mu_i(K) + 2\|f\| \varepsilon/(4\|f\|) < \varepsilon. \qquad \square$$

4. Properties of Positive Limit Sets

If $\gamma^+(\mu)$ is a precompact positive orbit, then we may apply Theorem 3.5 of Chapter II to see that the positive limit set $L^+(\mu)$ is nonempty, compact, connected, and weakly invariant. Thus we have an invariance theorem.

Theorem 4.1. Let $\mathscr{B}, \mathscr{A} \subset \mathscr{M}$ with the property that $U(t)\mu \to \mathscr{A}$ as $t \to \infty$ whenever $\mu \in \mathscr{B}$. Then $U(t)\mu \to \mathscr{S}$ as $t \to \infty$, where $\mathscr{S}$ is the largest weakly invariant subset of $\mathscr{M}$.

We will establish a stronger invariance in Section 6. In particular we will be able to say where the underlying Markov process $\{\xi(t): t \in \mathbb{R}^+\}$ goes as $t \to \infty$.

Some important Markov processes, such as diffusions, possess a kind of stability property which we now formulate. This condition (4.1), as well as conditions (3.2) and (3.3) will hold throughout this section.

(4.1) $\quad \bigcup_{t \in \mathbb{R}^+} T_t f$ is an equicontinuous family for each $f \in C$.

Theorem 4.2. The system $(\mathscr{M}, U)$ is Lyapunov stable.

Proof: It is sufficient to show that $\mu_i \to \mu$ in $\mathscr{M}$ implies for each $f \in C$ that $(f, U(t)\mu_i) \to (f, U(t)\mu)$ uniformly in $t \in \mathbb{R}^+$. Since $\|T_t f\| \le \|f\|$ for all $t \in \mathbb{R}^+$, then Proposition 2.1(ii) shows that $(f, U(t)\mu_i) = (T(t)f, \mu_i) \to (T(t)f, \mu) = (f, U(t)\mu)$, uniformly in $t \in \mathbb{R}^+$.

Theorem 4.3. Suppose $\gamma^+(\mu)$ is precompact for some $\mu \in \mathcal{M}$. Then

(i) $L^+(\mu)$ is a compact positively minimal set.

(ii) $L^+(\mu)$ is invariant, hence U extends to a Lyapunov stable dynamical system on the minimal set $L^+(\mu)$.

(iii) $L^+(\mu)$ is equi-almost periodic under U.

Proof: (i) and (ii) follow from Theorems 2.4 and 2.5 of Chapter III. For (iii), let $\varepsilon > 0$. Property (i) of Proposition 2.1 and Theorem 4.2 imply there exists $\delta > 0$ such that $\sup_{t \in \mathbb{R}^+} \rho(U(t)\nu, U(t+\tau)\nu) < \varepsilon$ whenever $\rho(\nu, U(\tau)\nu) < \delta$ and $\nu \in L^+(\mu)$. Since $L^+(\mu)$ is compact minimal, the trajectory through ν is recurrent. Thus the set $\{\tau \in \mathbb{R}: \rho(\nu, U(\tau)\nu) < \delta\}$ is relatively dense in $\mathbb{R}$. Consequently the set $\{\tau \in \mathbb{R}: \sup_{t \in \mathbb{R}^+} \rho(U(t)\nu, U(t+\tau)\nu) < \varepsilon\}$ is also relatively dense in $\mathbb{R}$. As $L^+(\mu) = \overline{\bigcup_{t \in \mathbb{R}} U(t)\nu}$, then U is equi-almost periodic in $L^+(\mu)$. (Note, (iii) following directly from Theorems 3.5, 4.10 and Corollary 4.9.) □

The reader should note the similarity between Theorem 4.3 and Theorem 6.1 of Chapter V. The contracting (nonlinear) semigroup on a Banach space has been replaced by a Lyapunov stable (linear) semigroup on a metric space. The equicontinuity condition (4.1) is the analogue to the (norm) contraction property.

A more remarkable property of the positive limit sets lies in the next theorem.

Theorem 4.4. The set $\mathcal{N} \overset{\text{def}}{=} \bigcup_{\mu \in \mathcal{M}} L^+(\mu)$ is closed in $\mathcal{M}$. Moreover, if every positive orbit $\gamma^+(\mu)$ is precompact, then $\mathcal{N}$ is convex.

Proof: We leave it to the reader to show $\mathcal{N}$ is closed. This is a general property of semidynamical systems. (See Exercise 7.3, Chapter III.) Now suppose every positive orbit in $\mathcal{M}$ is precompact. Let $\mu, \nu \in \mathcal{N}$. Then $\mu \in L^+(\mu)$ as each positive limit set must be minimal. There exists a sequence $\{t_n\} \subset \mathbb{R}^+$ with $t_n \to \infty$ so that $U(t_n)\mu \to \mu$. Also, $\nu \in L^+(\nu)$. As $L^+(\nu)$ is compact, we may assume (by choosing a subsequence if necessary) that $U(t_n)\nu \to \lambda$ for some $\lambda \in L^+(\nu)$. Then $U(t_n+t)\nu \to U(t)\lambda$ for every $t \in \mathbb{R}^+$. Let $0 < r < 1$, and consider the probability measure

$$\beta_r = r\mu + (1-r)U(t)\lambda.$$

Then $\beta_r \in \mathcal{N}$. Indeed,

$$\begin{aligned}\beta_r &= \lim_{n\to\infty} [rU(t_n)\mu + (1-r)U(t_n+t)\nu] \\ &= \lim_{n\to\infty} U(t_n)(r\mu + (1-r)U(t)\nu).\end{aligned}$$

Thus $\beta_r \in L^+(r\mu + (1-r)U(t)\nu) \subset \mathcal{N}$. Thus the "line segment" $\mathcal{L}_t$ joining μ and $U(t)\lambda$ lies in $\mathcal{N}$ for every $t \in \mathbb{R}^+$. As $\lambda \in L^+(\nu)$, then $\gamma^+(\lambda)$ is dense in $L^+(\nu)$. Therefore there exists a sequence $\{\tau_n\} \subset \mathbb{R}^+$ with $\tau_n \to \infty$ so that $\lim_{n\to\infty} U(\tau_n)\lambda = \nu$. Consequently, $\lim_{n\to\infty} \mathcal{L}_{\tau_n} = \mathcal{L}$, the line segment joining μ and ν. As $\mathcal{N}$ is closed we must have $\mathcal{L} \subset \mathcal{N}$. □

Remark 4.5. The proof of Theorem 4.4 only requires the space to be a linear space. Thus, the theorem must hold in a Banach space.

In the event there is an underlying Markov process $\{\xi(t): t \in \mathbb{R}^+\}$, we show in Section 7 that $\xi(t) \to \overline{\text{supp}}\ L^+(\mu)$ in some sense when $\xi(0)$ has probability measure μ. The next theorem makes it easier to find $\overline{\text{supp}}\ L^+(\mu)$. Denote by

δ_x the Dirac measure at $x \in \mathbb{R}^d$; i.e., $\delta_x(\{x\}) = 1$.

Theorem 4.6. If $\mu \in \mathcal{M}$ has compact support $K \subset \mathbb{R}^d$, then $\operatorname{supp} L^+(\mu) \subset \overline{\bigcup_{x\in K} \operatorname{supp} L^+(\delta_x)}$.

Proof: Fix $\mu \in \mathcal{M}$ with compact support K. According to Proposition 2.1(iv), $\mathcal{M}_K$ is a compact convex subset of $\mathcal{M}$. As the extreme points of $\mathcal{M}_K$ are Dirac measures, the collection of all sums $\sum_{i=1}^{m} a_i\delta_{x_i}$, $a_i \in \mathbb{R}^+$, $\sum_{i=1}^{m} a_i = 1$ is dense in $\mathcal{M}_K$ (Dunford and Schwartz [1], p. 440). There exists a sequence $\mu_m = \sum_{i=1}^{m} a_i\delta_{x_i}$ of the above sort which converges to μ as $m \to \infty$. First we establish that

$$\operatorname{supp} L^+(\mu_m) \subset \sum_{i=1}^{m} \operatorname{supp} L^+(\delta_{x_i}). \tag{4.2}$$

Then we show

$$\operatorname{supp} L^+(\mu) \subset \overline{\bigcup_{m=1}^{\infty} \operatorname{supp} L^+(\mu_m)}. \tag{4.3}$$

The conclusion of the theorem will follow readily.

So let μ_m be defined as above for some fixed m. Then $U(t)\mu_m = \sum_{i=1}^{m} a_iU(t)\delta_{x_i}$. If $\nu \in L^+(\mu)$, there exists $t_j \to \infty$ with $\sum_{i=1}^{m} a_iU(t_j)\delta_{x_i} \to \infty$ as $j \to \infty$. As the sequence $\{U(t_j)\delta_{x_1}\}_{j=1}^{\infty}$ is contained in the precompact set $\gamma^+(\delta_{x_1})$, we may assume (by choosing a subsequence if necessary) that $U(t_j)\delta_{x_1} \to \nu_1$ as $j \to \infty$. Continue in this manner so that we obtain $U(t_j)\delta_{x_i} \to \nu_i \in L^+(\delta_{x_i})$ as $j \to \infty$, $1 \le i \le m$. Thus $\sum_{i=1}^{m} a_iU(t_j)\delta_{x_i} \to \sum_{i=1}^{m} a_i\nu_i \overset{\text{def}}{=} \nu$. This means $\nu \in \bigcup_{i=1}^{m} a_iL^+(\delta_{x_i})$. Thus $L^+(\mu) \subset \bigcup_{i=1}^{m} a_iL^+(\delta_{x_i})$.

Now as $\nu \in L^+(\mu)$ implies $\nu = \sum_{i=1}^{m} a_i\nu_i$, $\nu_i \in L^+(\delta_{x_i})$,

$1 \leq i \leq m$, then $\operatorname{supp} \nu \not\subset \bigcup_{i=1}^{m} \operatorname{supp} \nu_i$ implies there exists $x \in \operatorname{supp} \nu$ and neighborhoods $N_i(x)$ of x, so that $\nu_i(N_i(x)) = 0$, $1 \leq i \leq m$. If $N(x) = \bigcap_{i=1}^{m} N_i(x)$, then $\nu(N(x)) = 0$ which contradicts the fact that $x \in \operatorname{supp} \nu$. Thus $\operatorname{supp} \nu \subset \bigcup_{i=1}^{m} \operatorname{supp} \nu_i$, and therefore $\operatorname{supp} \nu \subset \bigcup_{i=1}^{m} \operatorname{supp} L^+(\delta_{x_i})$. This proves the Inclusion (4.2).

Let $Q = \bigcup_{m=1}^{\infty} \operatorname{supp} L^+(\mu_m)$. We claim (4.3) holds. Otherwise, there exists $x \in \operatorname{supp} L^+(\mu)$ with $x \notin \overline{Q}$. Suppose G is a closed set containing $\overline{Q}$ but not x. Define $f \in C$ so that $f = 0$ on $\overline{Q}$, $f \geq 1$ outside of G. Now there exists $\nu \in L^+(\mu)$ so that $\nu(N(x)) > 0$ for every neighborhood $N(x)$ of x. As $\mu_m \to \mu$, we see that $L^+(\mu_m) \to L^+(\mu)$ (in the topology of the Hausdorff metric on the nonempty compact subsets of $\mathscr{M}$). Indeed, $(\mathscr{M}_K, U)$ is uniformly Lyapunov stable. So by Proposition 2.2 of Chapter III, $(\mathscr{M}_K, U)$ is a stable system. The diagram of Figure 7.4 of Chapter II shows that L^+ is continuous on $\mathscr{M}_K$. Thus there exists $\nu_m \in L^+(\mu_m)$ so that $\nu_m \to \nu$. The definition of Q implies $(f, \nu_m) = 0$ for all m, while $(f, \nu) > 0$. But this is a contradiction. Thus, the Inclusion (4.3) does hold.

Now (4.2) implies $L^+(\mu_m) \subset \bigcup_{x \in K} \operatorname{supp} L^+(\delta_x)$. Let $m \to \infty$ and apply (4.3) to conclude the proof of the theorem. □

5. Critical Points for Markov Processes

By a critical point for the semidynamical system $(\mathscr{M}, U)$ we mean a probability measure μ so that $\int \mathscr{P}(t,x,\cdot)\mu(dx) = \mu(\cdot)$ for all $t \in \mathbb{R}^+$. Such a measure is also called *invariant*. The following theorem provides for invariant probability measures under conditions which have already been assumed. In particular, we require conditions (3.2) and (3.3). Condition

(4.1) is not used here.

Theorem 5.1. If there exists a precompact positive orbit $\gamma^+(\mu)$ through $\mu \in \mathcal{M}$, then there exists an invariant probability measure β, unique within the closed convex hull of $\gamma^+(\mu)$. Moreover,

$$\beta = \lim_{t\to\infty} t^{-1} \int_0^t U(s)\mu ds.$$

Proof: Let $\gamma^+(\mu)$ be precompact, $\mu \in \mathcal{M}$. Then $\gamma^+(\mu)$ is uniformly tight, and therefore so is $\text{co}\ \gamma^+(\mu)$ by a simple calculation. Thus $\mathcal{K} = \overline{\text{co}}\ \gamma^+(\mu)$ is a compact, convex subset of $\mathcal{M}$. (Note that the closed convex hull of a compact subset of a locally convex space need not be compact.) We claim that $U(t)\mathcal{K} \subset \mathcal{K}$ for every $t \in \mathbb{R}^+$. In fact if $\nu \in \text{co}\ \gamma^+(\mu)$ so that $\nu = \sum_{i=1}^m a_i U(t_i)\mu$ for some $\{a_i\} \subset \mathbb{R}^+$, $\sum_{i=1}^m a_i = 1$, then $U(t)\nu = \sum_{i=1}^m a_i U(t_i+t)\mu \in \text{co}\ \gamma^+(\mu)$. Now extend $U(t)$ to $\mathcal{K}$ by continuity. According to the Markov-Kakutani fixed point theorem (see Appendix A), there exists $\beta \in \mathcal{K}$ such that $U(t)\beta = \beta$ for all $t \in \mathbb{R}^+$.

In order to establish the uniqueness of β within $\mathcal{K}$, we need to consider the strong topology on $\mathcal{M}$. By this we mean the topology defined by the variation norm $\|\mu\| = |\mu|(\mathbb{R}^d)$. $\mathcal{M}$ is a Banach space under this topology. Then $\mathcal{K}$ is strongly closed (Dunford and Schwartz [1], V. 3.13). If β' is another invariant probability measure in $\mathcal{K}$, then there must exist sequences $\beta_m \to \beta$, $\beta'_m \to \beta'$ (norm convergence) with $\beta_m, \beta'_m \in \text{co}\ \gamma^+(\mu)$; i.e., $\beta_m = P_m\mu$, $\beta'_m = P'_m\mu$ for some convex combinations P_m, P'_m of the operators $U(t)$. Clearly $P'_m\beta = P_m\beta = \beta$, $P'_m\beta' = P_m\beta' = \beta'$, $P_m P'_n = P'_n P_m$ for all positive integers n,m. Moreover $\|\beta'-\beta\| = \|P_m\beta' - P'_n\beta\| \leq \|P_m(\beta' - P'_n\mu)\| +$

$\|P_n'(P_m\mu-\beta)\| \leq \|\beta'-\beta_n'\| + \|\beta-\beta_m\| \to 0$ as $n,m \to \infty$. Thus the invariant measure in $\mathscr{K}$ is unique.

Define the measure $\beta_t = t^{-1}\int_0^t U(s)\mu ds$. Then $\beta_t \in \overline{co}\,\gamma^+(\mu)$ as can be seen from the proof of Theorem 6.2 (iv), Chapter V. Thus the set $\{\beta_t\colon t \in \mathbb{R}^+\}$ is contained in $\mathscr{K}$, and hence is precompact. Let $\hat{\beta}$ be any w*-limit point so that $\hat{\beta} = \lim_{n\to\infty} \beta_{t_n}$ for some $t_n \to \infty$. Then the same argument that was used in the proof of Theorem 6.2(iv) of Chapter V shows that $\hat{\beta}$ is invariant. Uniqueness implies $\hat{\beta} = \beta$. □

6. Stochastic Differential Equations

We examine a special, but important class of Markov processes: those generated by solutions to stochastic differential equations. From an intuitive point of view a stochastic differential equation is an ordinary differential equation with an added perturbation term whose values are distributed according to some probability distribution. Thus the perturbation term gives rise to random fluctuations in the "solution" to the differential equation. An important case of random perturbations is that called "white noise." Formally, a white noise stochastic process is the time derivative of a Brownian motion process. But as a Brownian motion process is not even of bounded variation, we cannot compute its derivative in the usual sense. This matter is dealt with by recourse to the Ito stochastic integral. A summary of needed results is provided in Appendix B.

The stochastic differential equation we shall consider is written in differential form

$$dx(t) = f(x(t))dt + G(x(t))dW(t). \tag{6.1}$$

If G is zero, then Equation (6.1) is an (autonomous) ordinary differential equation. The white noise term could be formally represented by $G(x)\dot{W}(t)$. The proper interpretation of Equation (6.1) is through the integral formulation

$$x(t) = x_0 + \int_0^t f(x(s))ds + \int_0^t G(x(s))dW(s), \tag{6.2}$$

where the second integral is the Ito stochastic integral.

Suppose $f\colon \mathbb{R}^d \to \mathbb{R}^d$ and $G\colon \mathbb{R}^d \to \mathbb{R}^{dp}$; i.e., $G(x)$ is an $d \times p$ matrix. $W = \{W(t)\colon t \in \mathbb{R}^+\}$ is normalized Brownian motion, and x_0 is an initial random variable, both relative to some probability space (Ω,Σ,P).

Definition 6.1. By a *solution* of Equation (6.2) we mean a family of $\mathbb{R}^d$-valued random variables $x = \{x(t)\colon t \in \mathbb{R}^+\}$ on (Ω,Σ,P) so that $x(t)$ satisfies (6.2) w.p.1.

Definition 6.2. A (homogeneous) *Markov process* is a family of $\mathbb{R}^d$-valued random variables $\{\xi(t)\colon t \in \mathbb{R}^+\}$ on (Ω,Σ,P) which satisfies (w.p.1)

$$P\{\xi(s+t) \in B|\Sigma_s\} = P\{\xi(s+t) \in B|\xi(s)\} \tag{6.3}$$

$$P\{\xi(s+t) \in B|\xi(s)\} = P\{\xi(t) \in B|\xi(0)\} \tag{6.4}$$

for all $s,t \in \mathbb{R}^+$ and $B \in \mathscr{B}^d$. $\Sigma_s \subset \Sigma$ is the minimal σ-algebra for which the family $\{\xi(u)\colon 0 \le u \le s\}$ is measurable. We can now define a Markov transition function (see Appendix B) so that for each fixed $t \in \mathbb{R}^+$ and $B \in \mathscr{B}^d$,

$$\mathscr{P}(t,x,B) = P\{\xi(t) \in B|\xi(0) = x\} \quad \text{w.p.1 on } \mathbb{R}^d,$$

and $\mathscr{P}$ satisfies Definition 3.1.

The following proposition insures the existence of solutions to Equation (6.2) which are Markov processes. We also see that it provides sufficient conditions for properties (3.2) and (3.3) to hold. The conditional probability P_x and expectation E_x are versions of $P\{\cdot|\xi(0) = x\}$ and $E\{\cdot|\xi(0) = x\}$ respectively. See Appendix B for details.

Proposition 6.3. Suppose f and G are locally Lipschitz continuous on $\mathbb{R}^d$. If $E|x_0|^2 < \infty$ and x_0 is independent of W, then there exists a solution ξ of Equation (6.2) which satisfies

(i) $\xi(0) = x_0$ w.p.1.

(ii) $\xi(t)$ is continuous on $\mathbb{R}^+$ w.p.1.

(iii) $\xi(t)$ is the unique solution of (6.2) w.p.1 which satisfies (i) and (ii).

(iv) ξ is a Markov process.

If the initial random variable x_0 is a constant $x \in \mathbb{R}^d$ w.p.1., then

(v) $E_x|\xi(t)|^2$ is bounded on bounded subsets of $\mathbb{R}^+$, uniformly for x in compact subsets of $\mathbb{R}^d$.

(vi) $P_x\{\sup_{0\le s\le t} |\xi(s)-x| > \varepsilon\} \to 0$ as $t \to \infty$ for each $\varepsilon > 0$.

(vii) $E_x h(\xi(t))$ is a bounded continuous functon of $x \in \mathbb{R}^d$ for every $t \in \mathbb{R}^+$, $h \in C$.

(viii) $\lim_{t\downarrow 0} E_x h(\xi(t)) = h(x)$ for every $h \in C$, uniformly for x in compact subsets of $\mathbb{R}^d$.

Proof: For a proof of parts (i) through (v), the reader is referred to Gihman and Skorohod [1], Chapters 1,2. The remainder of the proof can be found in Doob [1], Chapter VI. □

Remark 6.4. As ξ is a Markov process, $\mathscr{P}$ generates the Markov semigroup T on C. This is related to the conditional expectation operator E_x via

$$(6.5) \qquad E_x h(\xi(t)) = \int h(y)P(t,x,dy) = T(t)h(x).$$

Thus parts (vii) and (viii) of Proposition 6.3 ensure that conditions (3.2) and (3.3) are met. It is also seen from Chapter 3 of Gihman and Skorohod [1], that ξ is a diffusion with drift vector $f(x)$ and diffusion matrix $g(x) = G(x)G(x)'$, G' denotes the transpose of the matrix G.

If, in addition to the hypotheses on f and G in Prop. 6.3, that for some $\alpha > 0$ and any $y \in \mathbb{R}^d$, $\sum_{i,j=1}^{d} g_{ij}(x)y_i y_j \geq \alpha|y|^2$, then $h \in C$ implies $|T(t)h(x)| \to 0$ as $t \to \infty$, uniformly for x in compact subsets of $\mathbb{R}^d$ (c.f. Dynkin [1], Chapter V, §6). It then follows that the equicontinuity condition (4.1) is satisfied.

The following is a criteria for precompactness of positive orbits in $\mathscr{M}$.

Theorem 6.5. If for some $b > 0$ the Markov process satisfies $\limsup_{t\to\infty} E_x|\xi(t)|^2 \leq b$ for each $x \in \mathbb{R}^d$, then $\gamma^+(\mu)$ is precompact for every $\mu \in \mathscr{M}$.

Proof: It is sufficient to establish that $\gamma^+(\mu)$ is uniformly tight. There exists $t_0 \in \mathbb{R}^+$ such that $E_x|\xi(t)|^2 \leq b$ for all $t \geq t_0$. Let $\varepsilon > 0$ and choose $R_\varepsilon > 0$ so that $b/R_\varepsilon^2 < \varepsilon$. Chebychev's inequality implies $P_x\{|\xi(t)| > R_\varepsilon\} \leq E_x|\xi(t)|^2/R_\varepsilon^2 < \varepsilon$ whenever $t \geq t_0$. If $K_\varepsilon = \{x \in \mathbb{R}^d: |x| \leq R_\varepsilon\}$, then $U(t)\mu(\mathbb{R}^d \setminus K_\varepsilon) = \int \mathscr{P}(t,x,\mathbb{R}^d \setminus K_\varepsilon)\mu(dx) = P_x\{|\xi(t)| > R_\varepsilon\} < \varepsilon$ for $t \geq t_0$. Thus $\bigcup_{t\geq t_0} U(t)\mu$ is precompact. But as the mapping $t \to U(t)\mu$ of $[0,t_0]$ into $\mathcal{M}$ is continuous, we conclude that $\gamma^+(\mu) = \bigcup_{t \in \mathbb{R}^+} U(t)\mu$ is precompact. □

7. The Invariance Principle for Markov Processes

The type of invariance principle which we will establish here goes beyond Theorem 4.1. In fact, we will be able to say where the Markov process itself goes as $t \to \infty$ instead of just where its probability measures go. We shall assume that the Markov process is a solution to the stochastic differential equation (6.2).

Theorem 7.1. Suppose the Markov process ξ has initial probability measure μ so that $\gamma^+(\mu)$ is precompact. Then $\xi(t) \to \overline{\text{supp } L^+(\mu)}$ in pr. as $t \to \infty$.

Proof: Let N_ε be an ε-neighborhood of $Q = \overline{\text{supp } L^+(\mu)}$ for $\varepsilon > 0$. It is sufficient to show that

$$\limsup_{t\to\infty} P_x\{\xi(t) \in \mathbb{R}^d \setminus \overline{N}_\varepsilon\} = 0, \tag{7.1}$$

for then $P_x\{\sup_{y\in Q}|\xi(t)-y| > \varepsilon\} \to 0$ as $t \to \infty$.

If (7.1) were false there would exist $\varepsilon_0 > 0$ and a sequence $t_n \to \infty$ such that $P_x\{\xi(t_n) \in \mathbb{R}^d \setminus \overline{N}_\varepsilon\} \geq \varepsilon_0$. Choose $f \in C$ such that $0 \leq f \leq 1$, $f = 0$ on $\overline{N}_{\varepsilon/2}$, and $f = 1$ on $\mathbb{R}^d \setminus N_\varepsilon$. We can assume that $U(t_n)\mu \to \nu \in L^+(\mu)$. Thus $(f,U(t_n)\mu) \to (f,\nu) \geq \varepsilon_0$. This means that $(\mathbb{R}^d \setminus N_\varepsilon) \cap \text{supp } \nu \neq \emptyset$, a contradiction to the definition of Q. □

Theorem 7.2. (The Invariance Principle.) Suppose the Markov process ξ has initial probability measure μ so that $\gamma^+(\mu)$ is precompact. If $\xi(t) \to H$ in pr. as $t \to \infty$, then $\xi(t) \to Q = \text{supp } \mathcal{S}$ in pr. as $t \to \infty$, where $\mathcal{S}$ is the largest weakly invariant subset of $\mathscr{M}$ whose support lies in $\overline{H}$. Moreover if the Markov process ξ is also a solution of (6.2) with initial probability measure $\nu \in \mathscr{S}$, then $\xi(t) \in Q$ w.p.1 for all $t \in \mathbb{R}^+$.

Proof: Use Theorem 7.1 and the weak invariance of $L^+(\mu)$. □

We apply the invariance principle in the following way. Denote by A^W the weak infinitesimal generator of the semigroup T. That is (c.f. Definition 2.8 of Chapter V), if $\xi(t)$ is the Markov process which is given by Proposition 6.3, with initial value $x \in \mathbb{R}^d$, then $f \in B_b(\mathbb{R}^d; \mathbb{R})$ belongs to $\mathscr{D}(A^W)$ if and only if

$$(7.2) \qquad w - \lim_{t \downarrow 0} \frac{f(x) - T(t)f(x)}{t}$$

exists for every $x \in \mathbb{R}^d$. In this event we write $A^W f(x)$ for the limit in (7.2). The following facts will be needed in the sequel. They may be found in Dynkin [1], Chapter 5.

Definition 7.3. A nonnegative random variable τ on Ω is called a *Markov time* if $\{\tau \leq t\} \in \Sigma_t$ for every $t \in \mathbb{R}^+$.

$$(7.3) \qquad \text{Dynkin's Formula:} \quad T(\tau)f(x) = f(x) - \int_0^\tau T(s)A^W f(x)ds,$$

whenever $f \in \mathscr{D}(A^W)$, and τ is a Markov time.

$$(7.4) \qquad A^W = -\sum_{i=1}^n f_i(x)\frac{\partial}{\partial x_i} - \frac{1}{2}\sum_{i=1}^n \sum_{j=1}^n g_{ij}(x)\frac{\partial^2}{\partial x_i \partial x_j},$$

where g_{ij} is the (i,j)-th element of GG'.

Remark 7.4. Observe that Dynkin's formula is the integrated version of Corollary 2.10 of Chapter V, except that t has been replaced by the random variable .

Suppose $V \in C(\mathbb{R}^d; \mathbb{R})$ and $Q_\lambda = \{x \in \mathbb{R}^d: V(x) < \lambda\} \neq \emptyset$

for each $\lambda > 0$. Set $\tau_\lambda = \inf_{t \in \mathbb{R}^+} \{\xi(t) \notin Q_\lambda\}$. Then τ_λ is a Markov time. Now define the stopped process $\xi_\lambda = \{\xi_\lambda(t): t \in \mathbb{R}^+\}$ by $\xi_\lambda(t) = \xi(t \wedge \tau_\lambda)$ where $t \wedge s = \min\{t,s\}$. Then $\xi_\lambda(t)$ is a solution of (6.2) until $\xi(t)$ "hits the boundary" of Q_λ, where upon it remains constant w.p.1. ξ_λ satisfies all of the conclusions of Proposition 6.3. Let T_λ denote the corresponding Markov semigroup and A_λ^w its weak infinitesimal generator. Assume $V \in \mathcal{D}(A_\lambda^w)$. This will obtain, if, for example, Q_λ is bounded. (The domain of V must therefore be limited to the range of ξ_λ w.p.1.) Finally let $\Omega_\lambda = \{\omega \in \Omega: \xi(t,\omega) \in Q_\lambda$ for all $t \in \mathbb{R}^+\}$.

Under the foregoing conditions we have the following stability result. For the basic properties of martingales, see Appendix B.

<u>Theorem 7.5.</u> Suppose Q_λ is bounded and $A_\lambda^w V(x) \geq k(x) \geq 0$ for some $k \in C(Q_\lambda; \mathbb{R}^+)$. Then

(i) $V(\xi(t))$ converges for almost all $\omega \in \Omega_\lambda$ as $t \to \infty$.

(ii) $P_x\{\sup_{t \in \mathbb{R}^+} V(\xi(t)) \geq \lambda\} \leq \frac{1}{\lambda} V(x)$ for each $x \in Q_\lambda$.

(iii) $\xi(t) \to H_\lambda \overset{\text{def}}{=} Q_\lambda \cap \{y \in \mathbb{R}^d: k(y) = 0\}$ with probability at least $1 - \frac{1}{\lambda}V(x)$ as $t \to \infty$ for each $x \in Q_\lambda$.

(iv) If $V(x) \to \infty$ as $|x| \to \infty$, then $\xi(t) \to \{y \in \mathbb{R}^d: k(y) = 0\}$ w.p.1. as $t \to \infty$.

<u>Proof</u>: (i) By Dynkin's formula for the stopped process,

$$T_\lambda(t)V(x) - V(x) = \int_0^{t\wedge\tau_\lambda} T_\lambda(s)A_\lambda^w V(x)ds \leq 0.$$

Replace x by $\xi_\lambda(s)$ and use Equation (6.5) to obtain

$$E_{\xi_\lambda(s)}\{V(\xi_\lambda(t))\} \le V(\xi_\lambda(s)) \quad \text{w.p.1.}$$

As ξ_λ is Markov, we have, using Equation 6.4),

$$E\{V(\xi_\lambda(s+t))|\Sigma_s\} \le V(\xi_\lambda(s)) \quad \text{w.p.1.}$$

Thus $\{V(\xi_\lambda(t)): t \in \mathbb{R}^+\}$ is a nonnegative supermartingale. Consequently $V(\xi_\lambda(t))$ converge w.p.1. as $t \to \infty$ according to the martingale convergence theorem. Thus (i) is established.

(ii) As $V \in \mathscr{D}(A_\lambda^w)$ we must have that $\underset{t \downarrow 0}{w - \lim}\, E_x V(\xi_\lambda(t)) = V(x)$ from Dynkin's formula. Now use the supermartingale inequality to obtain (ii) whenever $x \in Q_\lambda$.

(iii) Since k is continuous on Q_λ there exists $\varepsilon_0 > 0$ so that whenever $0 < \varepsilon < \varepsilon_0$ there is a $\delta = \delta(\varepsilon) > 0$ with the property that $x \in P_{\lambda,\delta} \overset{\text{def}}{=} Q_\lambda \setminus B_\delta(H_\lambda)$ implies $k(x) \ge \varepsilon$. Choose $0 < \varepsilon_2 < \varepsilon_1 < \varepsilon_0$ and $\delta_2 = \delta(\varepsilon_2) < \delta_1 = \delta(\varepsilon_1)$ accordingly. Let $\eta_x(t,\delta_i)$ be the random variable which denotes the total time $\xi(s)$ spends in P_{λ,δ_i}, $i = 1,2$, during $[t \wedge \tau_\lambda, \tau_\lambda)$. Then $\eta_x(t,\delta_i) \to 0$ w.p.1 as $t \to \infty$, $i = 1,2$. Indeed, $E_x\eta_x(t,\delta_i)$ is finite since

$$T_\lambda(t)V(x) \ge V(x) \ge T_\lambda(\tau_\lambda)V(x) - T_\lambda(t)V(x)$$

$$= \int_{t\wedge\tau_\lambda}^{\tau_\lambda} T_\lambda(s)A_\lambda^w V(x)ds \ge E_x \int_{t\wedge\tau_\lambda}^{\tau_\lambda} k(\xi(s))ds \ge \delta_i E_x\eta_x(t,\delta_i).$$

Now suppose that there are paths $\xi(t,\omega)$ which remain in Q_λ for all $t \in \mathbb{R}^+$ but move back and forth between $B_{\delta_2}(H_\lambda)$ and $Q_\lambda \setminus B_{\delta_1}(H_\lambda)$ infinitely often in any interval $[t,\infty)$. These movements must occur within the total elapsed time of $\eta_x(t,\delta_2)$. We show the set of ω for which this occurs has probability zero.

Choose $h > 0$ so that $P_x\{\sup_{0 \le s \le h} |\xi(s)-x| < \delta_1 - \delta_2\} \ge \frac{1}{2}$ for every $x \in Q_\lambda \setminus B_{\delta_1}(H_\lambda)$. We may do this in view of Proposition 6.3(vi). In particular if $\xi(t) \in Q_\lambda \setminus B_{\delta_1}(H_\lambda)$, $\xi(s)$ remains outside of $B_{\delta_2}(H_\lambda)$ with probability $\ge 1/2$ whenever $s \in [t,t+h)$. But we may choose t so that $P_x\{\eta_x(t,\delta_2) > h\} < 1/2$, which obviously is a contradiction. We conclude that if $x \in Q_\lambda$, then $P_x\{k(\xi(s)) \ge \varepsilon_1$ for some $s \in [t,\infty)\} \to 0$ as $t \to \infty$. Thus whenever $\omega \in \Omega_\lambda$, $\xi(s) \in B_{\delta_1}(H_\lambda)$ for all $x \ge c$ for some random variable $c = c(\delta_1) < \infty$ w.p.1. As ε_1 was arbitrary and in view of the estimate in (ii), we conclude that (iii) is true.

(iv) The proof of this statement follows easily from parts (ii) and (iii) since $P_x(\Omega_\lambda) \to 1$ as $\lambda \to \infty$. □

We now demonstrate by way of an example how to apply the invariance principle (Theorem 7.2).

Example 7.6. Consider the system

$$\begin{aligned} dx_1 &= x_2 dt \\ dx_2 &= -g(x_1)dt - \alpha x_2 dt - \beta x_2 dW \end{aligned} \tag{7.5}$$

where $\int_0^x g(s)ds \to \infty$ as $x \to \infty$, $sg(s) > 0$ for $s > 0$, and $g(0) = 0$. We will prove that solutions of Equation (7.5) tend to the origin w.p.1. as $t \to \infty$.

Define

$$V(x_1,x_2) = x_2^2 + 2\int_0^{x_1} g(s)ds.$$

Then Q_λ is a bounded open subset of $\mathbb{R}^2$. The solution $\xi(t)$ of (7.5) can be defined up until the first exist from Q_λ; thereafter we consider the stopped process. The positive orbit of measures $\gamma_\lambda^+(\mu)$ of the stopped process is clearly precompact. Now $V \in \mathcal{D}(A_\lambda^W)$. In fact if $x \in Q_\lambda$ then

$$A_\lambda^w V(x) = x_2^2(\beta^2 - 2\alpha)$$

from Equation (7.4). When $\beta^2 \le 2\alpha$ then

$$P_x\{\sup_{t\in\mathbb{R}^+} V(\xi(t)) \ge \lambda\} \le \frac{V(x)}{\lambda} \to 0 \quad \text{as} \quad \lambda \to \infty.$$

Thus $\xi(t)$ is uniquely defined on $\mathbb{R}^+$ w.p.1.

If $\beta^2 < 2\alpha$, then in view of Theorem 7.5(iv) we find that $\xi(t) \to \{(x_1,x_2): x_2 = 0\}$ w.p.1. as $t \to \infty$. By considering the system (7.5) and the Invariance Principle we must have $\xi(t) \to (0,0)$ in pr. as $t \to \infty$. But we can obtain yet stronger convergence. Indeed, as $V(\xi(t))$ converges w.p.1. as $t \to \infty$, then $\xi(t) \to (0,0)$ w.p.1. as $t \to \infty$.

8. Exercises

8.1. Prove that $L^+(\mu)$ is nowhere dense for every $\mu \in \mathscr{M}$ for which $L^+(\mu) \neq \emptyset$.

8.2. Suppose the Markov process ξ remains in a compact set Q for all $t \in \mathbb{R}^+$. We can therefore assume that the semigroup T satisfies

$$T(t)C(Q;\mathbb{R}) \subset C(Q;\mathbb{R}), \quad t \in \mathbb{R}^+,$$

$$\lim_{t\downarrow 0} \|T(t)f - f\| = 0, \quad f \in C(Q;\mathbb{R}).$$

Let $\mathscr{M}_Q$ be the set of all probability measures on Q and $\mathscr{M}_Q^s$ the set $\mathscr{M}_Q$ with the strong topology: $\mathscr{M}_Q^s$ is a Banach space with the variation norm, $\|\mu\| = |\mu|(Q)|$. (The characterization, $\|\mu\| = \sup\{|\int f d\mu|: f \in C(Q;\mathbb{R}), \|f\| \le 1\}$ is useful.)

(a) Prove that the map $U: \mathscr{M}_Q^s \times \mathbb{R}^+ \to \mathscr{M}_Q^s$ given by $U(\mu,t) = U(t)\mu$ yields a semidynamical system $(U, \mathscr{M}_Q^s)$.

(b) Show that $U(t)$ is a contraction on $\mathscr{M}_Q^s$.

(c) Denote by $L_s^+(\mu)$ the strong limit set of $\mu \in \mathscr{M}_Q$. If $L_s^+(\mu) \neq \emptyset$, prove that $L_s^+(\mu)$ is positively minimal.

(d) Show that if $\mathscr{F} \subset \mathscr{M}_Q$ is positively minimal, then $U(t)$ is an isometry on $\mathscr{F}$.

(e) Prove that the restriction of U to $\Omega_s = \bigcup_{\mu \in \mathscr{M}_Q} L_s^+(\mu)$ yields a dynamical system (U, Ω_s). Show that $U(t)$ is an isometry on Ω_s.

(f) Consider $\mathscr{M}_Q$ with the w*-topology as in Section 2. Then every positive orbit in $\mathscr{M}_Q$ is asymptotically almost periodic.

8.3. Consider the semidynamical system $(\mathscr{M}, U)$ as defined in Section 3, and assume the following hold for some set $\mathscr{K} \subset \mathscr{M}$ and function $V: \mathbb{R}^d \to \mathbb{R}^+$:

(i) $\mathscr{K}$ is a compact invariant set,

(ii) $V \in \mathscr{D}(A^{w^2})$,

(iii) $(A^w V, \mu^2) > 0$ whenever $\mu \in \mathscr{M} - \mathscr{K}$ with $(A^w V, \mu) = 0$, and

(iv) $\mu \in \mathscr{K}$ implies $(V, \mu) = 0$.

(a) Show that if $\nu \in \mathscr{M} \setminus \mathscr{K}$, there exists at most one $t_0 > 0$ for which $(A^w V, U(t_0)\nu) = 0$.

(b) If such t_0 exists, prove that $(A^w V, U(t)\nu) > 0$ when $t > t_0$, and $(A^w V, U(t)\nu) < 0$ when $0 \leq t < t_0$.

(c) Let $\beta \in \mathscr{M}$. If $(V, U(t)\beta)$ is either non-increasing or non-decreasing for sufficiently large t, prove that (ν, μ) is constant for $\mu \in L^+(\beta)$.

(d) Prove that $L^+(\mu) \subset \mathscr{K}$ whenever $L^+(\mu) \neq \emptyset$, $\mu \in \mathscr{M} \setminus \mathscr{K}$.

(e) Suppose the semidynamical system $(U, \mathcal{M})$ is generated by a Markov process ξ which remains in a compact set Q. Assume (i) through (iv) are satisfied for $\mathcal{M}$ replaced by $\mathcal{M}_Q$. Show that if $\mu \in \mathcal{M} \setminus \mathcal{K}$, then $U(t)\mu \to \mathcal{K}$ as $t \to \infty$.

(f) Consider the scalar stochastic differential equation $dx(t) = f(x(t))dt + g(x(t))dW(t)$ where f and g satisfy the conditions of Proposition 6.3. Moreover let $f,g \in C^2$ with f'',g'' bounded. In addition let $f(0) = g(0) = 0$, $xf(x) > 0$ for $x \neq 0$ and $g(x) > 0$ for $x \neq 0$. Now suppose ξ_Q denotes the solution of the stochastic differential equation, stopped on the boundary of Q. Prove that the process ξ_Q satisfies conditions (i) through (iv) when $K = \{\delta_0\}$ (the Dirac measure at zero) and $V = x^2$ (on the interior of Q).

(g) Prove that the process $\xi_Q(t) \to 0$ in pr. as $t \to \infty$.

8.4. Let $u(t,x) = T(t)h(x)$, $h \in C^2(\mathbb{R}^d; \mathbb{R})$. Show that $h \in \mathcal{D}(A^W)$ and $\frac{\partial u}{\partial t} + A^W u = 0$.

8.5. Let $\mathcal{S} = \{1,2,\ldots,N\}$. Then $C(\mathcal{S}; \mathbb{R}) = \mathbb{R}^N$ and $\mathcal{M} = \{x \in \mathbb{R}^N: \sum_{i=1}^{N} x_i = 1, x_i \geq 0\}$. The strong and w^* topologies on $\mathcal{M}$ agree. Replace the parameter $\mathbb{R}^+$ by the nonnegative integers $\mathbb{Z}^+$. A finite state Markov chain is generated by a stochastic matrix $P = [p_{ij}]_{i,j=1}^N$; i.e., $p_{ij} \geq 0$ with $\sum_{j=1}^{N} p_{ij} = 1$ for each $i = 1,2,\ldots,N$. We can define a (discrete) parameter Markov transition operator $\mathcal{P}$ by $[\mathcal{P}(n,i,j)]_{i,j=1}^N$, $P^0 = I$. Assume the matrix P is irreducible. It is a basic property of Markov chains that $\lim_{k\to\infty} P^{kd}$ exists for some positive integer d. Prove that

(a) the positive trajectories for the Markov chain have form $\{\nu P^n\}_{n=0}^{\infty}$ for $\nu \in \mathcal{M}$.

(b) the positive limit sets are of the form $\{\mu, \mu P, \ldots, \mu P^{d-1}\}$, for some $\mu \in \mathcal{M}$, where $\mu P^d = \mu$ and d is minimal with this property.

9. Notes and Comments

Sections 3-5. Theorem 3.2 is from Saperstone [1]. Also see Kushner [1,2]. Theorem 4.4 is also due to Saperstone [1]. Theorem 4.6 is from Boyarski [1]. Theorem 5.1 is based on a result of Benes [1,2].

Sections 6-7. Theorem 6.5 is due to Miyahara [1]. Theorem 7.1 is from Kushner [2], as well as the asymptotic stability result of Theorem 7.5. Also see Boyarski [2] and Wonham [1]. Example 7.6 is also from Kushner [2].

CHAPTER VIII

WEAK SEMIDYNAMICAL SYSTEMS AND PROCESSES

1. Introduction

Many of the important properties of semidynamical systems which were developed in the first three chapters can essentially be obtained with a weaker continuity axiom (Definition 2.1(iii) of Chapter I); namely, assume that $\pi(x,t)$ is only continuous in $x \in X$. In particular, we still obtain weak invariance of compact positive limit sets. In addition, if the continuity in x is uniform with respect to $t \in \mathbb{R}^+$, then (X,π) extends to a weak dynamical system on the positive limit sets. Moreover, the positive limit sets will then be minimal with respect to this flow. Finally, we will still be able to show, as in Chapter III, that the positive limit sets are equi-almost periodic.

One motivation behind this development is to establish some general criteria for the stability of solutions of some evolutionary systems. By weakening the continuity axiom we are still able to establish stability in a wide variety of examples without requiring the positive motions to be continuous in $t \in \mathbb{R}^+$. This affects a considerable savings in the stability analysis of solutions.

Another motive can be found by considering nonautonomous evolutionary systems. It was seen in Chapter IV how to create a semidynamical system associated with the solution of a nonautonomous ordinary differential equation. One reason for this came from the fact that the positive limit set of such a solution through an initial value (x_0, t_0) need not be even weakly invariant. Consequently, a new type of invariance called quasi-invariance was established. This resulted in Theorem 4.17 and the Invariance Principle, Theorem 4.23, of Chapter IV. In order to do something similar for other nonautonomous systems, say partial differential equations, we present a general framework so that from the solutions of these systems we can create a weak semi-dynamical system. Quasi-invariance is obtained for which Theorem 4.17 of Chapter IV is a special case.

In Section 2 we develop the essential properties of weak semidynamical systems along the lines of Chapter I and II. Section 3 introduces the notion of a compact process, due to Dafermos. We show that a process generates a weak semidynamical system, and characterize the resulting limit sets. We do the same in Section 4 for uniform processes. In Section 5 we revisit the nonautonomous ordinary differential equations of Chapter IV and establish a weak invariance principle. Finally in Section 6 we study the stability of solutions of a wave equation.

2. Weak Semidynamical Systems

This presentation will closely follow the development of the first three chapters. Since our objective is only to establish certain properties positive limit sets, we define only those concepts which will be needed in the sequel. The reader may generalize other notions from the earlier chapters. One of the concepts we will require is Lyapunov stability. As we cannot expect our phase space to be metric, we will take it to be uniform. Lyapunov stability may be readily defined in such spaces. The reader who is not familiar with uniform spaces may consult Kelley [1], or the brief review in Appendix A. Additionally, all closure operations will be taken with respect to only sequential limits. Thus completeness, compactness, continuity, etc., will be in the sequential sense instead of the framework of arbitrary nets. In order to avoid confusion, the sequential closure of a set M will be denoted by $\overline{M}^s$.

Definition 2.1. Let $(Y,\mathscr{U})$ be a uniform space and π a mapping from $Y \times \mathbb{R}^+$ to Y where

(i) $\pi(y,0) = y$ for each $y \in Y$,

(ii) $\pi(y,s+t) = \pi(\pi(y,s),t)$, for each $y \in Y$, $s,t \in \mathbb{R}^+$,

(iii) $\pi^t\colon Y \to Y$ is sequentially continuous in the topology of the uniformity $\mathscr{U}$ for each $t \in \mathbb{R}^+$.

The triple $(Y,\mathscr{U},\pi)$ is called a *weak semidynamical system* on Y. If $\mathbb{R}^+$ is replaced by $\mathbb{R}$, the triple $(Y,\mathscr{U},\pi)$ is called a *weak dynamical system* on Y. At times we will just refer to the triple by π.

We employ the same notation for π as developed in Chapter I; that is, $\pi^t = \pi(\cdot,t)$, $yt = \pi(y,t)$. Positive orbits, hulls, and limit sets are also as defined in Chapters I and II. We denote the corresponding sets by $\gamma^+(y)$, $H^+(y)$, and $L^+(y)$ respectively. Again, we recall the closures are sequential. The concepts of invariance and minimality are also unchanged. In particular, an examination of the proofs of the results of Chapter I and II shows that many of the conclusions of these theorems are true for weak semidynamical systems. We record here those properties which will be needed later.

Proposition 2.2. For every $y \in Y$ we have $H^+(y) = \gamma^+(y) \cup L^+(y)$. Also, $H^+(y)$ and $L^+(y)$ are sequentially closed, positively invariant sets.

Proposition 2.3. If $H^+(y)$ is sequentially compact for some $y \in Y$, then $L^+(y)$ is nonempty, sequentially compact, and weakly invariant.

Definition 2.4. A weak semidynamical system $(Y, \mathcal{U}, \pi)$ is called *Lyapunov stable* if for every $V \in \mathcal{U}$ there exists $U \in \mathcal{U}$ such that $(x,y) \in U$ implies $(xt,yt) \in V$ for all $t \in \mathbb{R}^+$.

We introduce a new concept.

Definition 2.5. A nonempty subset $M \subset Y$ is called *strongly invariant* for $(Y, \mathcal{U}, \pi)$ if for each $t \in \mathbb{R}^+$, π^t is a (sequential) homeomorphism of M onto M.

Remark 2.6. If $M \subset Y$ is strongly invariant for $(Y, \mathcal{U}, \pi)$, it follows that $(\pi^t)^{-1} = \pi^{-t}$ on M for every $t \in \mathbb{R}^+$. Thus $(M, \mathcal{U}, \pi)$ is a weak dynamical system.

Theorem 2.7. Suppose $(Y, \mathcal{U}, \pi)$ is a Lyapunov stable weak semidynamical system. If $(Y, \mathcal{U})$ is sequentially complete, then every nonempty positive limit set $L^+(y)$, $y \in Y$, is strongly invariant and minimal.

Proof: Let $x \in L^+(y)$. Then $L^+(x) \subset L^+(y)$. Now let $z \in L^+(y)$. There exist sequences $\{t_n\}$, $\{s_n\} \subset \mathbb{R}^+$ with $t_n \to \infty$, $s_n \to \infty$ such that $yt_n \to x$, $ys_n \to z$. We may assume (by choosing subsequences if necessary) that $\tau_n \overset{\text{def}}{=} s_n - t_n \geq n$ for each $n \in \mathbb{N}$. We claim $x\tau_n \to z$. Let $V \in \mathcal{U}$ and pick $U \in \mathcal{U}$ so that $U^2 \subset V$. Then $((yt_n)\tau_n, z) \in U$ for all sufficiently large $n \in \mathbb{N}$. By Lyapunov stability of $(Y, \mathcal{U}, \pi)$ we can assume $(yt_n, x) \in U$ implies $((yt_n)\tau_n, x\tau_n) \in U$. Hence $(x\tau_n, z) \in U^2 \subset V$ for all sufficiently large $n \in \mathbb{N}$, so $z \in L^+(x)$. We have shown $L^+(x) = L^+(y)$.

Next we show that $zt_n \to z$ for every $z \in L^+(y)$, where $\{t_n\}$ is the sequence chosen earlier. Recall also that $x\tau_n \to z$. Let $V \in \mathcal{U}$ and choose $U \in \mathcal{U}$ such that $U^3 \subset V$. Again, by Lyapunov stability we can assume $(x\tau_n, z) \in U$ and $(xt_n, x) \in U$ imply $((x\tau_n)t_n, zt_n) \in U$ and $((xt_n)\tau_n, x\tau_n) \in U$. Therefore, $(zt_n, z) \in V$ for all sufficiently large $n \in \mathbb{N}$, so $zt_n \to z$.

Fix $t \in \mathbb{R}^+$. We show π^t is one-to-one on $L^+(y)$. Suppose $xt = zt$ for some $x, z \in L^+(y)$. Then $xt_n \to x$, $zt_n \to z$. As $x(t_n - t)$ is well defined for large enough n,

$$x = \lim_{n\to\infty} xt_n = \lim_{n\to\infty} (xt)(t_n - t) = \lim_{n\to\infty} (zt)(t_n - t) = \lim_{n\to\infty} zt_n = z.$$

Hence π^t is one-to-one on $L^+(y)$.

To see that π^t maps onto $L^+(y)$ fix $z \in L^+(y)$ and again choose large enough $m \in \mathbb{N}$ such that $z(t_m - t)$ is well defined. We claim $\{z(t_m-t)\}$ is Cauchy. Let $V \in \mathcal{U}$ and choose $U \in \mathcal{U}$ so that $U \subset V$. As $z(t_m-t) \in L^+(y)$, we have $((zt_n)(t_m-t), z(t_m-t)) \in U$ for all sufficiently large n. As $zt_m \to z$, then by Lyapunov stability we obtain $((zt_m)(t_n-t), z(t_n-t)) \in U$ for all $t_n \geq t$. Thus $(z(t_m-t), z(t_n-t)) \in U^2 \subset V$ for all sufficiently large m,n. Consequently by the sequential completeness of the uniformity $\mathcal{U}$, there exists $x \in L^+(y)$ such that $z(t_m-t) \to x$. Therefore $xt = \lim_{n\to\infty} z(t_n-t)t = \lim_{n\to\infty} zt_n = z$. Thus π^t maps $L^+(y)$ in a one-to-one fashion onto itself. We may write $\pi^{-1}(z,t) = \pi^{-t}(z) = z(-t)$.

Next we establish the sequential continuity of π^{-t} on $L^+(y)$. Let $\{z_m\} \subset L^+(y)$ converge to $z \in L^+(y)$. Then $z(t_n-t) \to z(-t)$ and $z_m(t_n-t) \to z_m(-t)$ as $n \to \infty$. Suppose $U \in \mathcal{U}$ and choose $V \in \mathcal{U}$ with $V^3 \subset U$. For each $m \in N$ and sufficiently large $n \in N$, $(z_m(t_n-t), z_m(-t)) \in V$, $(z(t_n-t), z(-t)) \in V$. Also for sufficiently large $m \in N$, $(z_m, z) \in V$. Again, by Lyapunov stability we obtain $(z_m(t_n-t), z(t_n-t)) \in V$ for $t_n \geq t$. Then $(z_m(-t), z(-t)) \in V^3 \subset U$, and so $z_m(-t) \to z(-t)$. This establishes that $L^+(y)$ is strongly invariant. In view of Definition 4.1 of Chapter II, then $L^+(y)$ must also be minimal. □

Definition 2.8. Suppose $M \subset Y$ is strongly invariant for the weak semidynamical system $(Y, \mathcal{U}, \pi)$. M is called *equi-almost periodic* if for every $V \in \mathcal{U}$, the set $\{\tau \in \mathbb{R}: (y\tau, y) \in V$ for each $y \in M\}$ is relatively dense in $\mathbb{R}$.

If $M = \gamma^+(y)$ for some $y \in Y$ so that M is only positively invariant, then Definition 2.8 suggests that we call the motion through y *(positively) almost periodic* provided for each $V \in \mathcal{U}$, the set $\{\tau \in \mathbb{R}^+: (y(t+\tau), yt) \in V$ for all $t \in \mathbb{R}^+\}$ is relatively dense in $\mathbb{R}^+$.

Theorem 2.9. Suppose $(Y, \mathcal{U}, \pi)$ is a Lyapunov stable weak semidynamical system. If for some $y \in Y$, $H^+(y)$ is sequentially compact, then $L^+(y)$ is equi-almost periodic.

Proof: Fix $z \in L^+(y)$. There exists a sequence $\{s_n\} \subset \mathbb{R}^+$ with $s_n \to \infty$ so that $ys_n \to z$. First we show that for every $U \in \mathcal{U}$ there exists $L > 0$ so that every interval of length L in $\mathbb{R}^+$ contains a point τ with $(x\tau, x) \in U$ for every $x \in \gamma^+(z)$. If not, there exists $U \in \mathcal{U}$ such that for every $n \in \mathbb{N}$ there exists $x_n \in \gamma^+(z)$ and a sequence $\{t_n\} \subset \mathbb{R}^+$ with $(x_n(t_n+\tau), x_n) \notin U$ for every $\tau \in [0,n]$. Write $y_n = z\tau_n$ and choose $V \in \mathcal{U}$ so that $V^3 \subset U$. The sequential compactness of $H^+(y)$ ensures that $\{y(s_n+\tau_n)\}$ has a cluster point $u \in L^+(y)$ and $\{y(s_n+\tau_n+t_n)\}$ has a cluster point $v \in L^+(y)$. Choose $W \in \mathcal{U}$ such that $W^2 \subset V$. There exists $m \in \mathbb{N}$ so that $(y(s_m+\tau_m), u) \in W$, $(y(s_m+\tau_m+t_m), v) \in W$, and $(ys_n, z) \in W$. By Lyapunov stability we may assume $(y(s_m+\tau_m), z\tau_m) = (y(s_m+\tau_m), y_m) \in W$ and $(y(s_m+\tau_m+t_m), y_mt_m) \in W$. Then $(y_m, u) \in V$ and $(y_mt_m, v) \in V$. Since $L^+(y)$ is positively minimal, then there exists $\tau \in \mathbb{R}^+$ with $(v\tau, u) \in V$. In view of the continuity of π^τ, we may assume $(y_m(t_m+\tau), v\tau) \in V$. Then $(y_m(t_m+\tau), y_m) \in V^3 \subset U$. As $y_m = z\tau_m \in \gamma^+(z)$, we have a contradiction. Thus the positive motion through each $x \in \gamma^+(z)$, $z \in L^+(y)$, is almost periodic.

Now let $x \in L^+(y)$, and choose a sequence $\{x_n\} \subset \gamma^+(z)$ so that $x_n \to x$. This is possible since $\overline{\gamma^+(z)}^s = L^+(y)$. We have already shown that for every $U \in \mathcal{U}$ there exists $L > 0$ such that every interval of length L contains a point τ with $(x_n\tau, x_n) \in U$ for every $n \in \mathbb{N}$. The continuity of π^τ implies $(x\tau, x) \in U$. This shows $L^+(y)$ is equi-almost periodic. □

3. Compact Processes

Definition 3.1. Suppose (X,d) is a complete metric space. A *process* on X is a mapping $u: \mathbb{R} \times X \times \mathbb{R}^+ \to X$ which satisfies

(i) $u(\tau,x,0) = x$ for each $\tau \in \mathbb{R}$, $x \in X$,

(ii) $u(\tau,x,s+t) = u(\tau+s,u(\tau,x,s),t)$ for each $\tau \in \mathbb{R}$, $x \in X$, and $s,t \in \mathbb{R}^+$,

(iii) for each fixed $t \in \mathbb{R}^+$ the family of mappings

$$u(\tau,\cdot,t): X \to X,$$

parametrized by $\tau \in \mathbb{R}$ is equicontinuous.

We may think of $u(\tau,x,t)$ as the solution of a nonautonomous equation of evolution with initial value x at time τ.

We endow the collection of functions $X^{\mathbb{R}\times X\times\mathbb{R}^+}$ with a uniform structure under which it is a complete topological space. Let $\mathcal{Q}$ denote the family of subsets of $\mathbb{R} \times X \times \mathbb{R}^+$ of the form $[\tau,\infty) \times \{x\} \times \{t\}$. Denote by $\mathcal{U}$ the uniformity of Y of uniform convergence on members of $\mathcal{Q}$; that is, $\mathcal{U}$ has a subbase the collection of all sets of the form

$$U_{r,Q} = \{(u,v): d(u(\tau,x,t),v(\tau,x,t)) < r \text{ for all } (\tau,x,t) \in Q\}$$

for $r > 0$ and $Q \in \mathcal{Q}$ (c.f. Kelley [1]). If $\mathcal{Q}$ just consists of the singletons of the form $Q = \{(\tau,x,t)\}$, then $\mathcal{U}$ is called the uniformity of pointwise convergence. The topology of the uniformity $\mathcal{U}$ on $X^{\mathbb{R}\times X\times\mathbb{R}^+}$ has as a subbase for the neighborhood system at each $v \in X^{\mathbb{R}\times X\times\mathbb{R}^+}$ the family of sets

$$U(v) = \{u \in X^{\mathbb{R}\times X\times\mathbb{R}^+} : (u,v) \in U\}$$

for every $U \in \mathscr{U}$. Consequently, the net $\{u_\alpha\} \subset X^{\mathbb{R}\times X\times\mathbb{R}^+}$ converges to $u \in X^{\mathbb{R}\times X\times\mathbb{R}^+}$ provided $u_\alpha(\tau,x,t) \to u(\tau,x,t)$, the convergence being uniform on the sets $Q \in \mathscr{Q}$. Since (X,d) is complete, then so is $(X^{\mathbb{R}\times X\times\mathbb{R}^+}, \mathscr{U})$. Unless otherwise indicated, we shall assume $\mathscr{Q}$ consists of the sets $[\tau,\infty) \times \{x\} \times \{t\}$.

For any process u and $s \in \mathbb{R}$ we call u_s the s-translate of u, where

$$u_s(\tau,x,t) = u(\tau+s,x,t).$$

Obviously, u_s is also a process.

Definition 3.2. The *hull* of a process u on X is the set

$$\mathscr{H}[u] = \{v \in X^{\mathbb{R}\times X\times\mathbb{R}^+} : v = \lim_{n\to\infty} u_{s_n} \text{ for some sequence } \{s_n\} \subset \mathbb{R}^+\}.$$

The *asymptotic hull* of u is the set

$$\mathscr{L}[u] = \{v \in \mathscr{H}[u] : v = \lim_{n\to\infty} u_{s_n} \text{ for some sequence } \{s_n\} \subset \mathbb{R}^+,\ s_n \to \infty\}.$$

Note that we are taking the sequential closure and limit in Definition 3.2. The space $(X^{\mathbb{R}\times X\times\mathbb{R}^+}, \mathscr{U})$ need not be first countable.

Lemma 3.3. Suppose u is a process on X. If $v \in \mathscr{H}[u]$, then v is also a process on X. Furthermore, for each fixed $t \in \mathbb{R}^+$, the family of mappings from X to X

$$\{v(\tau,\cdot,t) : v \in \mathscr{H}[u],\ \tau \in \mathbb{R}\}$$

is equicontinuous.

Proof: If $v \in \mathscr{A}[u]$, there exists a sequence $\{s_n\} \subset \mathbb{R}^+$ such that $u_{s_n} \to v$, uniformly on each $Q \in \mathscr{Q}$. We verify that v is a process. First, as $u_{s_n}(t,x,0) = x$ for all s_n, $t \in \mathbb{R}$, and $x \in X$, then $v(t,x,0) = \lim_{n\to\infty} u_{s_n}(t,x,0) = x$. Second, fix $(t,x,\tau) \in \mathbb{R} \times X \times \mathbb{R}^+$ and $s \in \mathbb{R}^+$. We have

$$\begin{aligned} d(v(t,x,s+\tau),v(t+s,v(t,x,s),\tau) \\ &\leq d(v(t,x,s+\tau),u_{s_n}(t,x,s+\tau)) \\ &\quad + d(u_{s_n}(t,x,s+\tau),u_{s_n}(t+s,v(t,x,s),\tau) \\ &\quad + d(u_{s_n}(t+s,v(t,x,s),\tau),v(t+s,v(t,x,s),\tau)). \end{aligned}$$

Since $u_{s_n} \to v$, the first and third terms on the right side of the last inequality tend to zero as $n \to \infty$. Using property (ii) of Definition 3.1 we have

$$\begin{aligned} u_{s_n}(t,x,s+\tau) &= u(t+s_n,x,s+\tau) \\ &= u(t+s+s_n,u(t+s_n,x,s),\tau) \\ &= u_{s_n}(t+s,u_{s_n}(t,x,s),\tau). \end{aligned}$$

The second term in the last inequality also tends to zero in view of the equicontinuity property (iii) of Definition (3.1). Thus we have shown v satisfies property (ii) of Definition 3.1.

Now fix $t \in R^+$. We establish the family $\{v(\tau,\cdot,t)\colon \tau \in \mathbb{R}\}$ is equicontinuous. Suppose $(x_0,t_0) \in X \times \mathbb{R}^+$ and $\varepsilon > 0$. Choose $\delta = \delta(x_0,t_0,\varepsilon) > 0$ so that

$$d(u(\tau,x,t),u(\tau,x_0,t_0)) < \varepsilon \quad \text{whenever} \quad d(x,x_0) + |t-t_0| < \delta,$$

uniformly in $\tau \in \mathbb{R}$. We have

$$d(v(\tau,x,t),v(\tau,x_0,t_0) \le d(v(\tau,x,t),u_{s_n}(\tau,x,t))$$
$$+ d(u(\tau+s_n,x,t),u(\tau+s_n,x_0,t_0))$$
$$+ d(u_{s_n}(\tau,x_0,t_0),v(\tau,x_0,t_0)).$$

Take the infimum over $n \in \mathbb{N}$ of the right side of the last inequality. As the estimate is independent of $\tau \in \mathbb{R}$, it cannot exceed ε. This finally establishes v is a process.

Next we observe that for each $(\tau,x,t) \in \mathbb{R} \times X \times \mathbb{R}^+$ and $s \in \mathbb{R}^+$,

$$\lim_{n\to\infty} u_{s+s_n}(\tau,x,t) = \lim_{n\to\infty} u_{s_n}(\tau+s,x,t) = v(\tau+s,x,t) = v_s(\tau,x,t).$$

Hence $v_s \in \mathscr{A}[u]$ whenever $v \in \mathscr{A}[u]$. Note that the choice of δ above is independent of v. Indeed, the sequence $\{s_n\}$ depends upon v, but the estimate derived from the last inequality does not depend upon $v \in \mathscr{A}[u]$ as well. Consequently we have shown that for each fixed $t \in \mathbb{R}^+$, the family of mappings $\{v(\tau,\cdot,t): v \in \mathscr{A}[u],\ \tau \in \mathbb{R}\}$ is equicontinuous. □

We take $\mathscr{A}[u]$ with the relativization of $\mathscr{U}$. Note that $\mathscr{A}[u]$ is not closed in $X^{\mathbb{R}\times X\times\mathbb{R}^+}$, as $\mathscr{A}[u]$ only contains the sequential limits of the set of translates $\{u_s: s \in \mathbb{R}^+\}$. But as $(X^{\mathbb{R}\times X\times\mathbb{R}^+},\mathscr{U})$ is complete, then $(\mathscr{A}[u],\mathscr{U})$ is sequentially complete. The following theorem is an immediate consequence of the last lemma. The proof is obvious.

<u>Theorem 3.4</u>. Suppose u is a process on X. Then the mapping $\pi^*: \mathscr{A}[u] \times \mathbb{R}^+ \to \mathscr{A}[u]$ given by $\pi^*(v,s) = v_s$ defines a weak semidynamical system $(\mathscr{A}[u],\mathscr{U},\pi^*)$.

<u>Corollary 3.5</u>. Suppose u is a process on X. Then $(\mathscr{A}[u],\mathscr{U},\pi^*)$ is Lyapunov stable.

Proof: Let $v,w \in \mathscr{A}[u]$. As we have

$$d(v_s(\tau,x,t),w_s(\tau,x,t)) = d(v(s+\tau,x,t),w(s+\tau,x,t)),$$

then $(v,w) \in U$ for some $U \in \mathscr{U}$ implies $(v_s,w_s) \in U$ for every $s \in \mathbb{R}^+$. □

Definition 3.6. Suppose u is a process on X. We call u *compact* if $\{u_s: s \in \mathbb{R}^+\}$ is sequentially relatively compact in $(\mathscr{A}[u],\mathscr{U})$.

Remark 3.7. Note that $\mathscr{L}[u] = L^+(u)$, where $L^+(u)$ is the (sequential) positive limit set of the positive motion through $u \in \mathscr{A}[u]$. Also observe that $(\mathscr{A}[u],\mathscr{U})$ is sequentially complete.

Lemma 3.8. Suppose u is a compact process on X.

(i) $\mathscr{L}[u]$ is minimal, strongly invariant, and equi-almost periodic in $(\mathscr{A}[u],\mathscr{U},\pi^*)$.

(ii) If $v \in \mathscr{L}[u]$, then v is also a compact process and $\mathscr{L}[v] = \mathscr{L}[u]$.

Proof: (i) These properties of $\mathscr{L}[u]$ follow directly from Theorems 2.7, 2.9, and Remark 3.7.

(ii) If $v \in \mathscr{L}[u]$, there exists a sequence $\{\tau_n\} \subset \mathbb{R}^+$ with $\tau_n \to \infty$ such that $u_{\tau_n} \to v$. Let $\{s_n\}$ be any sequence in $\mathbb{R}^+$, and set $r_n = \tau_n + s_n$. Since u is a compact process, we may assume (by choosing a subsequence if necessary) that $u_{r_n} \to w$. For any $(\tau,x,t) \in \mathbb{R} \times X \times \mathbb{R}^+$, we have

$$d(v_{s_n}(\tau,x,t),w(\tau,x,t))$$

$$\leq d(v(\tau+s_n,x,t),u_{\tau_n}(\tau+s_n,x,t)) + d(u_{r_n}(\tau,x,t),w(\tau,x,t)).$$

Consequently, $v_{s_n}(\tau,x,t) \to w(\tau,x,t)$, the convergence being uniform on sets $Q \in \mathscr{Q}$. Thus $v_{s_n} \to w$ in $(X^{\mathbb{R}\times X\times\mathbb{R}^+},\mathscr{U})$, so v is a compact process.

Finally $\mathscr{L}[v] = \mathscr{L}[u]$ as these are just positive limit sets which are minimal in $(\mathscr{A}[u],\mathscr{U},\pi^*)$. □

We may interpret Lemma 3.8 as follows. If u is a compact process, then for each $x \in X$, $t \in \mathbb{R}^+$ and $\varepsilon > 0$, the set $\Gamma_\varepsilon = \{s \in \mathbb{R}: d(v(0,x,t),v(s,x,t)) < \varepsilon$ for each $v \in \mathscr{L}[u]\}$ is relatively dense in $\mathbb{R}$. This enables us to establish a stronger convergence in $X^{\mathbb{R}\times X\times\mathbb{R}^+}$. Note that a process u is compact on X if and only if for every sequence $\{\tau_n\} \subset \mathbb{R}^+$, there exists a subsequence $\{\tau_{n_k}\}$ and some process v on X such that

$$u_{\tau_{n_k}}(\tau,x,t) \to v(\tau,x,t) \quad \text{for all} \quad (\tau,x,t) \in \mathbb{R} \times X \times \mathbb{R}^+,$$

uniformly in τ on every subset of $\mathbb{R}$ which is bounded from below.

Definition 3.9. A process u on X is called *uniformly compact* if for every sequence $\{\tau_n\} \subset \mathbb{R}^+$, there exists a subsequence $\{\tau_{n_k}\}$ and some process v on X such that

$$u_{\tau_{n_k}}(\tau,x,t) \to v(\tau,x,t) \quad \text{for all} \quad (\tau,x,t) \in \mathbb{R} \times X \times \mathbb{R}^+,$$

uniformly in τ on $\mathbb{R}$.

Lemma 3.10. Suppose u is a compact process on X. Then every $v \in \mathscr{L}[u]$ is uniformly compact.

Proof: Let $w \in \mathscr{L}[v] = \mathscr{L}[u]$, say $w = \lim_{n\to\infty} v_{s_n}$ for some sequence $\{s_n\} \subset \mathbb{R}^+$ with $s_n \to \infty$. Consequently, $d(v_{s_n}(\tau,x,t),w(\tau,x,t)) \to 0$, uniformly in τ belonging to subsets of $\mathbb{R}^+$ which are bounded from below. Now fix $x \in X$, $\tau \in \mathbb{R}$, and choose s in the relatively dense set $\Gamma_\varepsilon \subset \mathbb{R}$. We have

$$d(v_{s_n}(\tau,x,t),w(\tau,x,t)) \le d(v_{s_n+\tau}(0,x,t),v_{s_n+\tau+s}(0,x,t))$$

$$+ d(v_{s_n}(\tau+s,x,t),w(\tau+s,x,t)) + d(w_{\tau+s}(0,x,t),w_\tau(0,x,t)).$$

As $\mathscr{L}[u]$ is strongly invariant, then both $v_{s_n+\tau}$ and $w_t \in \mathscr{L}[u]$. Consequently, $v_{s_n}(\tau,x,t) \to w(\tau,x,t)$, where the convergence is uniform in τ on $\mathbb{R}$. □

We may associate with a process u on X another semidynamical system with phase space $\mathscr{A}[u] \times X$. Note that $\mathscr{A}[u] \times X$ is a uniform space under the product uniformity $\mathscr{V}$ from the relativization of $\mathscr{U}$ to $\mathscr{A}[u]$ and the metric space (X,d). Define a mapping $\pi\colon \mathscr{A}[u] \times X \times \mathbb{R}^+ \to \mathscr{A}[u] \times X$ by

$$\pi(v,x,t) = (v_t, v(0,x,t)).$$

Theorem 3.11. Suppose u is a process on X. Then the triple $(\mathscr{A}[u] \times X, \mathscr{V}, \pi)$ is a weak semidynamical system.

Proof: Property (i) of Definition 2.1 is obvious. Secondly, $\pi(v,x,s+t) = (v_{s+t}, v(0,x,s+t)) = (v_{s+t}, v(s,v(0,x,s),t) = (v_{s+t}, v_s(0,v(0,x,s),t)) = \pi(v_s, v(0,x,s),t) = \pi(\pi(v,x,s),t)$ holds in view of property (ii) of Definition 3.1. Thus property (ii) of Definition 2.1 is satisfied. Finally, let $\{(v_n,x_n)\}$ converge to (v,x) in $\mathscr{A}[u] \times X$. Then $\pi(v_n,x_n,t) = ((v_n)_t, v_n(0,x_n,t))$. We have

$$d(v_s(\tau,x,t),(v_n)_s(\tau,x,t)) = d(v(\tau+s,x,t),v_n(\tau+s,s,t) \to 0.$$

uniformly in $s \in \mathbb{R}^+$ according to convergence in $X^{\mathbb{R}\times X\times\mathbb{R}^+}$.

Consequently, $(v_n)_t \to (v)_t$. Also,

$$d(v_n(0,x_n,t),v(0,x,t)) \leq d(v_n(0,x_n,t),v_n(0,x,t)) + d(v_n(0,x,t),v(0,x,t)).$$

Now $d(v_n(0,x_n,t),v_n(0,x,t)) \to 0$ from Lemma 3.3. Also $d(v_n(0,x,t),v(0,x,t)) \to 0$ since $v_n \to v$. Therefore property (iii) of Definition 2.1 is confirmed. □

As in Chapter IV, our objective here is to obtain some characteristics of the process u from properties of the associated weak semidynamical system $(\mathscr{A}[u] \times X, \mathscr{V},\pi)$.

Definition 3.12. Suppose u is a process on X. The *positive motion* of u through $(\tau,x) \in \mathbb{R} \times X$ is the mapping $u(\tau,x,\cdot)$: $\mathbb{R}^+ \to X$. The *positive orbit* of u through (τ,x) is the set $\cup\{u(\tau,x,t)\colon t \in \mathbb{R}^+\}$. The *positive limit set* of u through (τ,x) is the set

$$\Omega_u^+(\tau,x) = \{y \in X\colon y = \lim_{n\to\infty} u(\tau,x,t_n) \text{ for some } \{t_n\} \subset \mathbb{R}^+, t_n \to \infty\}$$

We are now ready for the invariance principle for compact processes.

Theorem 3.13. Suppose u is a compact process on X, and let

$$y = \lim_{n\to\infty} u(0,x,t_n) \in \Omega_u^+(0,x),$$

$$v = \lim_{n\to\infty} u_{t_n} \in \mathscr{L}[u].$$

Then $v(0,y,t) \in \Omega_u^+(0,x)$ for all $t \in \mathbb{R}^+$. Moreover, if the positive orbit of u through $(0,x)$ is precompact in X, there is a mapping $\hat{v}(y,\cdot)\colon \mathbb{R} \to X$ such that $\hat{v}(y,0) = y$ and $\hat{v}(y,s+t) = v(s,\hat{v}(y,s),t)$ for every $y \in \Omega_w^+(0,x)$, $s \in \mathbb{R}$, and $t \in \mathbb{R}^+$.

Proof: $(v,y) \in \mathscr{L}[u] \times \Omega_u^+(0,x) \subset L^+(u,x)$, the positive limit set of (u,x) in the system $(\mathscr{U}[u] \times X, \mathscr{V}, \pi)$. According to Proposition 2.2, $L^+(u,x)$ is positively invariant. As $\pi(x,y,t) = (v_t, v(0,y,t))$, then $v(0,y,t) \in \Omega_u^+(0,x)$ for all $t \in \mathbb{R}^+$.

Now suppose the positive orbit of u through $(0,x)$ is precompact. Then $\Omega_u^+(0,x)$ is compact as (X,d) is complete. As u is a compact motion on X, then $L^+(u,x) = \mathscr{L}[u] \times \Omega_u^+(0,x)$ is also compact. (See Lemma 4.4 of Chapter IV.) Proposition 2.3 says that $L^+(u,x)$ is weakly invariant. So there must exist $\hat{\pi}: L^+(u,x) \to L^+(u,x)$ so that $\hat{\pi}(v,y,0) = (v,y)$ and $\hat{\pi}(v,y,s+t) = \pi(\hat{\pi}(v,y,s),t)$ for every $s \in \mathbb{R}$, $t \in \mathbb{R}^+$. Denote by $\mathscr{P}$ the projection of $\mathscr{U}[u] \times X$ onto X and define $\hat{v}(y,\cdot): \mathbb{R} \to X$ by

$$\hat{v}(y,t) = \mathscr{P}(\hat{\pi}(v,y,t)).$$

First, $\hat{v}(y,0) = \mathscr{P}(\hat{\pi}(v,y,0)) = \mathscr{P}(v,y) = y$. Second, as $\hat{\pi}(v,y,t) = (v_t, \hat{v}(y,t))$ for all $t \in \mathbb{R}$, then

$$\begin{aligned}\hat{v}(y,s+t) &= \mathscr{P}(\hat{\pi}(v,y,s+t)) = \mathscr{P}(\pi(\hat{\pi}(v,y,s),t)) \\ &= \mathscr{P}(\pi(v_s,\hat{v}(y,s),t)) = \mathscr{P}(v_{s+t}, v_s(0,\hat{v}(y,s),t) \\ &= v_s(0,\hat{v}(y,s),t)) = v(s,\hat{v}(y,s),t)).\end{aligned}$$ □

Remark 3.14. Theorem 3.13 postulates a property which was called quasi-invariance in Chapter IV (Definition 4.10). Also observe that $\hat{v}(y,\cdot)$ is an extension of the motion $v(0,y,\cdot)$ to all of $\mathbb{R}$.

Definition 3.15. Let u be a compact process on X. A

mapping $V: X \times \mathbb{R} \to \mathbb{R}$ is called a *Lyapunov function for* u if

(i) $V(u(\tau,x,t),\tau+t) \leq V(x,\tau)$ for all $\tau \in \mathbb{R}$, $x \in X$, $t \in \mathbb{R}^+$,

(ii) for any sequence $\{\tau_n\} \subset \mathbb{R}^+$ so that $\{u_{\tau_n}\}$ is convergent, then $\{V(x,\tau_n)\}$ is convergent for every $x \in X$.

Now suppose V is a Lyapunov function for a compact process u on X. If $v \in \mathscr{A}[u]$ with $v = \lim_{n\to\infty} u_{\tau_n}$ for some sequence $\{\tau_n\} \subset \mathbb{R}^+$, and $x \in X$, define

$$W(v,x) = \lim_{n\to\infty} V(x,\tau_n).$$

We have

$$W(\pi(v,x,t)) = W(v_t,v(0,x,t)) = \lim_{n\to\infty} V(v(0,x,t),t+\tau_n)$$

$$\leq \lim_{n\to\infty} V(x,\tau_n) = W(v,x).$$

In view of the definition of W we have that W is a (sequentially) continuous Lyapunov function on $(\mathscr{A}[u] \times X, \mathscr{V}, \pi)$. (See Definition 8.1 of Chapter II.) The next theorem says what we expect; W is constant on positive limit sets. The proof is left for the reader.

Theorem 3.16. Suppose u is a compact process on X and V is a Lyapunov function for u. In addition, suppose the family of mappings $\{V(\tau,\cdot): \tau \in R\}$ of X to $\mathbb{R}$ is equicontinuous. Let $y = \lim_{n\to\infty} u(0,x,t_n) \in \Omega_u^+(0,x)$ for some $x \in X$, and $v = \lim_{n\to\infty} u_{t_n} \in \mathscr{L}[u]$. Then

$$W(v_t,v(0,y,t)) = W(v,y) \quad \text{for all } t \in \mathbb{R}^+.$$

4. Uniform Processes

We can strengthen the notion of convergence in $X^{\mathbb{R}\times X\times\mathbb{R}^+}$, to obtain a new class of processes.

Definition 4.1. Equip $X^{\mathbb{R}\times X\times\mathbb{R}^+}$ with the topology of uniform convergence on members of $\mathcal{Q}$, where $Q \in \mathcal{Q}$ has the form $[\tau,\infty) \times \{x\} \times \mathbb{R}^+$, $\tau \in \mathbb{R}$, $x \in X$. The corresponding uniformity will also be denoted by $\mathcal{U}$. Then a process u on X is called *uniform* if the family of mappings from X to X

$$\{u(\tau,\cdot,t)\colon \tau \in \mathbb{R},\ t \in \mathbb{R}^+\}$$

is equicontinuous.

We leave it to the reader to show the following. The proofs are similar to the case for compact processes.

Lemma 4.2. Suppose u is a uniform process on X. Then

(i) If $v \in \mathscr{A}[u]$, then v is a uniform process.

(ii) $(\mathscr{A}[u] \times X, \mathscr{V}, \pi)$ is Lyapunov stable.

Finally, we obtain an invariance principle for uniform processes. The proof is left to the reader.

Theorem 4.3. Suppose u is a uniform process on X, and let

$$y = \lim_{n\to\infty} u(0,x,t_n) \in \Omega_u^+(0,x),$$

$$v = \lim_{n\to\infty} u_{t_n} \in \mathscr{L}[u].$$

Then $\Omega_u^+(0,x) = \Omega_v^+(0,y)$. If, in addition, the positive orbit of u through $(0,x)$ is precompact in X, then the motion $\hat{v}(y,\cdot)\colon \mathbb{R} \to \Omega_u^+(0,x)$ obtained in Theorem 3.13 is almost periodic. If $u(\tau,x,t)$ is continuous in $t \in \mathbb{R}^+$, the collection of all $\hat{v}$ defined on $\Omega_u^+(0,x)$ is equi-almost periodic.

5. Solutions of Nonautonomous Ordinary Differential Equations Revisited - A Compact Process

The solutions of the nonautonomous ordinary differential equation of Chapter IV

$$(5.1) \qquad \dot{x} = f(x,t), \quad x(0) = x_0 \in \mathbb{R}^d,$$

provide us with our first example of a process. We will show that many of the results of Chapter IV can be obtained via the framework of this chapter. We will assume that the solutions of Equation (5.1), denoted by $\phi(f,x;t)$, with $\phi(f,x;0) = x$, are unique and are defined for all $t \in \mathbb{R}^+$. In particular, f fulfills the Carathéory conditions and satisfies $\breve{\mathcal{F}}_1$ and $\mathcal{F}_2$ of Chapter IV, Section 2.

Let $X = W$, an open subset of $\mathbb{R}^d$, be endowed with the usual Euclidean metric. Fix f as above, and define $u \in W^{\mathbb{R}\times W\times\mathbb{R}^+}$ by

$$u(\tau,x,t) = \phi(f_\tau,x;t).$$

It was seen in Chapter IV that $u(\tau,x,t)$ is the solution of $\dot{x} = f(x,t)$ through (x,τ) at time $\tau + t$. Since ϕ is continuous on $\mathcal{F} \times W \times \mathbb{R}^+$ with $\mathcal{F}$ a compact metric space, then as $f_\tau \in \mathcal{F}$ for all $\tau \in \mathbb{R}^+$, we must have that $u(\tau,x,t)$ is jointly continuous in (τ,x,t), uniformly with respect to $\tau \in \mathbb{R}^+$. This, combined with Lemma 2.10 of Chapter IV ensures that u is a process.

We claim u is a compact process. Indeed, $\mathcal{A}[u] = \{\phi(f^*,\cdot;\cdot): f^* \in H^+(f)\}$. Define $R: H^+(f) \to \mathcal{A}[u]$ by $R(f^*)(\tau,x,t) = \phi(f^*_\tau,x;t)$. Since $H^+(f)$ is compact, and R maps onto $\mathcal{A}[u]$, we can establish sequential compactness of $\mathcal{A}[u]$ by showing R is continuous. So let $g^*_n \to g$ in $H^+(f)$, and fix $(\tau,x,t) \in \mathbb{R}^+ \times W \times \mathbb{R}^+$. The properties of ϕ ensure that $R(g_n^*)(\tau,x,t) \to R(g^*)(\tau,x,t)$, uniformly in $\tau \in \mathbb{R}^+$. Thus R is continuous. Note that $H^+(f)$ is metric, so it was sufficient to show R is sequentially continuous.

From the definition of $\Omega_u^+(0,x)$ we see that

$$\Omega_u(0,x) = \Omega_f^+(x),$$

the positive limit set of the compact motion $\phi(f,x;\cdot)$. Now suppose $y \in \Omega_u^+(0,x)$. In view of Theorem 3.13 there exists $f^* \in L^+(f)$ so that $\phi(f^*,y;t) \in \Omega_u^+(0,x)$ for all $t \in \mathbb{R}^+$. Furthermore, there exists a motion $\hat{\phi}(f^*,y;\cdot): \mathbb{R} \to \Omega_u^+(0,x)$ which agrees with $\phi(f^*,y;\cdot)$ on $\mathbb{R}^+$, and that $\hat{\phi}(f^*,y;s+t) = \phi(f_s^*,\hat{\phi}(f^*,y;s);t)$ for all $s \in \mathbb{R}$, $t \in \mathbb{R}^+$. Regrettably, Theorem 3.13 does not give us as strong a result as Theorem 5.13 of Chapter IV. Since the properties of a process were developed without regard to the originating evolutionary system, we cannot claim, just on the basis of Theorem 3.13 that $\hat{\phi}(f^*,y;t)$, $t \in \mathbb{R}$, is the solution of $\dot{x} = f^*(x,t)$, $x(0) = y$.

6. Solutions of a Wave Equation - A Uniform Process

We apply the techniques of this chapter to a nonautonomous, yet linear version of the wave equation with weak damping as given in Example 5.12 of Chapter V, namely

$$\frac{\partial^2 u}{\partial t^2} = \frac{\partial^2 u}{\partial x^2} - a(x,t)\frac{\partial u}{\partial t} + f(x,t), \quad x \in [0,1]. \tag{6.1}$$

Assume $a(x,t) > 0$ on $[0,1] \times \mathbb{R}$ with both

$$\int_\tau^\infty\int_0^1 |a(x,\xi)|^2 dx d\xi < \infty, \quad \int_\tau^\infty\int_0^1 |f(x,\xi)|^2 dx d\xi < \infty$$

for every $\tau \in \mathbb{R}$. Finally, let there exist a function $a^*(x)$

so that

$$\int_0^1 |a^*(x)|^2 dx < \infty$$

and

$$\int_\tau^\infty \int_0^1 |a(x,\xi) - a^*(x)|^2 dx d\xi < \infty$$

for every $\tau \in \mathbb{R}$. For boundary conditions take

$$u(0,t) = u(1,t) = 0, \quad t > \tau,$$

and for initial values take

$$u(x,\tau) = u_0(x), \quad \frac{\partial u}{\partial t}(x,\tau) = u_1(x), \quad x \in [0,1].$$

As in Example 4.2 of Chapter V, we can view the solution $u(x,t)$ of the damped wave equation as a map $t \mapsto u(t) \stackrel{\text{def}}{=} u(\cdot,t)$ from $[\tau,\infty)$ to $L^2 = L^2([0,1];\mathbb{R})$. In this fashion we obtain the abstract equation in L^2

$$\ddot{u}(t) + a(t)\dot{u}(t) - \Delta u(t) = f(t), \quad t \geq \tau \tag{6.2}$$

and initial values

$$u(\tau) = u_0, \quad \dot{u}(\tau) = u_1. \tag{6.3}$$

(The dot $\dot{}$ here indicates differentiation with respect to t; a prime $'$ will indicate differentiation with respect to x.) The assumptions on a, a^* and f become

$$a,f \in L^1([\tau,\infty);L^2), \quad a^* \in L^2, \quad a-a^* \in L^1([\tau,\infty);L^2).$$

Let $X = H_0^1 \times L^2$ and endow it with the norm

$$\|(u,v)\| = \sqrt{\int_0^1 [(u')^2 + v^2] dx}, \quad (u,v) \in X,$$

so that

$$\|(u,v)\|^2 = \|u'\|_2^2 + \|v\|_2^2,$$

where $\|\cdot\|_2$ is the L^2 norm arising from the usual inner product

$$\langle u,v\rangle_2 = \int_0^1 uv\,dx.$$

Observe that the boundary conditions are built into the space $X = H_0^1 \times L^2$. We are ready to define the process. Set

$$\phi(\tau,u_0,u_1,t) = (u(\tau+t),\dot{u}(\tau+t)), \tag{6.4}$$

where u is the solutions of Equations (6.2) and (6.3) on $[\tau,\infty)$. We will assume henceforth that Equation (6.2) admits a unique solution through the initial value (6.3).

<u>Lemma 6.1</u>. $\|\phi(\tau,u_0,u_1,t)\| \leq \|(u_0,u_1)\| + \int_0^t \|f(\tau+\xi)\|_2 d\xi.$

<u>Proof</u>: Apply the operator $\langle\cdot,\dot{u}(t)\rangle_2$ to Equation (6.2) and integrate over $[\tau,\tau+t]$. We get

$$\int_\tau^{\tau+t} \langle\ddot{u}(\xi),\dot{u}(\xi)\rangle_2 d\xi + \int_\tau^{\tau+t} \langle a(\xi)\dot{u}(\xi),\dot{u}(\xi)\rangle_2 d\xi$$
$$- \int_\tau^{\tau+t} \langle\Delta u(\xi),\dot{u}(\xi)\rangle_2 d\xi = \int_\tau^{\tau+t} \langle f(\xi),\dot{u}(\xi)\rangle_2 d\xi.$$

Upon integrating by parts and noting that $\Delta u = u''$, we have

$$\langle\Delta u(\xi),\dot{u}(\xi)\rangle_2 = \int_0^1 u''(\xi)\dot{u}(\xi)dx = u'(\xi)\dot{u}(\xi)\Big|_0^1 - \int_0^1 u'(\xi)\dot{u}'(\xi)dx$$
$$= -\langle u'(\xi),\dot{u}'(\xi)\rangle_2 = -\frac{1}{2}\frac{d}{d\xi}\langle u'(\xi),u'(\xi)\rangle_2.$$

Therefore

$$2\int_\tau^{\tau+t} \langle\Delta u(\xi),\dot{u}(\xi)\rangle_2 d\xi = -\|u'(\tau+t)\|_2^2 + \|u'(\tau)\|_2^2.$$

Likewise,

$$2\int_\tau^{\tau+t} \langle\ddot{u}(\xi),\dot{u}(\xi)\rangle_2 d\xi = \int_\tau^{\tau+t} \frac{d}{d\xi}\langle\dot{u}(\xi),\dot{u}(\xi)\rangle_2 d\xi$$
$$= \|\dot{u}(\tau+t)\|_2^2 - \|\dot{u}(\tau)\|_2^2.$$

Therefore,

$$\|\dot{u}(\tau+t)\|_2^2 - \|u_1\|_2^2 + 2\int_\tau^{\tau+t} \langle a(\xi)\dot{u}(\xi),\dot{u}(\xi)\rangle_2 d\xi$$

$$+ \|u'(\tau+t)\|_2^2 - \|u'(\tau)\|_2^2 = 2\int_\tau^{\tau+t} \langle f(\xi),\dot{u}(\xi)\rangle_2 d\xi.$$

Then

$$\|u'(\tau+t)\|_2^2 + \|\dot{u}(\tau+t)\|_2^2 = \|u'(\tau)\|_2^2 + \|u_1\|_2^2$$

$$(6.5) \qquad - 2\int_\tau^{\tau+t} \langle a(\xi)\dot{u}(\xi),\dot{u}(\xi)\rangle_2 d\xi + 2\int_\tau^{\tau+t} \langle f(\xi),\dot{u}(\xi)\rangle_2 d\xi.$$

Since $\|\phi(\tau,u_0,u_1,t)\|^2 = \|u'(\tau+t)\|_2^2 + \|\dot{u}(\tau+t)\|_2^2$ and $a(\xi)$ is a positive operator, then

$$\|\phi(\tau,u_0,u_1,t)\|^2 \le \|(u_0,u_1)\|^2 + 2\int_0^t \|f(\tau+\xi)\|_2 \|\phi(\tau,u_0,u_1,\xi)\| \, d\xi.$$

An application of Gronwall's Inequality (Appendix A) yields

$$\|\phi(\tau,u_0,u_1,t)\| \le \|(u_0,u_1)\| + \int_0^t \|f(\tau+\xi)\|_2 d\xi. \qquad \square$$

<u>Proposition 6.2</u>. The mapping $\phi\colon \mathbb{R} \times H_0^1 \times L^2 \times \mathbb{R}^+ \to H_0^1 \times L^2$ given by Equation (6.4) is a uniform process on $H_0^1 \times L^2$.

<u>Proof</u>: The definition of ϕ and the uniqueness of solutions to the linear equation (6.2) ensure that parts (i) and (ii) of Definition 3.1 are satisfied. Thus, we need only show that the family of maps $\{\phi(\tau,\cdot,\cdot,t)\colon \tau \in \mathbb{R},\ t \in \mathbb{R}^+\}$ is equicontinuous. Now the linearity of Equation (6.2) implies that

$$\phi(\tau,u_0,u_1,t) - \phi(\tau,v_0,v_1,t)$$

is a solution of

$$\ddot{u}(t) + a(t)\dot{u}(t) - \Delta u(t) = 0, \quad (u(\tau),\dot{u}(\tau)) = (u_0-v_0,u_1-v_1).$$

It follows from Lemma 6.1 that

$$\|\phi(\tau,u_0,u_1,t) - \phi(\tau,v_0,v_1,t)\| \leq \|(u_0-v_0,u_1-v_1)\|$$

for every $\tau \in \mathbb{R}$, (u_0,u_1) and $(v_0,v_1) \in H_0^1 \times L^2$, $t \in \mathbb{R}^+$. □

Now consider the equation

$$\frac{\partial^2 v}{\partial t^2} = \frac{\partial^2 v}{\partial x^2} - a^*(x)\frac{\partial v}{\partial t}, \quad x \in [0,1].$$

$a^*(x)$ was defined earlier. In L^2 this equation becomes

$$\ddot{v}(t) + a^*\dot{v}(t) - \Delta v(t) = 0. \tag{6.6}$$

Take for initial values

$$v(0) = u_0, \quad \dot{v}(0) = u_1. \tag{6.7}$$

Given $(u_0,u_1) \in H_0^1 \times L^2$, $\tau \in \mathbb{R}$ and $t \in \mathbb{R}^+$, set

$$\phi^*(\tau,u_0,u_1,t) = (v(t),\dot{v}(t)) \tag{6.8}$$

where v is the solution of Equation (6.6) with initial values (6.7). Equation (6.6) is a special case of Equation (6.2), so ϕ^* is well defined and independent of τ. Also,

$$\|\phi^*(\tau,u_0,u_1,t)\| \leq \|(u_0,u_1)\|$$

follows from Lemma 6.1.

<u>Proposition 6.3</u>. $\mathscr{L}[\phi] = \{\phi^*\}$; that is, ϕ is a compact uniform process on $H_0^1 \times L^2$.

<u>Proof</u>: It will be sufficient to prove for any $\tau_0 \in \mathbb{R}$, $(u_0,u_1) \in H_0^1 \times L^2$ that

$$\lim_{s\to\infty} \phi(\tau+s,u_0,u_1,t) = \phi^*(\tau,u_0,u_1,t)$$

uniformly in $\tau \in [\tau_0,\infty)$ and $t \in \mathbb{R}^+$. From the definitions of ϕ and ϕ^* we can write

$$\phi(\tau+s,u_0,u_1,t) - \phi^*(\tau,u_0,u_1,t) = (z(\tau+s+t),\dot{z}(\tau+s+t)),$$

where

$$\ddot{z}(\xi) + a(\xi)\dot{z}(\xi) - \Delta z(\xi) = f(\xi) - [a(\xi)-a^*]\dot{v}(\xi),$$

(6.9)

$$z(\tau+s) = 0, \quad \dot{z}(\tau+s) = 0.$$

Now apply the operator $<\cdot,\dot{z}(\xi)>_2$ to Equation (6.9) and integrate over $[\tau+s,\tau+s+t]$ to obtain (as was done in the proof of Lemma 6.1)

$$\|\dot{z}(\tau+s+t)\|_2^2 + \|z'(\tau+s+t)\|_2^2 = -2\int_{\tau+s}^{\tau+s+t} <a(\xi)\dot{z}(\xi),\dot{z}(\xi)>_2 d\xi$$

$$+ 2\int_{\tau+s}^{\tau+s+t} <f(\xi),\dot{z}(\xi)>_2 d\xi - 2\int_{\tau+s}^{\tau+s+t} <[a(\xi)-a^*]\dot{v}(\xi),\dot{z}(\xi)>_2 d\xi.$$

Thus,

$$\|(z(\tau+s+t),\dot{z}(\tau+s+t))\|^2 \le 2\int_{\tau+s}^{\tau+s+t} \|f(\xi)\|_2 \|\dot{z}(\xi)\|_2 d\xi$$

$$+ 2\int_{\tau+s}^{\tau+s+t} \|a(\xi)-a^*\|_2 \|\dot{v}(\xi)\|_2 \; \|\dot{z}(\xi)\|_2 d\xi$$

$$\le 2\int_{\tau+t}^{\tau+t+s} \{\|f(\xi)\|_2 + \|a(\xi)-a^*\|_2 \|(u_0,u_1)\|\} \|(z(\xi),\dot{z}(\xi))\| d\xi.$$

Once again we see from Gronwall's inequality (Appendix A) that

$$\|\phi(\tau+s,u_0,u_1,t) - \phi^*(\tau,u_0,u_1,t)\|$$

$$\le \int_{\tau+s}^{\tau+s+t} \|f(\xi)\|_2 d\xi + \|(u_0,u_1)\| \int_{\tau+s}^{\tau+s+t} \|a(\xi)-a^*\|_2 d\xi.$$

The integrability assumptions on f and $a-a^*$ ensure that for any fixed $\tau_0 \in \mathbb{R}$, the last two integrals tend to zero as

$s \to \infty$, uniformly in $\tau \in [\tau_0,\infty)$ and $t \in \mathbb{R}^+$. □

We want to establish the precompactness of the positive orbits of the process. First we need an estimate along the lines of Lemma 6.1.

Lemma 6.4. Suppose for each $\tau \in \mathbb{R}$ that $a(t)$ and $f(t)$ are absolutely continuous in $t \in [\tau,\infty)$ with both $\dot{a},\dot{f} \in L^1([\tau,\infty);L^2)$. Then

$$
\begin{aligned}
&\{\|\dot{u}'(\tau+t)\|_2^2 + \|\ddot{u}(\tau+t)\|_2^2\}^{\frac{1}{2}} \\
(6.10)\quad &\le \{\|\dot{u}'(\tau)\|_2^2 + \|\ddot{u}(\tau)\|_2^2\}^{\frac{1}{2}} + 2\int_\tau^{\tau+t}\{\|\dot{a}(\xi)\|_2\,\|\dot{u}(\xi)\|_2 \\
&\qquad + \|\dot{f}(\xi)\|_2\}d\xi.
\end{aligned}
$$

Proof: Apply the operator $\langle\cdot,\ddot{u}(t)\rangle_2$ to the (time) derivative of Equation (6.2) and integrate over $[\tau,\tau+t]$. We get

$$
\begin{aligned}
&\int_\tau^{\tau+t}\langle\dddot{u}(\xi),\ddot{u}(\xi)\rangle_2 d\xi + \int_\tau^{\tau+t}\langle\dot{a}(\xi)\dot{u}(\xi),\ddot{u}(\xi)\rangle_2 d\xi \\
&\quad + \int_\tau^{\tau+t}\langle a(\xi)\ddot{u}(\xi),\ddot{u}(\xi)\rangle_2 d\xi - \int_\tau^{\tau+t}\langle\Delta\dot{u}(\xi),\ddot{u}(\xi)\rangle_2 d\xi \\
&\quad = \int_\tau^{\tau+t}\langle\dot{f}(\xi),\ddot{u}(\xi)\rangle_2 d\xi.
\end{aligned}
$$

According to the proof of Lemma 6.1 the first and fourth integrals above become

$$
\int_\tau^{\tau+t}\langle\dddot{u}(\xi),\ddot{u}(\xi)\rangle_2 d\xi = \frac{1}{2}\|\ddot{u}(\tau+t)\|_2^2 - \frac{1}{2}\|\ddot{u}(\tau)\|_2^2
$$

$$
-\int_\tau^{\tau+t}\langle\Delta\dot{u}(\xi),\ddot{u}(\xi)\rangle_2 d\xi = \frac{1}{2}\|\dot{u}'(\tau+t)\|_2^2 - \frac{1}{2}\|\dot{u}'(\tau)\|_2^2.
$$

Consequently since $a(\xi)$ is a positive operator, we have

$$\|\dot{u}'(\tau+t)\|_2^2 + \|\ddot{u}(\tau+t)\|_2^2 \le \|\dot{u}'(\tau)\|_2^2 + \|\ddot{u}(\tau)\|_2^2$$

$$- \int_\tau^{\tau+t} \langle\dot{a}(\xi)\dot{u}(\xi),\ddot{u}(\xi)\rangle_2 d\xi + \int_\tau^{\tau+t} \langle\dot{f}(\xi),\ddot{u}(\xi)\rangle_2 d\xi$$

$$\le \|\dot{u}'(\tau)\|_2^2 + \|\ddot{u}(\tau)\|_2^2 + 2\int_\tau^{\tau+t} \{\|\dot{a}(\xi)\|_2\|\dot{u}(\xi)\|_2$$

$$+ \|\dot{f}(\xi)\|_2\} \|\ddot{u}(\xi)\|_2 d\xi.$$

From Gronwall's inequality (Appendix A) we obtain

$$\{\|\dot{u}'(\tau+t)\|_2^2 + \|\ddot{u}(\tau+t)\|_2^2\}^{\frac{1}{2}} \le \{\|\dot{u}'(\tau)\|_2^2 + \|\ddot{u}(\tau)\|_2^2\}^{\frac{1}{2}}$$

$$+ 2\int_\tau^{\tau+t} \{\|\dot{a}(\xi)\|_2\|\dot{u}(\xi)\|_2 + \|\dot{f}(\xi)\|_2\} d\xi. \quad \square$$

Proposition 6.5. For each $(u_0,u_1) \in H_0^1 \times L^2$ the positive orbit through (τ,u_0,u_1), namely $\gamma^+(\tau,u_0,u_1) = \bigcup_{t\in\mathbb{R}^+} \phi(\tau,u_0,u_1,t)$, is precompact.

Proof: Recall from Example 4.2 of Chapter V that $(H_0^1 \cap H^2) \times H_0^1$ is dense in $H_0^1 \times L^2$. Also the Sobolev embedding theorem (Appendix A) says that $W^{1,1}([\tau,\infty);L^2)$ is dense in $L^1([\tau,\infty);L^2)$. It was shown in the proof of Theorem 4.3 of Chapter V that the set $\{(u_0,u_1) \in H_0^1 \times L^2\colon \gamma^+(\tau,u_0,u_1)$ is precompact$\}$ is closed in $H_0^1 \times L^2$. Therefore it will be sufficient to establish the precompactness of $\gamma^+(\tau,u_0,u_1)$ when $(u_0,u_1) \in (H_0^1 \cap H^2) \times H_0^1$ and $a,f \in W^{1,1}([\tau,\infty);L^2)$; that is, $a(\xi)$ and $f(\xi)$ are absolutely continuous in $\xi \in [\tau,\infty)$ with $\dot{a},\dot{f} \in L^1([\tau,\infty);L^2)$. The linearity of Equation (6.2) and the estimate given by Lemma 6.1 ensures that we may extend to the case when $(u_0,u_1) \in H_0^1 \times L^2$ and $a,f \in L^1([\tau,\infty);L^2)$.

As $\phi(\tau,u_0,u_1,t) = (u(\tau+t),\dot{u}(\tau+t))$, then

$\frac{d}{dt}\,\phi(\tau,u_0,u_1,t) = (\dot{u}(\tau+t),\ddot{u}(\tau+t))$. Hence

$$\left\|\frac{d}{dt}\,\phi(\tau,u_0,u_1,t)\right\| = \{\|\dot{u}'(\tau+t)\|_2^2 + \|\ddot{u}(\tau+t)\|_2^2\}^{\frac{1}{2}}.$$

We see from Lemma 6.1 that $\|\phi(\tau,u_0,u_1,t)\|$ is uniformly bounded in $t \in \mathbb{R}^+$, hence so must $\|\dot{u}(\tau+t)\|_2$. We have already assumed $\|\dot{a}(\xi)\|_2$ and $\|\dot{f}(\xi)\|_2$ are integrable on $[\tau,\infty)$. Then it follows from Lemma 6.4 that $\left\|\frac{d}{dt}\,\phi(\tau,u_0,u_1,t)\right\|$ is also uniformly bounded in $t \in \mathbb{R}^+$. Consequently, the positive orbit $\gamma^+(\tau,u_0,u_1)$ is precompact. □

<u>Remark 6.6</u>. At this point the reader should compare this result with Examples 5.12 and 6.4 of Chapter V. The problem considered there was of the form $\ddot{v}(t) + aq(\dot{v}(t)) - \Delta v(t) = 0$. Equation (6.6) above is just a special case of this when $q(y) = y$. The importance of Proposition 6.5 though is the establishment of precompactness for the nonautonomous version of Example 6.4 of Chapter V.

We conclude this example with an explicit representation of ϕ^*. Recall that $\phi^*(0,u_0,u_1,t) = (v(t),\dot{v}(t))$, where $v(t)$ is the solution of $\ddot{v}(t) + a^*\dot{v}(t) - \Delta v(t) = 0$, $v(0) = u_0$, $\dot{v}(0) = u_1$. The mapping $V\colon H_0^1 \times L^2 \times \mathbb{R} \to \mathbb{R}$ defined by

$$V(u_0,u_1,\tau) = \|(u_0,u_1)\| + \int_\tau^\infty \|f(\xi)\|_2 d\xi \tag{6.11}$$

is a Lyapunov function for the process ϕ on $H_0^1 \times L^2$ as can be verified from Lemma 6.1. Now define $W^*\colon H_0^1 \times L^2 \to \mathbb{R}$ by

$$W^*(u_0,u_1) = W(\phi^*,u_0,u_1),$$

where W is the Lyapunov function on $\mathscr{A}[\phi] \times H_0^1 \times L^2$ generated from V as constructed following Definition 3.15 of Section 3. In view of the definition of W we have

$$(6.12)\qquad W^*(u_0,u_1) = \lim_{\tau\to\infty} V(u_0,u_1,\tau) = \|(u_0,u_1)\|.$$

Since all positive limit sets in $\mathscr{A}[\phi] \times H_0^1 \times L^2$ are of the form $\{\phi^*\} \times \Omega_\phi^+(\tau,u_0,u_1)$, it follows that W^* is constant on $\Omega_\phi^+(\tau,u_0,u_1)$. In particular, if $(v_0,v_1) \in \Omega_\phi^+(\tau,u_0,u_1)$, then

$$(6.13)\qquad W^*(\phi^*(0,v_0,v_1,t)) = W^*(v_0,v_1) = \|(v_0,v_1)\|$$

for all $t \in \mathbb{R}^+$. According to Equation (6.5) we obtain

$$(6.14)\qquad \|v'(t)\|_2^2 + \|\dot v(t)\|_2^2 = \|v'(0)\|_2^2 + \|\dot v(0)\|_2^2 - 2\int_0^t \langle a^*\dot v(\xi),\dot v(\xi)\rangle_2 d\xi$$

for the special case of Equation (6.6). We can rewrite this as

$$(6.15)\qquad \|\phi^*(0,v_0,v_1,t)\|^2 = \|(v_0,v_1)\|^2 - 2\int_0^t \langle a^*\dot v(\xi),\dot v(\xi)\rangle_2 d\xi.$$

But from Equations (6.11) and (6.12) we have

$$W^*(\phi^*(0,v_0,v_1,t)) = \lim_{\tau\to\infty} V(\phi^*(0,v_0,v_1,t),\tau) = \|\phi^*(0,v_0,v_1,t)\|,$$

so in view of Equation (6.13) we conclude that

$$\|(v_0,v_1)\|^2 = \|(v_0,v_1)\|^2 - 2\int_0^t \langle a^*\dot v(\xi),\dot v(\xi)\rangle_2 d\xi$$

for all $t \in \mathbb{R}^+$. Thus $\int_0^t \langle a^*\dot v(\xi),\dot v(\xi)\rangle_2 d\xi = 0$ for all $t \in \mathbb{R}^+$; hence $\dot v(t) = 0$ for all $t \in \mathbb{R}^+$ since a^* is positive. It follows that $v(t)$ must satisfy

$$\ddot v(t) - \Delta v(t) = 0,$$

the undamped wave equation treated in Example 6.4 of Chapter V. In particular, the analysis there shows that $v(t) = 0$ for every $t \in \mathbb{R}^+$. Consequently, $\Omega_\phi^+(\tau,u_0,u_1) = \{(0,0)\}$. We conclude under the hypotheses set forth the following Equation (6.1) that

<u>Proposition 6.7</u>. If $u(x,t)$ is the solution of Equation (6.1) with initial values $u(\cdot,\tau) = u_0 \in H_0^1$ and $\dot{u}(\cdot,\tau) = u_1 \in L^2$, then

$$\lim_{t\to\infty} u(x,t) = 0 \quad (\text{in } H_0^1),$$

$$\lim_{t\to\infty} \dot{u}(x,t) = 0 \quad (\text{in } L^2).$$

7. Exercises

7.1. Suppose $(Y, \mathscr{U}, \pi)$ is a Lyapunov stable weak semidynamical system. Denote by $\mathscr{L}$ the set $\{y \in Y: \gamma^+(y)$ is sequentially relatively compact$\}$. Prove that $\mathscr{L}$ is positively invariant and sequentially closed.

7.2. Suppose $(Y, \mathscr{U}, \pi)$ is a Lyapunov stable weak semidynamical system for which $(Y, \mathscr{U})$ is sequentially complete. Let $V: Y \to \mathbb{R}$ be a sequentially lower semi-continuous function with $V(yt) \leq V(y)$ for every $(y,t) \in Y \times \mathbb{R}^+$. If $L^+(x) \neq \emptyset$ for some $x \in Y$, then $V(yt) = V(y)$ for every $y \in L^+(x)$ and $t \in \mathbb{R}$.

7.3. Verify Theorem 3.16.

7.4. Verify Lemma 4.2.

7.5. Verify Theorem 4.3.

7.6. Suppose u is a uniform process on X and V a Lyapunov function for u. Furthermore, assume that the family of mappings of X to $\mathbb{R}$, $\{V(\tau,\cdot): \tau \in \mathbb{R}^+\}$ is lower equi-semicontinuous. If for some $x \in X$ we have $y = \lim_{n\to\infty} u(0,x,t_n) \in \Omega_u^+(0,x)$ and $v = \lim_{n\to\infty} u_{t_n} \in \mathscr{L}[u]$, prove that $W(v_\tau, v(0,y,\tau)) = W(v,y)$ for all $\tau \in \mathbb{R}^+$.

7.7. Show that if (X,π) is a semidynamical system with X a metric space, then π is a compact process on X.

7.8. Suppose u is a process on X for which $u_\tau = u$ for some $\tau > 0$ and that $u(s,x,t)$ is continuous in t. Prove that $\mathscr{L}[u] = \{u_s: 0 \leq s \leq \tau\}$.

7.9. Suppose u is a uniform process on X and $\tau \in \mathbb{R}$. Denote by $\mathscr{K}_\tau$ the set of $x \in X$ so that the positive orbit of u through (τ,x) is precompact in X. Show that $\mathscr{K}_\tau$ is closed.

7.10. Suppose u is a uniform process on X and $\tau \in \mathbb{R}$. Denote by $\mathscr{C}_\tau$ the set of $x \in X$ so that the motion through (τ,x) is continuous; i.e., $u(\tau,x,\cdot): \mathbb{R}^+ \to X$ is continuous. Show that $\mathscr{C}_\tau$ is closed.

7.11. Let u be a uniform process on X and suppose $v \in \mathscr{L}[u]$. Prove that $\mathscr{L}[v] = \mathscr{L}[u]$.

8. Notes and Comments

Sections 1-4. This chapter is based on the work of Dafermos [1-7,9]. The theory of processes is developed in Dafermos [3,6]. Our exposition is based on the function space setting of Dafermos [7]. The relationship to almost periodic solutions of evolution equations is taken up in Dafermos [8]. Slemrod [1] has also developed a framework suitable for handling solutions of abstract equations of evolution in Banach space.

Section 6. This example is based upon Dafermos [4]. There are numerous other papers which characterize the asymptotic behavior of nonautonomous partial differential equations in terms of the dynamical system framework; for example, Dafermos [1,2,5], Slemrod [1], Slemrod and Infante [1], and Ball and Peletier [1].

APPENDIX A

0. Preliminaries

We assume the reader is familiar with analysis at the level presented in Royden [1], *Real Analysis*. All terms not defined below can be found in Royden. Additional references for this appendix are as follows: Sections 2 and 3, see Kelley [1]; Sections 4-10, see Yoshida [1]; Section 12, see Corduneanu [1]; Section 13, see Brezis [2].

1. Commonly Used Symbols

$\emptyset$	Empty set
$\mathbb{Z}$	Integers
$\mathbb{Z}^+$	Nonnegative integers
$\mathbb{N}$	Positive integers
$\mathbb{R}$	Real numbers
$\mathbb{R}^+$	Nonnegative real numbers
$\mathbb{R}^-$	Nonpositive real numbers
$\mathbb{C}$	Complex numbers
$\mathbb{Q}$	Rational numbers
$\bar{A}$	Closure of set A
∂A	Boundary of set A

Int A	Interior of set A
$f\|_A$	Restriction of the function f to the set A
$A \setminus B$	Compliment of B with respect to A
$\mathbb{R}^d$	All d-tuples, $(x_1, x_2, \ldots, x_d)$, $x_i \in \mathbb{R}$, $1 \leq i \leq d$
$\|\cdot\|$	Absolute value

2. Nets

A relation $\geq$ *directs* a set Λ if Λ is nonempty and

(i) if $\alpha, \beta, \gamma \in \Lambda$ with $\alpha \geq \beta$ and $\beta \geq \gamma$, then $\alpha \geq \gamma$,
(ii) if $\alpha \in \Lambda$, then $\alpha \geq \alpha$, and
(iii) if $\alpha, \beta \in \Lambda$, then there is $\gamma \in \Lambda$ with $\gamma \geq \alpha$ and $\gamma \geq \beta$.

A *directed set* is a pair $(\Lambda, \geq)$ such that $\geq$ directs Λ. A *net* is a pair $(x, \geq)$ where x is a function with domain Λ and $\geq$ directs Λ. If $\Gamma \subset \Lambda$ is also directed by $\geq$, then $(x|_\Gamma, \geq)$ is a *subnet* of the net $(x, \geq)$. $(x|_\Gamma, \geq)$ is also a net. The value of a net $(x, \geq)$ at a point $\alpha \in \Lambda$ is denoted by x_α. For convenience we usually write $\{x_\alpha\}$ instead of $(x, \geq)$. Let x have range in X. The net $\{x_\alpha\}$ is *in* the set $U \subset X$ if $x_\alpha \in U$ for each $\alpha \in \Lambda$. For convenience we usually write $x_\alpha \in U$. The net $\{x_\alpha\}$ is *eventually* in $U \subset X$ if there is some $\beta \in \Lambda$ so that $x_\alpha \in U$ for every $\alpha \geq \beta$ in Λ. The net $\{x_\alpha\}$ is *frequently* in U if for each $\alpha \in \Lambda$, there is some $\beta \in \Lambda$ so that $\beta \geq \alpha$ and $x_\beta \in U$. If X is a topological space, the net $\{x_\alpha\}$ *converges* to a point $y \in X$ if $\{x_\alpha\}$ is eventually in every neighborhood of y. We write this as $\lim_\alpha x_\alpha = y$ or $x_\alpha \to y$.

Prop: A point $y \in \overline{U} \subset X$ (a topological space) if and only if there exists a net $\{x_\alpha\}$ in X which converges to y.

Suppose $\{x_\alpha\}$ is a net in a topological space X. A point $y \in X$ is called a *cluster point* of $\{x_\alpha\}$ if $\{x_\alpha\}$ is frequently in every neighborhood of y. X is *compact* if every net in X has a cluster point in X.

Prop: The following are equivalent for a topological space X:

(i) X is compact.

(ii) Every net in X has a subnet which converges to some point of X.

(iii) Every collection of open sets which covers X contains a finite subcollection which also covers X.

Suppose $f: X \to Y$. If $\{x_\alpha\}$ is a net in X, then $\{f(x_\alpha)\}$ is a net in Y, directed by the same set which directs $\{x_\alpha\}$. When X and Y are topological spaces, f is continuous at $y \in X$ if and only if for each net $\{x_\alpha\}$ converging to y, the net $\{f(x_\alpha)\}$ converges to $f(y)$.

A mapping $d(\cdot,\cdot): X \times X \to \mathbb{R}$ is a *metric* for a set X if

(i) $d(x,y) \geq 0$ for all $x,y \in X$,

(ii) $d(x,y) = 0$ if and only if $x = y$,

(iii) $d(x,y) = d(y,x)$ for all $x,y \in X$, and

(iv) $d(x,z) \leq d(x,y) + d(y,z)$ for all $x,y,z \in X$.

The pair (X,d) is a *metric space*. When appropriate, a metric space will just be referred to by the set X. In the event X is a metric space, the preceeding remarks about nets may be replaced by sequences.

3. Uniform Topologies

Let X be a nonempty set and U a relation (subset) of $X \times X$. Denote by U^{-1} the set $\{(x,y) \in X \times X: (y,x) \in U\}$. U is *symmetric* if $U^{-1} = U$. If U and V are relations, denote by UV the set $\{(x,z) \in X \times X$: there exists $y \in X$ such that $(x,y) \in V, (y,z) \in U\}$. By U^2, U^3, etc., we mean UU, UUU, etc. The relation $\{(x,x): x \in X\}$ is called the *identity (diagonal)* and is denoted by Δ. A *uniformity* for a set X is a nonempty family $\mathscr{U}$ of subsets of $X \times X$ so that

(i) if $U \in \mathscr{U}$, then $\Delta \subset U$,

(ii) if $U \in \mathscr{U}$, then $U^{-1} \in \mathscr{U}$,

(iii) if $U \in \mathscr{U}$, then $V^2 \subset U$ for some $V \in \mathscr{U}$,

(iv) if $U,V \in \mathscr{U}$, then $U \cap V \in \mathscr{U}$, and

(v) if $U \in \mathscr{U}$ and $U \subset V \subset X \times X$, then $V \in \mathscr{U}$.

The pair $(X, \mathscr{U})$ is called a *uniform space*. A subfamily $\mathscr{B}$ of $\mathscr{U}$ is called a *base* for $\mathscr{U}$ if each member of $\mathscr{U}$ contains a member of $\mathscr{B}$. A subfamily $\mathscr{S}$ of $\mathscr{U}$ is called a *subbase* for $\mathscr{U}$ if the finite intersections of members of $\mathscr{S}$ is a base for $\mathscr{U}$.

Prop: A family $\mathscr{S}$ of subsets of $X \times X$ is a subbase for some uniformity for X if

(i) whenever $U \in \mathscr{S}$, then $\Delta \subset U$,

(ii) whenever $U \in \mathscr{S}$, then U^{-1} contains a member of $\mathscr{S}$, and

(iii) whenever $U \in \mathscr{S}$, there exists $V \in \mathscr{S}$ with $V^2 \subset U$.

For each $x \in X$ denote by $U[x]$ the set $\{y \in X: (x,y) \in U\}$. If $(X, \mathscr{U})$ is a uniform space then X can be given the *uniform topology (the topology of the uniformity* $\mathscr{U}$*)*:

a set $V \subset X$ is open if for each $x \in V$ there exists some $U \in \mathscr{U}$ so that $U[x] \subset V$. In the event X admits a metric d, then the sets $U[x]$ are just the open balls of the form $B_r(x) = \{y \in X: d(x,y) < r\}$ for some $r > 0$. The metric space (X,d) then is a uniform space.

A net $\{x_\alpha\}$ in the uniform space $(X,\mathscr{U})$ is a *Cauchy net* if for each $U \in \mathscr{U}$, there exists $\gamma \in \Lambda$ = domain x such that $(x_\alpha, x_\beta) \in U$ whenever $\alpha,\beta \in \Lambda$ with $\alpha \geq \beta$, $\beta \geq \gamma$. A uniform space $(X,\mathscr{U})$ is *complete* if every Cauchy net in X converges to a point of X.

Let F be a family of functions on a set Y to some metric space (X,d) and $\mathscr{Q}$ be a family of subsets of Y. The sets $\{g \in F: d(g(x),f(x)) < r$ for every $x \in Q\}$ for all $f \in F$, $r > 0$ and $Q \in \mathscr{Q}$ form a subbase for the uniformity of *uniform convergence on members of* $\mathscr{Q}$, $\mathscr{U}_{\mathscr{Q}}$.

Prop: Let F be the family of all functions on a set Y to a metric space (X,d). Suppose $\mathscr{Q}$ is a family of subsets of X which covers X. If (X,d) is complete, then $(F,\mathscr{U}_{\mathscr{Q}})$ is complete.

4. Compactness

A topological space X is *compact* if every collection of open sets of X which covers X contains a finite subcollection which also covers X. A topological space X is *sequentially compact* if every sequence $\{x_n\}$ in X has a cluster point in X. If (X,d) is a metric space, then compactness and sequential compactness are equivalent. A topological space X is *locally compact* if each point $x \in X$ is contained in an open set which has compact closure. A subset S of a topological space X is *precompact* if $\overline{S}$ is compact (in the relative topology). A family F of functions from a topological space X to a metric space (Y,d) is *equicontinuous* on X if for each $x \in X$ and each $\varepsilon > 0$, there is an open set U containing x so that $d(f(x),f(y)) < \varepsilon$ for all $y \in U$ and all $f \in F$.

Prop (Ascoli's Theorem): Let F be an equicontinuous family of functions from a separable metric space X to a metric space Y. Let $\{f_n\}$ be a sequence in F such that for each $x \in X$, $\bigcup_{n=1}^{\infty} f_n(x)$ is precompact. Then there is a subsequence $\{f_{n_k}\}$ of $\{f_n\}$ which converges pointwise to a continuous function f on X, and the convergence is uniform on compact subsets of X.

Let X be a metric space and $C_b(X; \mathbb{R})$ be the set of all bounded continuous real valued functions on X. For $f \in C_b(X; \mathbb{R})$, $\|f\| = \sup_{x \in X} |f(x)|$ defines a norm on the linear space $C_b(X; \mathbb{R})$. Under this norm, $C_b(X; \mathbb{R})$ is a *Banach space*; that is, a complete normed linear space. If X is compact, Ascoli's theorem becomes

Prop: If X is a compact metric space, than a subset of $C_b(X; \mathbb{R})$ is precompact if and only if it is bounded and equicontinuous.

5. Linear Spaces

All linear spaces will be taken over the scalar field $\mathbb{R}$. Let X,Y be linear spaces. The domain of a mapping T from X to Y is denoted by $\mathscr{D}(T)$; the range by $\mathscr{R}(T)$. When appropriate, we write $T: \mathscr{D}(T) \subset X \to Y$ to denote a mapping or *operator* from X to Y with domain $\mathscr{D}(T)$; otherwise we just write $T: X \to Y$. T is a *linear operator* if

$T(\alpha x+\beta y) = \alpha T(x) + \beta T(y)$ for all $\alpha,\beta \in \mathbb{R}$ and all $x,y \in \mathscr{D}(T)$. We usually write Tx for $T(x)$. If $Y = \mathbb{R}$, T is a *linear functional*. If X and Y are normed linear spaces, the linear operator $T\colon X \to Y$ is *bounded* if there exists a constant $M > 0$ such that $\|Tx\|_Y \leq M\|x\|_X$. (We subscript the norm $\|\cdot\|$ by the appropriate space so as to avoid confusion.) The least such M is the *norm* of the linear operator T, and is denoted by $\|T\|$. Then $\|T\| = \sup\{\|Tx\|_Y\colon \|x\|_X \leq 1\}$. Bounded linear operators are uniformly continuous. If a linear operator is continuous at a point $x \in X$, it is bounded. The space of all bounded linear operators from a normed linear space X to a Banach space Y is itself a Banach space with norm $\|T\|$ as given above.

Prop (Hahn-Banach Theorem): Let p be a real-valued function defined on the linear space X and which satisfies

(i) $p(x+y) \leq p(x) + p(y)$ for all $x,y \in X$, and
(ii) $p(\alpha x) = \alpha p(x)$ for all $\alpha \in \mathbb{R}^+$, $x \in X$.

If f is a linear functional defined on a linear subspace $S \subset X$ with $f(x) \leq p(x)$ for each $x \in S$, then there exists a linear functional $\hat{f}$ with domain X such that

(i) $\hat{f}(x) \leq p(x)$ for all $x \in X$, and
(ii) $\hat{f}(x) = f(x)$ for all $x \in S$.

Prop: Let X be a normed linear space and suppose $x \in X$. There exists a bounded linear functional f on X such that $f(x) = \|f\|\,\|x\|$.

6. Duality

Given a Banach space X, the *dual* of X is the space X^* consisting of all continuous linear functionals on X. For any $x \in X$ and $x^* \in X^*$, denote by (x,x^*) the value of the functional x^* at the point x. X^* is a Banach space with norm $\|x^*\| = \sup_{\|x\|_X \leq 1} |(x,x^*)|$. The dual of X^*, X^{**}, is the *bidual* of X. X is isometrically isomorphic to a subspace of X^{**} as follows. For $x \in X$, $F_x(x^*) \overset{\text{def}}{=} (x,x^*)$ is a linear functional on X^*. $\|F_x\|_{X^{**}} = \sup_{\|x^*\|_{X^*} \leq 1} |(x,x^*)| = \|x\|_X$. If this isomorphism maps X onto X^{**}, then X is *reflexive* and is identified with its bidual.

The *weak topology* on a Banach space X is the weakest topology on X so that every $x^* \in X^*$ is continuous on X. The topology on X induced by the norm is the *strong topology*. A net $\{x_\alpha\}$ in X *converges weakly* to some $x \in X$ if and only if $(x_\alpha,x^*) \to (x,x^*)$ for every $x^* \in X^*$. We write $x_\alpha \overset{w}{\to} x$ or $\text{w-lim}\, x_\alpha = x$. *Strong convergence* of a net $\{x_\alpha\}$ to x is written $x_\alpha \overset{s}{\to} x$ or $\text{s-lim}\, x_\alpha = x$. A strongly convergent net (sequence) is weakly convergent. A set $K \subset X$ is *weakly compact* if K is compact in the weak topology of X; $K \subset X$ is *weakly sequentially compact* if every sequence $\{x_n\} \subset K$ contains a subsequence which converges weakly to a point in K.

Prop: A set in a reflexive Banach space is weakly sequentially compact if and only if it is bounded.

For a normed linear space the *weak* topology* on X^* is the weakest topology on X^* so that each $x \in X$ when considered as an element of X^{**} is continuous on X^*. A net $\{x^*_\alpha\}$ in X^* *converges weak** to some $x^* \in X^*$ if and only if

$(x,x_\alpha^*) \to (x,x^*)$ for every $x \in X$. If X is reflexive then the weak* topology on X (considered as the dual of X^*) and the weak topology on X coincide.

Prop (Alaoglu): If X is a Banach space, the closed unit ball $\{x^* \in X^*: \|x^*\|_{X^*} \leq 1\}$ of X^* is weak* compact.

7. Hilbert Spaces

Let X be a linear space (over $\mathbb{R}$). An *inner product* on X is a mapping $\langle\cdot,\cdot\rangle: X \times X \to \mathbb{R}$ which satisfies

(i) $\langle \alpha x+\beta y,z\rangle = \alpha\langle x,z\rangle + \beta\langle y,z\rangle$ for all $\alpha,\beta \in \mathbb{R}$ and $x,y,z \in X$,

(ii) $\langle x,y\rangle = \langle y,x\rangle$ for all $x,y \in X$,

(iii) $\langle x,x\rangle \geq 0$ for all $x \in X$, and

(iv) $\langle x,x\rangle = 0$ if and only if $x = 0$.

A linear space with an inner product is called an *inner product space* or a *pre-Hilbert space*. Two elements x and y of an inner product space are *orthogonal* if $\langle x,y\rangle = 0$. If $\langle x,y\rangle = 0$ for every $x \in X$, then $y = 0$.

Prop (Cauchy Schwarz Inequality): If X is an inner product space, then

$$|\langle x,y\rangle| \leq \sqrt{\langle x,x\rangle\langle y,y\rangle}$$

for every $x,y \in X$. Every inner product space is a normed linear space with norm $\|x\| = \sqrt{\langle x,x\rangle}$. If an inner product space X is complete under the norm given by $\sqrt{\langle x,x\rangle}$, then X is a *Hilbert space*. The dual X^* of a Hilbert space may be identified with X itself; in particular X is reflexive.

Prop: A necessary and sufficient condition for a sequence $\{x_n\}$ in a Hilbert space X to converge strongly to some

$x \in X$ is that (i) $\sup_{n \geq 1} \|x_n\| < \infty$ and (ii) $\langle x_n, y\rangle \to \langle x, y\rangle$ for all $y \in X$.

8. Vector Valued Integration

Let J be an interval in $\mathbb{R}$. For $1 \leq p < \infty$ denote by $L^p(J; \mathbb{R})$ the set of all real valued Lebesgue measurable functions f on J such that $|f(t)|^p$ is Lebesgue integrable over J. Identify two functions f and g if $f = g$ a.e. in J (that is, $f(t) = g(t)$ for all $t \in J$, except perhaps on a set of Lebesgue measure zero). Denote the Lebesgue measure on $\mathbb{R}$ by m. As all integrals will be with respect to Lebesgue measure, henceforth we shall drop the modifier Lebesgue. $L^p(J; \mathbb{R})$ is a Banach space with norm $\{\int_J |f(t)|^p dt\}^{1/p}$. Now suppose X is a Banach space with norm $\|\cdot\|$ and X^* its dual. A function $f: J \to X$ is a *step function* if $\mathcal{R}(f)$ is a finite set. The step function f is *measurable* if $f^{-1}(x)$ is a measurable set for each $x \in X$. f is *integrable* if it is both measurable and $m(f^{-1}(x)) < \infty$. Define $\int_J f(t)dt = \sum_{x \in X} m(f^{-1}(x))x$. In general, $f: J \to X$ is *measurable* if there exists a sequence $\{f_n\}$ of measurable step functions $f_n: J \to X$ such that $f_n \to f$ a.e. in J. Finally, $f: J \to X$ is *integrable* if there exists a sequence $\{f_n\}$ of integrable step functions $f_n: J \to X$ such that for each $n \in \mathbb{N}$ the mapping $t \to \|f_n(t) - f(t)\|$ is integrable (that is, $\|f_n(\cdot) - f(\cdot)\| \in L^1(J; \mathbb{R})$) and $\lim_{n \to \infty} \int_J \|f_n(t) - f(t)\| dt = 0$. Then $\int_J f_n dt$ converges in X, and its limit is independent of the choice of sequence $\{f_n\}$. Denote the limit by $\int_J f\, dt$.

Prop (Bochner): Let X be a Banach space. Then $f: J \to X$

is integrable if and only if f is measurable and $\|f(\cdot)\|$ is integrable.

Prop (Lebesgue Dominated Convergence): Let $\{f_n\}$ be a sequence of integrable functions from J into the Banach space X such that $f_n \to f$ a.e. in J. Moreover, suppose $\phi: J \to \mathbb{R}$ is integrable with $\|f_n(t)\| \leq \phi(t)$ for all $n \in \mathbb{N}$, a.e. in J. Then

(i) f is integrable,

(ii) $\int_J \|f_n(t)-f(t)\| dt \to 0$, and

(iii) $\int_J f_n dt \to \int_J f\, dt$.

For $1 \leq p < \infty$ denote by $L^p(J;X)$ the set of measurable functions f such that $\|f(\cdot)\| \in L^p(J;\mathbb{R})$. These are Banach spaces with norms $\|\cdot\|_p$ given by $\|f\|_p = \{\int_J \|f(t)\|^p dt\}^{1/p}$. For $p = \infty$, $L^\infty(J;X)$ is also a Banach space with norm

$$\|f\|_\infty = \operatorname{ess\,sup} \|f(t)\| \overset{\text{def}}{=} \inf\{M: m\{t \in J: \|f(t)\| > M\} = 0\}.$$

If X is reflexive and $1 < p < \infty$, the dual of $L^p(J;X)$ can be identified with $L^q(J;X)$ where $\frac{1}{q} + \frac{1}{p} = 1$. In this event, the spaces $L^p(J;X)$ are reflexive. If J is a semi-infinite or infinite interval, then $L^p_{loc}(J;X)$ consists of those measurable functions f for which $\int_B \|f(t)\|^p dt < \infty$ for every bounded measurable subset B of J. Such an f is *locally integrable*. If X is a Hilbert space, $L^2(J;X)$ is also a Hilbert space with inner product $\langle f,g\rangle = \int_J \langle f(t),g(t)\rangle_X dt$.

Let $J = [a,b]$, $-\infty < a < b < \infty$ and X be a Banach space. $f: J \to X$ is *absolutely continuous* if for each $\varepsilon > 0$ there exists $\delta > 0$ such that if $\mathcal{P} = \{a = t_0 < t_1 < \ldots < t_n = b\}$ is any partition of J with $\sum_{k=1}^n |t_k - t_{k-1}| < \delta$, then

$\sum_{k=1}^{n} \| f(t_k) - f(t_{k-1}) \| < \varepsilon$. Denote by $AC(J;X)$ the set of all absolutely continuous functions from J to X. Denote by $W^{1,p}(J;X)$ for $1 \leq p \leq \infty$ those functions $f: J \to X$ for which there exists a function $g \in L^p(J;X)$ so that $f(t) = f(a) + \int_a^t g(s)ds$ for every $t \in J$. Consequently, $\frac{df}{dt} = g$, a.e. in J.

Prop (Kōmura): If X is reflexive, then $AC([a,b]; X) \subset W^{1,p}([a,b]; X)$, $1 \leq p \leq \infty$.

9. Sobolev Spaces

Let $\Omega \subset \mathbb{R}^d$ be open. $C^k(\Omega; \mathbb{R})$ is the collection of all functions $u: \Omega \to \mathbb{R}$ which are continuously differentiable up to order k. $C(\Omega; \mathbb{R}) \stackrel{def}{=} C^0(\Omega; \mathbb{R})$ is just the set of continuous functions from Ω into $\mathbb{R}$. Let $\alpha = (\alpha_1, \ldots, \alpha_n)$, where α_i is a nonnegative integer and $|\alpha| = \sum_{i=1}^{d} \alpha_i$. For $u \in C^k(\Omega; \mathbb{R})$,

$$D^\alpha u(x) \stackrel{def}{=} \frac{\partial^{|\alpha|} u(x)}{\partial x_1^{\alpha_1} \cdots \partial x_d^{\alpha_d}}$$

exists and is continuous for $|\alpha| \leq k$. Under the norms

$$\|u\|_{k,p} = \left\{ \int_\Omega \sum_{|\alpha| \leq k} |D^\alpha u(x)|^p dx \right\}^{1/p}, \quad 0 \leq k < \infty, \ 1 \leq p < \infty,$$

$C^k(\Omega; \mathbb{R})$ is a normed linear space. The completion of this space in $L^p(\Omega; \mathbb{R})$ under the norm $\|u\|_{k,p}$ is the Banach space $W^{k,p}(\Omega; \mathbb{R})$. For $r > k$, $W^{r,p}(\Omega; \mathbb{R}) \subset W^{k,p}(\Omega; \mathbb{R})$. Observe that $W^{0,p}(\Omega; \mathbb{R}) = L^p(\Omega; \mathbb{R})$.

Prop (Sobolev Embedding Theorem): If $\Omega \subset \mathbb{R}^d$ is a bounded open set with sufficiently smooth boundary and $\frac{1}{q} > \frac{1}{p} - \frac{k}{d}$ where $1 \leq p < \infty$, $1 \leq q < \infty$, $k \geq 1$, then $W^{k,p}(\Omega; \mathbb{R}) \subset L^q(\Omega; \mathbb{R})$. This inclusion is both dense and compact; by the latter it is meant that the inclusion map is compact.

$C_0^k(\Omega;\mathbb{R})$ is the subspace of $C^k(\Omega;\mathbb{R})$ consisting of all u with *compact support*; that is, u is zero outside a compact subset of Ω. The closure of $C_0^k(\Omega;\mathbb{R})$ in the $\|\cdot\|_{k,p}$ norm is a subspace of $W^{k,p}(\Omega;\mathbb{R})$ denoted by $W_0^{k,p}(\Omega;\mathbb{R})$. It consists of all $u \in W^{k,p}(\Omega;\mathbb{R})$ for which $D^\alpha u = 0$ on the boundary of Ω for all $|\alpha| \leq k-1$. The spaces $W^{k,p}(\Omega;\mathbb{R})$ and $W_0^{k,p}(\Omega;\mathbb{R})$ are reflexive for $1 < p < \infty$.

The spaces $H^k(\Omega;\mathbb{R}) \stackrel{\text{def}}{=} W^{k,2}(\Omega;\mathbb{R})$ and $H_0^k(\Omega;\mathbb{R}) \stackrel{\text{def}}{=} W_0^{k,2}(\Omega;\mathbb{R})$, $k \geq 0$, are *Sobolev spaces*. Under the inner product

$$\langle u,v\rangle_{H^k} = \int_\Omega \sum_{|\alpha|\leq k} D^\alpha u D^\alpha v \, dx,$$

the space $H^k(\Omega;\mathbb{R})$ is a Hilbert space with corresponding norm $\|u\|_{H^k} = \sqrt{\langle u,u\rangle_{H^k}} = \|u\|_{k,2}$. The inclusions

$$H^k(\Omega;\mathbb{R}) \subset H^{k-1}(\Omega;\mathbb{R}) \subset \cdots \subset H^1(\Omega;\mathbb{R}) \subset L^2(\Omega;\mathbb{R})$$

are dense and compact. In case $\Omega = [a,b] \subset \mathbb{R}$, then

$$H^1([a,b];\mathbb{R}) = \{u \in L^2([a,b];\mathbb{R}) : u \in AC([a,b];\mathbb{R}), \frac{du}{dx} \in L^2([a,b];\mathbb{R})\},$$

$$H_0^1([a,b];\mathbb{R}) = \{u \in H^1([a,b];\mathbb{R}) : u(a) = u(b) = 0\}.$$

10. Convexity

A subset K of a linear space X is *convex* if for each $x,y \in K$, $\lambda x + (1-\lambda)y \in K$ for all λ satisfying $0 \leq \lambda \leq 1$. The *convex hull* of a subset S of a linear space X is the intersection of all convex sets containing S, and it is denoted by $\text{co}\, S$. If X is a normed linear space, the *closed convex hull* of S is the intersection of all closed convex sets containing S, and it is denoted by $\overline{co}\ S$. $co\ S$ consists of all linear combinations $\sum_{i=1}^{n} \lambda_i x_i$ of elements $\{x_i\}_{i=1}^{n} \subset S$ for which $0 \leq \lambda_i \leq 1$ and $\sum_{i=1}^{n} \lambda_i = 1$, $n \in \mathbb{N}$. A convex subset of a Banach space is closed if and only if it is weakly closed.

11. Fixed Point Theorems

Suppose X,Y are Banach spaces. An operator $T: \mathscr{D}(T) \subset X \to Y$ is *closed* if for each sequence $\{x_n\} \subset \mathscr{D}(T)$ with $x_n \to x$ and $Tx_n \to y$, then $x \in \mathscr{D}(T)$ and $Tx = y$. T is *compact* if it maps bounded sets in X into compact sets in Y. T is *weakly compact* if it maps bounded sets in X into weakly sequentially compact sets in Y. T is *locally Lipschitz* if for each compact set $K \subset X$, there exists a constant $\ell_K > 0$ so that $\|Tx-Ty\|_Y \leq \ell_K \|x-y\|_X$ for all $x,y \in K$. T is *(globally) Lipschitz* if there exists a constant $\ell > 0$ so that $\|Tx-Ty\|_Y \leq \ell \|x-y\|_X$ for all $x,y \in X$. ℓ is the *Lipschitz constant* for T and T is said to be *ℓ-Lipschitz*. T is *contracting* if $\|Tx-Ty\|_Y \leq \|x-y\|_X$ for all $x,y \in X$.

Prop (Brouwer): Let B denote the closed unit ball in $\mathbb{R}^d$. If $T: B \to B$ is continuous, then T has a fixed point in B.

Prop (Contraction Mapping Principle): If U is a closed subset of a Banach space and $T: U \to U$ is an ℓ-Lipschitz mapping with $\ell < 1$, then T has a unique fixed point in U.

Prop (Schauder): If K is a compact, convex subset of a Banach space and $T: K \to K$ is continuous, then T has a fixed point in K.

Prop (Krasnoselskii): Let K be a closed, bounded, convex subset of a Banach space. Suppose $T: K \to K$ is an ℓ-Lipschitz mapping with $\ell < 1$ and $S: K \to K$ is a compact mapping. Then $T+S$ has a fixed point in K.

Prop (Markov-Kakutani): Suppose K is a compact, convex subset of a Banach space X and $\mathcal{F}$ is a commuting family of continuous affine mappings of K into itself. Then $\mathcal{F}$ has a common fixed point in K.

12. Almost Periodicity

Let X be a Banach space. A continuous function $f: \mathbb{R} \to X$ is *almost periodic* if for each $\varepsilon > 0$ there exists $L = L(\varepsilon) > 0$ such that any interval in $\mathbb{R}$ of length L contains a point τ with $\|f(t+\tau) - f(t)\| < \varepsilon$ for every $t \in \mathbb{R}$. If $f: \mathbb{R} \to X$ is almost periodic, then for each $\theta \in \mathbb{R}$,

$$A(\theta) \overset{\text{def}}{=} \lim_{t\to\infty} \frac{1}{t} \int_0^t f(s)e^{i\theta s}ds$$

exists in X and is nonzero for at most a countable set $\{\theta_k\}$. Let $a_k = A(\theta_k)$. The sum $\sum_{k=1}^{\infty} e^{i\theta_k t} a_k$ is the *Fourier series* for f and we write $f \sim \sum_{k=1}^{\infty} e^{i\theta_k t} a_k$. $\sum_{k=1}^{\infty} \|a_k\|^2 < \infty$ and this representation is unique. There also exists a sequence $\{\sigma_m(t)\} \subset X$ defined by $\sigma_m(t) = \sum_{k=1}^{n} r_{k,m} e^{i\theta_k t} a_k$, where $n = n(m)$, $r_{k,m}$ are rational numbers and do not depend upon a_k and $\sigma_m(t)$ converges to $f(t)$ as $m \to \infty$, uniformly in $t \in \mathbb{R}$.

13. Differential Inequalities

Prop (Gronwall):

(i) Suppose $g \in L^1([a,b]; \mathbb{R})$ with $g(t) \geq 0$ a.e. on $[a,b]$ and $\alpha \in \mathbb{R}^+$. If $\phi \in C([a,b]; \mathbb{R})$ satisfies $\phi(t) \leq \alpha + \int_a^t g(s)\phi(s)ds$ for every $t \in [a,b]$, then $\phi(t) \leq \alpha e^{\int_a^t g(s)ds}$

(ii) Suppose $g \in L^1([a,b];\mathbb{R})$ with $g(t) \geq 0$ a.e. on $[a,b]$. If $\phi,\psi \in C([a,b];\mathbb{R})$ satisfies $\phi(t) \leq \psi(t) + \int_a^t g(s)\phi(s)ds$ for every $t \in [a,b]$, then
$$\phi(t) \leq \psi(t) + \int_a^t g(s)\psi(s)e^{\int_s^t g(\xi)d\xi}ds.$$

(iii) Suppose $g \in L^1([a,b];\mathbb{R})$ with $g(t) \geq 0$ a.e. on $[a,b]$ and $\alpha \in \mathbb{R}^+$. If $\phi \in C([a,b];\mathbb{R})$ satisfies $\phi^2(t) \leq \alpha^2 + 2\int_a^t g(s)\phi(s)ds$ for every $t \in [a,b]$, then $|\phi(t)| \leq \alpha + \int_a^t g(s)ds$.

APPENDIX B

1. Probability Spaces and Random Variables

Given a set Ω and a σ-algebra Σ of subsets of Ω, the pair (Ω,Σ) is called a *measurable space*. If Λ is a family of subsets of Ω, there exists on Ω a minimal σ-algebra $\Sigma(\Lambda)$ which contains Λ. $\Sigma(\Lambda)$ is called the *σ-algebra generated* by Λ. If $\Omega = \mathbb{R}^d$ and Λ is the collection of open subsets of $\mathbb{R}^d$, $\mathscr{B}^d \overset{\text{def}}{=} \Sigma(\Lambda)$ is called the family of *Borel* sets of $\mathbb{R}^d$.

A function P with domain a σ-algebra Σ and range $[0,1]$ is called a *probability measure* on (Ω,Σ) provided (i) $P(\Omega) = 1$ and (ii) $P(\bigcup_{i=1}^{\infty} A_i) = \sum_{i=1}^{\infty} P(A_i)$ whenever $A_i \in \Sigma$, $i = 1,2,\ldots$, with the $\{A_i\}$ mutually disjoint. The triplet (Ω,Σ,P) is called a *probability space*. We can assume without loss of generality that Σ contains all subsets of Ω whose P-measure is zero.

Let (Ω,Σ) be a measurable space. A mapping $\xi\colon \Omega \to \mathbb{R}^d$ for which $\xi^{-1}(B) \in \Sigma$ whenever $B \in \mathscr{B}^d$ is called a *random variable* (or *Σ-measurable*) on (Ω,Σ). The set $\xi^{-1}(B)$ is often denoted by $\{\xi \in B\}$. If $\zeta\colon \Omega \to \mathbb{R}^d$, the family of sets

$\{\zeta \in B\}_{B \in \mathscr{B}^d}$ is a σ-algebra and is denoted by $\Sigma(\zeta)$. It is called the *σ-algebra generated* by ζ. We do not distinguish between random variables ξ and ζ on (Ω,Σ) when $P\{\omega \in \Omega: \xi(\omega) = \zeta(\omega)\} = 1$. In this case we write $\xi = \zeta$ w.p.1 (i.e., with probability one). In fact if a property holds on $\Omega \setminus N$ with $P(N) = 0$, we say this property holds w.p.1.

We say the random variables ξ and ζ on (Ω,Σ,P) are *independent* if

$$(1.1) \qquad P\{\xi \in A, \zeta \in B\} = P\{\xi \in A\}P\{\zeta \in B\}$$

for every $A,B \in \mathscr{B}^d$. This definition extends to an arbitrary collection of random variables if (1.1) holds for every finite combination of the random variables.

For a random variable ξ on (Ω,Σ,P), define the probability measure P_ξ on $(\mathbb{R}^d,\mathscr{B}^d)$ by $P_\xi(B) = P\{\xi \in B\}$, $B \in \mathscr{B}^d$. The function P_ξ is called the *distribution* of ξ.

2. Expectation

Fix a probability space (Ω,Σ,P), and let ξ be a random vairable on (Ω,Σ). The *expectation* of ξ, denoted by $E\xi$, is given by $\int_\Omega \xi(\omega)P(d\omega)$. This is well defined if $E|\xi| < \infty$. For brevity we write $E\xi$ as $\int \xi dP$. If $\phi: \mathbb{R}^d \to \mathbb{R}$ is $\mathscr{B}^d$-measurable with $E|\phi(\xi)| < \infty$, then

$$(2.1) \qquad \int_\Omega \phi(\xi)dP = \int_{\mathbb{R}^d} \phi(x)P_\xi(dx).$$

Let Σ' be a σ-algebra, $\Sigma' \subset \Sigma$. If ξ is a random variable on (Ω,Σ) for which $E|\xi| < \infty$, it follows from the Radon-Nikodym Theorem (c.f. Royden [1], p. 238) that there exists a random variable g on (Ω,Σ') so that for each $A \in \Sigma'$, $\int_A g\, dP = \int_A \xi\, dP$. (Note, ξ need not be Σ'-measurable.) Moreover g is unique w.p.1. that is, any two versions differ on a set of P-measure zero. We write $E(\xi|\Sigma')$ for g and call it the *conditional expectation of* ξ *given* Σ'.

For each $\omega \in \Omega$ and $A \in \Sigma$ define $P(A|\Sigma')(\omega) = E(\chi_A|\Sigma')(\omega)$. Regrettably, $P(\cdot|\Sigma')(\omega)$ need not be a probability on Σ. We can obtain a probability by transferring over to $(\mathbb{R}^d, \mathscr{B}^d)$ via a random variable ξ on (Ω,Σ). Then (c.f. Breiman [1], pp. 77-78) there exists a function p on $\mathscr{B}^d \times \Omega$, called a *regular conditional distribution for* ξ *given* Σ', so that for each fixed $B \in \mathscr{B}^d$, $P(B,\cdot)$ is a version of $P\{\xi \in B|\Sigma'\}$, and for each fixed $\omega \in \Omega$, $p(\cdot,\omega)$ is a probability on $(\mathbb{R}^d, \mathscr{B}^d)$.

If ζ be a random variable on (Ω,Σ), and $\Sigma' = \Sigma(\zeta)$ is the σ-algebra generated by ζ, we define $E(\xi|\zeta)$ to be $E(\xi|\Sigma')$. There exists a $\mathscr{B}^d$-measurable function ϕ so that $E(\xi|\zeta) = \phi(\zeta)$. ϕ is uniquely defined up to a set $N \in \mathscr{B}^d$ of images of ζ for which $P\{\zeta \in N\} = 0$ (c.f. Breiman [1], p. 73). Then

$$(2.2) \quad \int_{\{\zeta\in B\}} \xi dP = \int_{\{\zeta\in B\}} E(\xi|\zeta)dP = \int_B \phi(x)P_\xi(dx), \; B \in \mathscr{B}^d.$$

As $E(\xi|\zeta)(\omega)$ is constant w.p.1. on each subset of Ω for which $\zeta(\omega)$ is constant, we write suggestively $E(\xi|\zeta=x) = \phi(x)$.

For each $B \in \mathscr{B}^d$ consider the $\Sigma' = \Sigma(\zeta)$-measurable function $p(B,\cdot)$. There is a $\mathscr{B}^d$-measurable function $q(B,\cdot)$ defined on $\mathbb{R}^d$ (unique up to a set of P_ζ-measure zero) so that $p(B,\omega) = q(B, \zeta(\omega))$, $\omega \in \Omega$. If $A \in \Sigma$, define $P(A|\zeta=x) = E(\chi_A|\zeta=x)$. For fixed $B \in \mathscr{B}^d$, $q(B,x)$ is a version of $P\{\xi\in B|\zeta=x\}$, and for fixed $x \in \mathbb{R}^d$, $q(\cdot,x)$ is a probability measure on $(\mathbb{R}^d, \mathscr{B}^d)$. We retain the notation $P\{\xi \in B | \zeta = x\}$ for $q(B,x)$. Then for $B \in \mathscr{B}^d$,

$$(2.3) \qquad P\{\xi \in B\} = \int_{\mathbb{R}^d} P\{\xi \in B | \zeta = x\} P_\zeta(dx).$$

When ϕ is any $\mathscr{B}^d$-measurable and integrable function,

$$(2.4) \qquad E\{\phi(\xi) | \zeta = x\} = \int_{\mathbb{R}^d} f(y) q(dy, x).$$

See Breiman [1], p. 79 for details.

3. Convergence of Random Variables

A sequence of random variables $\{\xi_n\}$ on (Ω, Σ, P) can converge to a random variable ξ in each of the following modes:

(i) ξ_n converges to ξ *with probability one* if $P\{\omega: \lim_{n\to\infty} \xi_n(\omega) = \xi(\omega)\} = 1$. In this case we write $\lim_{n\to\infty} \xi_n = \xi$ w.p.1.

(ii) ξ_n converges to ξ *in* p^{th} *mean* $(p > 0)$ if $\lim_{n\to\infty} E|\xi_n - \xi|^p = 0$. If $p = 2$, we say $\lim_{n\to\infty} \xi_n = \xi$ in *mean square*.

(iii) ξ_n converges to ξ *in probability* if for every $\varepsilon > 0$, $\lim_{n\to\infty} P\{\omega: |\xi_n(\omega) - \xi(\omega)| > \varepsilon\} = 0$. In this case write $\lim_{n\to\infty} \xi_n = \xi$ in pr.

4. Stochastic Processes; Martingales and Markov Processes

Fix a probability space (Ω, Σ, P). A *stochastic process* is an indexed collection $\xi = \{\xi(t)\}_{t \in I}$ of $\mathbb{R}^d$-valued random variables on (Ω, Σ) with index set an interval $I \subset \mathbb{R}$ and *state space* $\mathbb{R}^d$. The dependence of $\xi(t)$ on $\omega \in \Omega$, namely $\xi(t,\omega)$ is usually suppressed. For each fixed $\omega \in \Omega$,

$\xi(\cdot,\omega): I \to \mathbb{R}^d$ and $\xi(\cdot,\omega)$ is called a *sample path* or *realization* of the stochastic process.

Let $\{\xi(t)\}_{t\in I}$ be a stochastic process, and suppose $\{\Sigma_t\}_{t\in I}$ is an increasing family of σ-algebras with $\Sigma_t \subset \Sigma$ for all $t \in I$, I an interval in $\mathbb{R}$. If $\xi(t)$ is Σ_t-measurable and $E|\xi(t)| < \infty$ for each $t \in I$, the pair $\{\xi(t),\Sigma_t\}_{t\in I}$ is called a *supermartingale* provided $s \leq t$, $s,t \in I$ implies

$$E(\xi(t)|\Sigma_s) \leq \xi(s) \quad \text{w.p.1.}$$

We have the following properties (c.f. Doob [1]). Let $I = \mathbb{R}^+$.

(i) Convergence: If $\sup_{t\in\mathbb{R}^+} E|\xi(t)| < \infty$, then there exists a random variable ξ_0 such that $\lim_{t\to\infty} \xi(t) = \xi_0$ w.p.1. Moreover, $E|\xi_0| < \infty$.

(ii) Inequality: If $\xi(t) \geq 0$ w.p.1. then

$$P\{\sup_{0\leq t\leq t_0} \xi(t) \geq \lambda\} \leq \lambda^{-1}E|\xi(0)| \tag{4.1}$$

for every $t_0 \in \mathbb{R}^+$ and $\lambda > 0$.

A stochastic process $\{\xi(t)\}_{t\in\mathbb{R}^+}$ on (Ω,Σ,P) is called a *Markov* process if for every $0 \leq s \leq t$ and $B \in \mathscr{B}^d$,

$$P\{\xi(t) \in B|\Sigma_s\} = P\{\xi(t) \in B|\xi(s)\} \quad \text{w.p.1.} \tag{4.2}$$

$\Sigma_t \subset \Sigma$ is the minimal σ-algebra for which the random variables $\{\xi(s): 0 \leq s \leq t\}$ are measurable. If $P\{\xi(t) \in B|\xi(s)\} = P\{\xi(t-s) \in B|\xi(0)\}$ w.p.1. for all $0 \leq s \leq t$, then the Markov process is called *homogeneous*.

Now suppose $\{\xi(t)\}_{t\in\mathbb{R}^+}$ is a homogeneous Markov process. For each $t \in \mathbb{R}^+$ we may use the construction of

Section 2 to find a conditional distribution $\mathscr{P}(t,x,B)$ corresponding to $P\{\xi(t) \in B | \xi(0) = x\}$. The function $\mathscr{P}$ on $\mathbb{R}^+ \times \mathbb{R}^d \times \mathscr{B}^d$ satisfies

(i) for each $t \in \mathbb{R}^+$ and $B \in \mathscr{B}^d$, $\mathscr{P}(t,x,B) = P\{\xi(t) \in B | \xi(0) = x\}$ on $\mathbb{R}^d$ w.p.1,

(ii) for each $t \in \mathbb{R}^+$ and $x \in \mathbb{R}^d$, $\mathscr{P}(t,x,\cdot)$ is a probability measure on $\mathscr{B}^d$,

(iii) for each $t \in \mathbb{R}^+$ and $B \in \mathscr{B}^d$, $\mathscr{P}(t,\cdot,B)$ is $\mathscr{B}^d$-measurable,

(iv) for each $x \in \mathbb{R}^d$ and $B \in \mathscr{B}^d$, $\mathscr{P}(0,x,B) = \chi_B(x)$,

(v) for each $s,t \in \mathbb{R}^+$, $x \in \mathbb{R}^d$, and $B \in \mathscr{B}^d$,
$\mathscr{P}(s+t,x,B) = \int_{\mathbb{R}^d} \mathscr{P}(s,y,B)\, \mathscr{P}(t,x,dy)$.

$\mathscr{P}(t,x,B)$ is called the *Markov transition function*. In a rather obvious way the finite dimensional distributions $P\{\xi(t_1) \in B_1,\ldots,\xi(t_k) \in B_k\}$, $0 \le t_1 < \ldots < t_k$, $B_i \in \mathscr{B}^d$, can be obtained from the Markov transition function. In particular, if $\xi(0)$ has distribution P_0, then

$$P\{\xi(t) \in B\} = \int_{\mathbb{R}^d} \mathscr{P}(t,x,B) P_0(dx).$$

Brownian motion $W = \{W(t)\}_{t \in \mathbb{R}^+}$ is a homogeneous Markov process with

$$\mathscr{P}(t,x,B) = \int_B (2\pi t)^{-d/2} \exp[-|x-y|^2/2t]\, dy. \tag{4.3}$$

Brownian motion enjoys the following properties:

(i) $E\xi(t) = 0$ for all $t \in \mathbb{R}^+$,

(ii) $E|\xi(t) - \xi(s)|^2 = |t-s|$,

(iii) For any sequence $0 \le t_1 \le t_2 < \ldots < t_k$, the random variables $\xi(t_1) - \xi(0)$, $\xi(t_2) - \xi(t_1)$, $\xi(t_k) - \xi(t_{k-1})$ are independent.

5. The Ito Stochastic Integral

Suppose $W = \{W(t): t \in \mathbb{R}^+\}$ is $\mathbb{R}^p$-valued Brownian motion on (Ω,Σ,P), and let $\Sigma_t \subset \Sigma$ be the minimal σ-algebra for which the family $\{W(t): 0 \leq s \leq t\}$ is measurable. Consider a function $f: \mathbb{R}^+ \times \Omega \to \mathbb{R}^{dp}$ (i.e., $f(t,\omega)$ is an $d \times p$ matrix) so that $f(t,\omega)$ has at most finitely many different values for each fixed $\omega \in \Omega$. Moreover let $f(t,\cdot)$ be Σ_t-measurable for each $t \in \mathbb{R}^+$. For any partition $\{t_0 < t_1 < \ldots < t_N = t\}$ of $[t_0,t]$, define

$$I(f) = \int_{t_0}^{t} f(s)dW(s) = \sum_{k=0}^{N-1} f(t_k)[W(t_{k+1})-W(t_k)]$$

where $f(s) = f(t_k)$ for $t_k \leq s < t_{k+1}$. We have suppressed the dependence of f upon $\omega \in \Omega$. Then

(i) $E\{I(f)\} = 0$

(ii) $I(\alpha f+\beta g) = \alpha I(f) + \beta I(g),\ \alpha,\beta \in \mathbb{R}$,

(iii) $E|I(f)|^2 = \int_{t_0}^{t} E|f(s)|^2 ds$.

Now suppose $f(t,\cdot)$ is Σ_t-measurable for each $t \in \mathbb{R}^+$, and that $f \in \mathscr{L}^2 \overset{\text{def}}{=} L^2([t_0,t] \times \Omega,\ m\times P)$. m is Lebesgue measure. f has norm $\{\int_{t_0}^{t} E|f(s)|^2 ds\}^{1/2}$. There exists step functions f_j of the sort defined above so that $\lim_{j\to\infty} f_j = f$ in $\mathscr{L}^2$. As the sequence $\{I(f_j)\} \subset L^2(\Omega,P)$ can be shown to be Cauchy, we define $I^*(f) = \lim_{j\to\infty} I(f_j)$; that is, $\lim_{j\to\infty} E|I^*(f) - I(f_j)|^2 = 0$. Thus I^* is an extension of I to all of $L^2(\Omega \times [t_0,t],\ P \times \lambda)$ and satisfies properties (i), (ii), and (iii) of $I(f)$. For further details, see Doob [1], Chapter VIII.

REFERENCES

Artstein, Z. [1], Continuous dependence of solutions of Volterra integral equations, SIAM J. Math. Anal. 6 (1975), 446-456.

__________ [2], Topological dynamics of an ordinary differential equation, J. Differential Equations 23 (1977), 216-223.

__________ [3], Topological dynamics of ordinary differential equations and Kurzweil equations, J. Differential Equations 23 (1977), 224-243.

__________ [4], The limiting equations of nonautonomous ordinary differential equations, J. Differential Equations 25 (1977), 184-202.

__________ [5], Uniform asymptotic stability via the limiting equations, J. Differential Equations 27 (1978), 172-189.

Artstein, Z.; Infante, E. F. [1], On the asymptotic stability of oscillators with unbounded damping, Quart. J. Appl. Math. 34 (1976), 195-199.

Baillon, J. B. [1], Générateurs et semi-groupes dans les espaces de Banach uniformement lisses, J. Funct. Anal. 29 (1978), 199-213.

Baillon, J. B.; Brezis, H. [1], Une remarque sur le comportement asymptotic des semigroupes non linéaires, Houston J. Math. 2 (1976), 5-7.

Baillon, J. B.; Bruck, R. E.; Reich, S. [1], On the asymptotic behavior of nonexpansive mappings and semigroups in Banach spaces, Houston J. Math. 4 (1978), 1-9.

Ball, J. M. [1], Continuity properties of nonlinear semigroups, J. Funct. Anal. 17 (1974), 91-103.

__________ [2], Strongly continuous semigroups, weak solutions, and the variation of constants formula, Proc. Amer. Math. Soc. 63 (1977), 370-373.

__________ [3], On the asymptotic behavior of generalized processes with applications to nonlinear evolution equations, J. Differential Equations 27 (1978), 224-265.

Ball, J. M.; Peletier, L. A. [1], Global attraction for the one dimensional heat equation with nonlinear time-dependent boundary conditions, Arch. Rat. Mech. Anal. 65 (1977), 193-201.

Barbu, V. [1], Nonlinear Semigroups and Differential Equations in Banach Spaces, Noordhoff International Publishing, Leyden, The Netherlands, 1976.

Bender, P. R. [1], Recurrent solutions to systems of ordinary differential equations, J. Differential Equations 5 (1969), 271-282.

Benes, V. E. [1], Existence of finite invariant measures for Markov processes, Proc. Amer. Math. Soc. 18 (1967).

__________ [2], Finite regular invariant measures for Feller processes, SIAM J. Control, 5 (1968), 203-209.

Bénilan, P. [1], Une remark sur la convergence des semi-groupes non linéaires, Acad. Sci. Paris, Sér. A. 272 (1971), 1182-1184.

__________ [2], Solutions intégrales d'equations d'évolution dans un espace de Banach, Acad. Sci. Paris, Sér. A. 274 (1972), 47-50.

Bénilan, P.; Brezis, H. [1], Solutions faibles d'équations d'évolution dans les espacés de Hilbert, Ann. Inst. Fourier, Grenoble 22 (1972), 311-329.

Bhatia, N. P. [1], Weak attractors in dynamical systems, Bol. Soc. Mat. Mexicana 11 (1966), 56-64.

__________ [2], Semidynamical systems. In "Mathematical Systems Theory and Economics II," W. H. Kuhn and G. P. Szego (eds.), Lecture Notes in Operations Research and Mathematical Economics, Vol. 12, Springer-Verlag, Berlin-Heidelberg-New York, 1969, pp. 303-318.

__________ [3], Attraction and nonsaddle sets in dynamical systems, J. Differential Equations 8 (1970), 229-249.

__________ [4], Characteristic properties of stable sets and attractors in dynamical systems, in "Symposia Mathematics," Vol. VI, Academic Press, New York, 1971, pp. 155-166.

__________ [5], Semi-Dynamical Systems, unpublished notes, 1971.

Bhatia, N. P.; Hajek, O. [1], "Local Semi-Dynamical Systems," Lecture Notes in Mathematics, Vol. 90, Springer-Verlag, Berlin-Heidelberg-New York, 1969.

________ [2], "Theory of Dynamical Systems, I," Technical Note BN-599, University of Maryland, IFDAM, 1969.

________ [3], "Theory of Dynamical Systems, IV," Technical Note BN-610, University of Maryland, IFDAM, 1969.

Bhatia, N. P.; Szego, G. P. [1], "Stability Theory of Dynamical Systems," Grundleheren d. math. Wiss. 161, Springer-Verlag, Berlin-Heidelberg-New York, 1970.

Bhatia, N. P.; Chow, S. N. [1], Weak attraction, minimality, recurrence, and almost periodicity in semi-systems, Funkcial, Ekvac. 15 (1972), 39-60.

Billotti, J. E.; LaSalle, J. P. [1], Dissipative periodic processes, Bull. Am. Math. Soc. 17 (1971), 1082-1088.

Birkhoff, G. D. [1], "Dynamical Systems," Amer. Math. Soc. Colloquium Publications, Vol. 9, New York, 1927.

Bochner, S. [1], Beitrage zur Theorie der Fastperiodischen Funktionen, I. Math. Ann. 96 (1926), 119-147.

Bondi, P.; Moauro, V.; Visentin, F. [1], Limiting equttions in the stability problem, Nonlinear Analysis 1 (1977), 123-128.

Boyarsky, A. [1], Limit sets of dynamical systems on the space of probability measures, J. Differential Equations 14 (1973), 559-567.

________ [2], Stochastic stability theory using the second-order infinitesimal generator, Int. J. Control 20 (1974), 857-863.

Breiman, L. [1], "Probability," Addison-Wesley, Reading, Mass., 1968.

Brezis, H. [1], On a problem of T. Kato, Comm. Pure Appl. Math. 24 (1971), 1-6.

________ [2], "Operateurs Maximaux Monotones," North Holland, Amsterdam, 1973.

Brezis, H; Pazy, A. [1], Accretive sets and differential equations in Banach spaces, Israel J. Math. 8 (1970), 367-383.

Bruck, R. E. [1], Asymptotic convergence of nonlinear contraction semigroups in Hilbert space, J. Functional Analysis 18 (1975), 15-26.

Carlson, D. H. [1], A generalization of Vinograd's theorem for dynamical systems, J. Differential Equations 11 (1972), 193-201.

Cellina, A. [1], On the nonexistence of solutions of differential equations in nonreflexive spaces, Bull. Amer. Math. Soc. 78 (1972), 1069-1072.

Chafee, N. [1], A stability analysis for a semilinear parabolic partial differential equation, J. Differential Equations 15 (1974), 522-540.

_________ [2], Asymptotic behavior for solutions of a one-dimensional parabolic equation with homogeneous Neumann boundary condition, J. Differential Equations 18 (1975), 111-134.

Chafee, N.; Infante, E. F. [1], Bifurcation and stability for a nonlinear parabolic partial differential equation, Bull. Amer. Math. Soc. 80 (1974), 49-52.

_________ [2], A bifurcation problem for a nonlinear partial differential equation of parabolic type, Applicable Anal. 4 (1974), 17-37.

Chow, S. N. [1], Perturbing almost periodic differential equations, Bull. Amer. Math. Soc. 76 (1970), 421-424.

_________ [2], Remarks on one-dimensional delay-differential equations, J. Math. Anal. Appl. 41 (1973), 426-429.

_________ [3], Existence of periodic solutions of autonomous functional differential equations, J. Differential Equations 15 (1974), 350-378.

Chow, S. N.; Hale, J. K. [1], Strongly limit-compact maps, Funkcial. Ekvac. 17 (1974), 31-38.

_________ [2], Periodic solutions of autonomous equations (to appear).

Chow, S. N.; Schur, J. D. [1], An existence theorem for ordinary differential equations in Banach spaces, Bull. Amer. Math. Soc. 77 (1971), 1018-1020.

Coddington, E.; Levinson, N. [1], "Theory of Ordinary Differential Equations," McGraw-Hill, New York, 1955.

Coleman, B. D.; Mizel, V. J. [1], Norms and semi-groups in the theory of fading memory, Arch. Rat. Mech. Anal. 23 (1966), 87-123.

_________ [2], A general theory of dissipation in materials with memory, Arch. Rat. Mech. Anal. 27 (1967), 255-274.

_________ 3 , On the general theory of fading memory, Arch. Rat. Mech. Anal. 29 (1968), 18-31.

_________ 4 , On the stability of solutions of functional-differential equations, Arch. Rat. Mech. Anal. 30 (1968), 173-196.

Cooke, K.; Yorke, J. A. [1], Some equations modelling growth processes and Gonorrhea Epidemics, Math. Biosci. 16 (1973), 75-101.

Corduneanu, C. [1], "Almost Periodic Functions," Interscience Publishers, New York, 1968.

_______ [2], Some differential equations with delay, Proc. EQUADIFF III (Czechoslovak Conference on Differential Equations and their Applications), J. E. Purkyne Univ., Brno, 1973, 139-143.

_______ [3], Functional equations with infinite delay, Boll. U.M.I. (4) 11 (1975), 173-181.

Costello, T. [1], On the fundamental theory of functional differential equations, Funkcial. Ekvac. 14 (1971), 177-190.

Crandall, M. G. [1], Differential equations on convex sets, J. Math. Soc. Japan 22 (1970), 443-455.

_______ [2], Semigroups of nonlinear transformations, in "Contributions to Nonlinear Functional Analysis" (E. Zarantonello, Ed.), Academic Press, New York, 1971, 157-179.

_______ [3], The semigroup approach to first order quasilinear equations in several space variables, Israel J. Math. 12 (1972), 108-132.

_______ [4], A generalized domain for semigroup generators, Proc. Amer. Math. Soc. 37 (1973), 434-440.

_______ [5], An introduction to evolution governed by accretive operators, "Dynamical Systems, An International Symposium," L. Cesari; J. K. Hale; J. P. LaSalle, eds., Academic Press, New York, 1976.

Crandall, M. G.; Liggett, T. M. [1], Generation of semigroups of nonlinear transformations on general Banach spaces, Amer. J. Math. 93 (1971), 265-293.

Crandall, M. G.; Pazy [1], Semigroups of nonlinear contractions and dissipative sets, J. Functional Analysis 3 (1969), 376-418.

_______ [2], Nonlinear evolution equations in Banach spaces, Israel J. of Math. 11 (1972), 57-94.

Cruz, M. A.; Hale, J. K. [1], Stability of functional differential equations of neutral type, J. Differential Equations 7 (1970), 334-355.

Cunningham, W. J. [1], A nonlinear differential-difference equation of growth, Proc. Nat. Acad. Sci. U.S.A. 40 (1954), 709-713.

Dafermos, C. M. [1], Wave equations with weak damping, SIAM J. A-pl. Math. 18 (1970), 759-767.

_________ [2], Asymptotic stability in viscoelasticity, Arch. Rat. Mech. Anal. 37 (1970), 297-308.

_________ [3], An invariance principle for compact processes, J. Differential Equations 9 (1971), 239-252. Erratum, J. Differential Equations 10 (1971), 179-180.

_________ [4], Applications of the invariance principle for compact processes, I. Asymptotically dynamical systems, J. Differential Equations 9 (1971), 291-299.

_________ [5], Applications of the invariance principle for compact processes, II. Asymptotic behavior of solutions of a hyperbolic conservation law, J. Differential Equations 11 (1972), 416-424.

_________ [6], Uniform processes and semicontinuous Liapunov functionals, J. Differential Equations 11 (1972), 401-415.

_________ [7], Semiflows associated with compact and uniform processes, Math. Systems Theory 8 (1974), 142-149.

_________ [8], Contraction semigroups and trend to equilibrium in continuum mechanics, in "Applications of Methods of Functional Analysis to Problems in Mechanics," Lecture Notes in Mathematics 503, Springer-Verlag, Berlin, 1976, 295-306.

_________ [9], Almost periodic processes and almost periodic solutions of evolution equations, "Dynamical Systems," A. Bednarek & L. Cesari eds., Academic Press, New York, 1977, 43-57.

_________ [10], Asymptotic behavior of solutions of evolution equations, "Nonlinear Evolution Equations," M. Crandall, ed., Academic Press, New York, 1978, 103-121.

Dafermos, C. M.; Slemrod, M. [1], Asymptotic behavior of nonlinear contraction semigroups, J. Functional Analysis 13 (1973), 97-106.

Datko, R. [1], Uniform asymptotic stability of evolutionary processes in a Banach space, SIAM J. Math. Anal. 3 (1972), 428-445.

Delfour, M. C.; Mitter, S. K. [1], Hereditary differential systems with constant delays. I. General case, J. Differential Equations 12 (1972), 213-235.

_________ [2], Hereditary differential systems with constant delays. II. A class of affine systems and the adjoint problem, J. Differential Equations, 18 (1975), 18-28.

Della Riccia, G. [1], Equicontinuous semi-flows (one-parameter semi-groups) on locally compact of complete metric spaces, Math. Systems Theory 4 (1970), 29-34.

Deysach, L. G.; Sell, G. R. [1], On the existence of almost periodic motions, Mich. Math. J. 12 (1965), 87-95.

Dieudonné, J. [1], Deux examples singuliers d'équations différentielles, Acta Sci. Math. Szeged (Leopoldo Fejér Frederico Riesz LXX annus natis dedicatus, pars B) 12 (1950), 38-40.

DiPasquantonio, F.; Kappel, F. [1], Applications to nuclear reactor kinetics of an extension of Lyapunov's direct method to functional differential equations, Energ. Nucl. 15 (1968), 761-770.

Doob, J. L. [1], "Stochastic Processes," John Wiley & Sons, New York, 1953.

Dunford, N.; Schwartz, J. T. [1], "Linear Operators" Part I, Interscience Publishers, New York, 1967.

Dynkin, E. B. [1], "Markov Processes," Vol. I, Springer-Verlag, New York, 1965.

Edelstein, M. [1], On non-expansive mappings of Banach spaces, Proc. Camb. Phil. Soc. 60 (1964), 439-447.

Egawa, J. [1], A remark on the flow near a compact invariant set, Proc. Japan Acad. 49 (1973), 247-251.

Evans, L. C. [1], Differentiability of a nonlinear semigroup in L^1, J. Math. Anal. Appl. 60 (1977), 703-715.

Evans, L. C.; Massey III, F. J. [1], A remark on the construction of nonlinear evolution operators, Houston J. Math 4 (1978), 35-41.

Feller, W. [1], "An Introduction to Probability Theory and Its Applications," Vol. II, John Wiley & Sons, New York, 1966.

Fennell, R. E. [1], Periodic solutions of functional differential equations, J. Math. Anal. Appl. 39 (1972), 198-201.

Fitzgibbon, W. E. [1], Weakly continuous nonlinear accretive operators in reflexive Banach spaces, Proc. Amer. Math. Soc. 41 (1973), 229-236.

__________ [2], Stability for abstract nonlinear Volterra equations involving finite delay, J. Math. Anal. Appl. 60 (1977), 429-434.

Flaschka, H.; Leitman, M. J. [1], On semigroups of nonlinear operators and the solution of the functional differential equation $\dot{x}(t) = F(x_t)$, J. Math. Anal. Appl. 49 (1975), 649-658.

Flugge Lots, I. [1], "Discontinuous Automatic Control," Princeton Univ. Press, Princeton, 1953.

Friedman, A. [1], Small random perturbations of dynamical systems and applications to parabolic equations. Indiana University Math. J., 24 (1974), 533-553.

Gihman, I. I.; Skorohod, A. V. [1], "Stochastic Differential Equations," Springer-Verlag, New York, 1972.

Hajek, O. [1], Structure of dynamical systems, Comment. Math. Univ. Carolinae 6 (1965), 53-72; Correction, Ibid. 6 (1965), 211-212.

Halanay, A.; Yorke, J. A. [1], Some new results and problems in the theory of differential-delay equations, SIAM Rev. 13 (1971), 55-80.

Hale, J. K. [1], A stability theorem for functional-differential equations, Proc. Nat. Acad. Sci. U.S.A. 50 (1963), 942-946.

________ [2], Sufficient conditions for stability and instability of autonomous functional-differential equations, J. Differential Equations 1 (1965), 452-482.

________ [3], Dynamical systems and stability, J. Math. Anal. Appl. 26 (1969), 39-59.

________ [4], "Ordinary Differential Equations," Wiley-Interscience, New York, 1969.

________ [5], A class of neutral equations with the fixed point property, Proc. Nat. Acad. Sci. U.S.A. 67 (1970), 136-137.

________ [6], Forward and backward continuation for neutral functional differential equations, J. Differential Equations 9 (1971), 168-181.

________ [7], Critical cases for neutral functional differential equations, J. Differential Equations 10 (1971), 59-82.

________ [8], Functional differential equations, in "Appl. Math. Sci.," Vol. 3, Springer-Verlag, New York, 1971.

________ [9], Local behavior of autonomous neutral functional differential equations, "Ordinary Differential Equations," pp. 95-107, Academic Press, New York, 1972.

________ [10], Some infinite-dimensional dynamical systems, "Dynamical Systems," Proc. Symp. at Univ. of Bahia, Salvador, Brazil, July 26 - Aug. 14, 1971, Academic Press, New York, 1973.

________ [11], Smoothing properties of neutral equations, An. Acad. Brazil Cienc. 45 (1973), 49-50.

________ [12], Behavior near constant solutions of functional differential equations, J. Differential Equations, 15 (1974), 278-294.

________ [13], Functional differential equations with infinite delays, J. Math. Anal. Appl. 48 (1974), 276-283.

Hale, J. K.; Cruz, M. A. [1], Asymptotic behavior of neutral functional differential equations, Arch. Rational Mech. Anal. 34 (1969), 331-353.

________ [2], Existence, uniqueness, and continuous dependence for hereditary systems, Ann. Mat. Pura Appl. (4) 85 (1970), 63-81.

Hale, J. K.; Imaz, C. [1], Existence, uniqueness, continuity, and continuation of solutions for retarded differential equations, Bol. Soc. Mat. Mex. (2) 11 (1966), 29-37.

Hale, J. K.; Infante, E. F. [1], Extended dynamical systems and stability theory, Proc. Nat. Acad. Sci. U.S.A. 58 (1967), 405-409.

Hale, J. K.; LaSalle, J. P.; Slemrod, M. [1], Theory of a general class of dissipative processes, J. Math. Anal. Appl. 39 (1972), 171-191.

Hale, J. K.; Lopes, O. [1], Fixed point theorems and dissipative processes, J. Differential Equations 13 (1973), 391-402.

Hale, J. K.; Meyer, K. [1], A class of functional differential equations of neutral type, Mem. Amer. Math. Soc. 76 (1967).

Hale, J. K.; Oliva, W. M. [1], One-to-oneness for linear retarded functional differential equations, J. Differential Equations 20 (1976), 28-36.

Hastings, S. P. [1], Backward existence and uniqueness for retarded functional differential equations, J. Differential Equations 5 (1969), 441-451.

Hénon, M. [1], A two-dimensional mapping with a strange attraction, Comm. Math. Phys. 50 (1976), 69-77.

Henry, D. [1], Small solutions of linear autonomous functional differential equations, J. Differential Equations 8 (1970), 494-501.

Hino, Y. [1], Asymptotic behavior of solutions of some functional differential equations, Tohoku Math. J. 22 (1970), 98-108.

________ [2], Continuous dependence for some functional differential equations, Tohoku Math. J. 23 (1971), 565-571.

_____ [3], On stability of the solution of some functional differential equations, Funkcial. Ekvac. 14 (1971), 47-60.

Hirsch, M. W.; Smale, S. [1], "Differential Equations, Dynamical Systems, and Linear Algebra," Academic Press, New York, 1974.

Jamison, B. [1], Asymptotic behavior of successive iterates of continuous functions under a Markov operator, J. Math. Anal. Appl. 9 (1964), 203-214.

Jones, G. S. [1], Hereditary structure in differential equations, Math. Systems Theory 1 (1967), 263-278.

_____ [2], Stability of compactness for functional differential equations, "Ordinary Differential Equations," pp. 433-458, Academic Press, New York, 1972.

_____ [3], Basic frequency periodic motions in noncompact dynamical processes, "Delay and Functional Differential Equations and their Applications," pp. 185-196, Academic Press, New York, 1972.

Kamke, E. [1], "Differentialgleichungen: Lösungsmethoden und Lösungen," Band 1, 3rd ed., Akademische Verlagsgesellschaft, Leipzig, 1944.

Kaplan, J. L.; Yorke, J. A. [1], On the stability of a periodic solution of a differential delay equation, SIAM J. Math. Anal. 6 (1975), 268-282.

_____ [2], Toward a unification of ordinary differential equations with nonlinear semi-group theory, in "International Conf. Differential Equations," H. A. Antosiewicz, ed., Academic Press, New York, 1975, 424-433.

Kappel, F. [1], The invariance of limit sets for autonomous functional-differential equations, SIAM J. Appl. Math. 19 (1970), 408-419.

Kappel, F.; DiPasquantonio, F. [1], Stability criteria for kinetic reactor equations, Arch. Rational Mech. Anal. 58 (1975), 317-338.

Kato, J. [1], Uniformly asymptotic stability and total stability, Tohoku Math. J. 22 (1970), 254-269.

Kato, J.; Yoshizawa, T. [1], A relationship between uniformly asymptotic stability and total stability, Funkcial. Ekvac. 12 (1969), 233-238.

Kato, T. [1], Nonlinear semigroups and evolution equations, J. Math. Soc. Japan 19 (1967), 508-520.

_____ [2], Accretive operators and nonlinear evolution equations in Banach spaces, in "Proc. Symp. Nonlinear Functional Analysis," Chicago, Amer. Math. Soc., 1968.

Kelley, J. L. [1], "General Topology," Van Nostrand, Princeton, 1955.

Kōmura, Y. [1], Nonlinear semigroups in Hilbert space, J. Math. Soc. Japan 19 (1967), 493-507.

Kushner, H. J. [1], The concept of invariant set for stochastic dynamical systems and applications to stochastic stability, in "Stochastic Optimization and Control," Ed. H. F. Karreman, John Wiley & Sons, 1968, 47-57.

__________ [2], Stochastic stability, in "Stability of Stochastic Dynamical Systems," Lecture Notes in Mathematics 294, Springer-Verlag, Berlin, 1972.

Ladas, G. E.; Lakshmikantham, V. [1], "Differential Equations in Abstract Spaces," Academic Press, New York, 1972.

Landau, L. D.; Lifschitz, E. M. [1], "Fluid Mechanics," Pergamon Press, Oxford, 1959.

LaSalle, J. P. [1], The extent of asymptotic stability, Proc. Nat. Acad. Sci. U.S.A. 46 (1960), 363-365.

__________ [2], Some extensions of Lyapunov's second method, IRE Trans. Circuit Theory CT-7 (1960); 520-527.

__________ [3], Asymptotic stability criteria, Hydrodynamic Instability in "Proc. of Symposia in Applied Mathematics" 13, American Mathematical Society, Providence, R.I., 1962, 299-307.

__________ [4], An invariance principle in the theory of stability, Differential Equations and Dynamical Systems, in "Proc. International Symposium," Puerto Rico, Academic Press, New York, 1967, 277-286.

__________ [5], Stability theory for ordinary differential equations, J. Differential Equations 4 (1968), 57-65.

__________ [6], Generalized invariance principles and the theory of stability in "INDAM, Symposia Mathematica," Vol. VI, Academic Press, New York, 1971, 355-360.

__________ [7], Dissipative systems, in "Ordinary Differential Equations," Academic Press, New York, 1972, 165-174.

__________ [8], "The Stability of Dynamical Systems," Vol. 25, Soc. Indust. Appl. Math., Phila., 1975.

__________ [9], Stability theory and invariance principles, "Dynamical Systems, An International Symposium," Academic Press, New York, 1976.

__________ [10], Stability of nonautonomous systems, Nonlinear Analysis 1 (1976), 83-91.

Levin, J. J.; Nohel, J. A. [1], Global asymptotic stability for nonlinear systems of differential equations and applications to reactor dynamics, Arch. Rat. Mech. Anal. 5 (1960), 194-211.

________ [2], On a nonlinear delay equation, J. Math. Anal. Appl. 8 (1964), 31-44.

Li, T.-Y. [1], Existence of solutions for ordinary differential equations in Banach spaces, J. Differential Equations 18 (1975), 29-40.

________ [2], Bounds for the periods of periodic solutions of differential delay equations, J. Math. Anal. Appl. 49 (1975), 124-129.

Li, T.-Y.; York, J. A. [1], The "simplest" dynamical system, in "Dynamical Systems, An International Symposium," Vol. 2, L. Cesari, J. K. Hale, and J. P. LaSalle (eds.), Academic Press, New York, 1976, pp. 203-206.

________ [2], Period three implies chaos, Am. Math. Monthly 82 (1975), 985-992.

Lillo, J. C. [1], Backward continuation of retarded functional differential equations, J. Differential Equations, 17 (1975), 349-360.

Lions, J. L. [1], "Équations Differentielles-Opérationnelles et Problèms aux Limites," Springer-Verlag, Berlin, 1961.

Lopes, O. [1], Periodic solutions of perturbed neutral differential equations, J. Differential Equations 15 (1974), 70-76.

Lyapunov, A. M. [1], Probleme general de la stabilite du mouvement, Annals of Math. Studies, No. 17, Princeton, Princeton Univ. Press, 1947.

MacCamy, R. C. [1], Exponential stability for a class of functional differential equations, Arch. Rational Mech. Anal. 40 (1970/71), 120-138.

Markov, A. A. [1], Sur une propriete generale des ensembles minimaux de M. Birkhoff, C. R. Acad. Sci. Paris 193 (1931), 823-825.

Markus, L. [1], Asymptotically autonomous differential systems, Ann. of Math. Studies, No. 36, Princeton, 1956, 17-29.

________ [2], "Lectures in Differential Dynamics," CBMS No. 3, Amer. Math. Soc., Providence, 1971.

Marotto, F. R. [1], Snap-back repellers imply chaos in $\mathbb{R}^n$, J. Math. Anal. Appl. 63 (1978), 199-223.

Martin, R. H., Jr. [1], Differential equations on closed subsets of a Banach space, Trans. Amer. Math. Soc. 179 (1973), 399-414.

May, R. E. [1], Biological populations with nonoverlapping generations: stable points, stable cycles, and chaos, Science 186 (1974), 645-647.

________ [2], Biological populations obeying difference equations: stable points, stable cycles, and chaos, J. Theor. Biol. 51 (1975), 511-524.

________ [3], Simple mathematical models with very complicated dynamics, Nature 261 (1976), 459-467.

May, R. E.; Oster, G. F. [1], Bifurcations and dynamic complexity in simple ecological models, The American Naturalist 110 (1976), 373-379.

May, R. [1], "Model Ecosystems," Princeton University Press, Princeton, N.J., 1973.

McCann, R. C. [1], An embedding theorem for semidynamical systems, Funkcial. Ekvac. 18 (1975), 23-34.

Melvin, W. R. [1], A class of neutral functional differential equations, J. Differential Equations 12 (1972), 524-534.

________ [2], Topologies for neutral functional differential equations, J. Differential Equations 13 (1973), 24-31.

________ [3], Stability properties of functional difference equations, J. Math. Anal. Appl. 48 (1974), 749-763.

________ [4], Liapunov's direct method applied to neutral functional differential equations, J. Math. Anal. Appl. 49 (1975), 47-58.

Miller, R. K. [1], Asymptotic behavior of solutions of nonlinear differential equations, Trans. Amer. Math. Soc. 115 (1965), 400-416.

________ [2], Asymptotic behavior of nonlinear delay-differential equations, J. Differential Equations 1 (1965), 293-305.

________ [3], Almost periodic differential equations as dynamical systems with applications to the existence of a.p. solutions, J. Differential Equations 1 (1965), 337-345.

________ [4], The topological dynamics of Volterra integral equations, Studies in Appl. Math. 5; Advances in Diff. & Integ. Eqns. SIAM, 1969, 82-87.

________ [5], Linear Volterra integrodifferential equations as semigroups, Funkcial. Ekvac. 17 (1974), 39-55.

Miller, R. K.; Sell, G. [1], A note on Volterra integral equations and topological dynamics, Bull. Amer. Math. Soc. 74 (1968), 904-908.

__________ [2], Existence, uniqueness and continuity of solutions of integral equations, Ann. Math. Pura Appl. (4) 8 (1968), 135-152.

__________ [3], Existence, uniqueness and continuity of solutions of integral equations. An addendum, Ann. Math. Pura Appl. (4) 84 (1970), 281-286.

__________ [4],"Volterra Integral Equations and Topological Dynamics," Mem. Amer. Math. Soc. 102 (1970).

__________ [5], Topological dynamics and its relation to integral equations and nonautonomous systems, in "Dynamical Systems, An International Symposium," L. Cesari; J. K. Hale; J. P. LaSalle, eds., Academic Press, New York, 1976.

Miyadera, I. [1], Some remarks on semigroups of nonlinear operators, Tôhoku Math. J. 23 (1971), 245-258.

__________ [2], Generation of semigroups of nonlinear contractions, J. Math. Soc. Japan 26 (1974), 389-404.

Miyadera, I.; Oharu, Shinnosuke [1], Approximation of semigroups of nonlinear operators, Tôhoku Math. J. 22 (1970), 24-47.

Miyahara, Y. [1], Invariant measures of ultimately bounded stochastic processes, Nagoya Math. J. 49 (1973), 149-153.

Morgan, A. P.; Narenda, K. S. [1], On the uniform asymptotic stability of certain linear nonautonomous differential systems, SIAM J. Control 15 (1977), 5-24.

Nemytski, V. V.; Stepanov, V. V. [1], "Qualitative Theory of Differential Equations," Princeton Univ. Press, Princeton, 1960.

Nussbaum, R. D. [1], A global bifurcation theorem with applications to functional differential equations, J. Functional Analysis 19 (1975), 1-20.

Onuchic, N. [1], Stability properties of a second order differential equation, Acta Mexicana de Ciencia y Tecnologia III (1969), 6-11.

__________ [2], Invariance properties in the theory of ordinary differential equations with applications to stability problems, SIAM J. Control 9 (1971), 97-104.

Pao, C. V.; Vogt, W. G. [1], On the stability of nonlinear operator differential equations, and applications, Arch. Rat. Mech. Anal. 35 (1969), 16-29.

Pao, C. V. [1], Semigroups and asymptotic stability of nonlinear differential equations, SIAM J. Math. Anal. 3 (1972), 371-379.

_________ [2], On the asymptotic stability of differential equations in Banach spaces, Math. Systems Theory 7 (1973), 25-31.

Papini, P. L. [1], Un 'osservazione sui semi-gruppi di isometric in certi-spazi di Banach, Boll. U.M.I. (4) 2 (1969), 682-685.

Parthasarathy, K. R. [1], "Probability Measures on Metric Spaces," Academic Press, New York, 1967.

Pazy, A. [1], On the applicability of Lyapunov's theorem in Hilbert space, SIAM J. Math. Anal. 3 (1972), 291-294.

_________ [2], A class of semi-linear equations of evolution, Israel J. Math. 20 (1975), 23-36.

_________ [3], On the asymptotic behavior of semigroups of nonlinear contractions in Hilbert spaces, J. Functional Analysis 27 (1978), 292-307.

Peng, T. K. C. [1], Invariance and stability for bounded uncertain systems, SIAM J. Control 10 (1972), 679-690.

Reich, S. [1], Asymptotic behavior of semigroups of nonlinear contractions in Hilbert spaces, Atti Accad. Naz. Lincei Rend. Cl. Sci. Fis. Mat. Natur. 56 (1974), 864-872.

_________ [2], Asymptotic behavior of semigroups of nonlinear contractions in Banach spaces, J. Math. Anal. Appl. 53 (1976), 277-290.

Rouché, N. [1], The invariance principle applied to noncompact limit sets, Boll. U.M.I. (4) 11 (1975), 306-315.

_________ [2], The invariance of limit sets for retarded differential equations, Ann. Mat. Pura Appl. (4) 108 (1976), 125-135.

Royden, H. L. [1], "Real Analysis," 2nd ed., MacMillan, New York, 1968.

Saperstone, S. H. [1], Semi-flows on spaces of probability measures, Math. Systems Theory 8 (1975), 256-265.

Saperstone, S.; Nishihama, M. [1], Continuity of the limit set maps in semidynamical systems, J. Differential Equations 23 (1977), 183-199.

Seifert, G. [1], A condition for almost periodicity with some applications to functional-differential equations, J. Differential Equations 1 (1965), 393-408.

_________ [2], Recurrence and almost periodicity in ordinary differential equations, Studies in Applied Mathematics 5, SIAM, Philadelphia, 1969.

Sell, G. R. [1], Periodic solutions and asymptotic stability, J. Differential Equations, 2 (1966), 143-157.

__________ [2], Nonautonomous differential equations and topological dynamics, I, II, Trans. Amer. Math. Soc. 127 (1967), 241-283.

__________ [3], "Topological Dynamics and Ordinary Differential Equations," Van Nostrand, London, 1971.

__________ [4], Topological dynamical techniques for differential and integral equations, in "Ordinary Differential Equations," 1971 NRL-MRC Conf., Academic Press, 1972, 287-304.

__________ 5 , Differential equations without uniqueness and classical topological dynamics, J. Differential Equations 14 (1973), 42-56.

__________ [6], The geometric theory of Volterra integral equations - a preliminary report, "Proc. Equadiff III (Czechoslovak Conference on Differential Equations and their Applications)," J. E. Purkyne Univ., Brno, 1973, 139-143.

Slemrod, M. [1], Asymptotic behavior of periodic, dynamical systems on Banach spaces, Annali Math. Pura Appl. (4) 86 (1970), 325-330. Erratum, Ann. Mat. Pura Appl. (4) 88 (1971).

__________ [2], Asymptotic behavior of a class of abstract dynamical systems, J. Differential Equations (1970), 584-600.

__________ [3], Nonexistence of oscillations in a nonlinear distributed network, J. Math. Anal. Appl. 36 (1971), 22-40.

Slemrod, M.; Infante, E. F. [1], An invariance principle for dynamical systems on Banach space: Application to the general problem of thermoelastic stability, in "Instability of Continuous Systems, IUTAM Sympos.," Herrenalb, 1969, Springer-Verlag, Berlin, 1971.

Stephan, B. H. [1], Hereditary differential equations, Trans. N.Y. Acad. Sci. (2) 33 (1971), 671-679.

Strauss, A.; Yorke, J. A. [1], Perturbation theorems for ordinary differential equations, J. Differential Equations 3 (1967), 15-30.

__________ [2], On asymptotically autonomous differential equations, Math. Systems Theory, 1 (1967), 175-182.

__________ [3], Perturbing uniform asymptotically stable nonlinear systems, J. Differential Equations 6 (1969), 452-483.

Ura, T. [1], Isomorphism and local characterization of local dynamical systems, Funkcial. Ekvac. 12 (1969), 99-122.

________ [2], Local isomorphism and local parallelizability in dynamical system theory, Math. Systems Theory 3 (1969), 1-16.

Vinograd, R. E. [1], On the limiting behavior of unbounded integral curves, Dokl. Akad. Nauk. SSSR 66 (1949), 5-8.

________ [2], On the limit behavior of an unbounded integral curve, Uchen. Zap. Moskov. Gos. Univ. 155, Mat., No. 5 (1952), 94-136.

Volterra, M. V. [1], Sur la théorie mathématique des phénomènes héréditaires, J. de Math. 7 (1928), 250-298.

________ [2], "Lecons sur la Théorie Mathématique de la Lutte pour la Vie," Gauthier-Villars, Paris, 1931.

Walker, J. A. [1], Some results on Liapunov functions and generated dynamical systems, J. Differential Equations 30 (1978), 424-440.

Walker, J. A.; Infante, E. F. [1], Some results on the precompactness of orbits of dynamical systems, J. Math. Anal. Appl. 51 (1975), 56-67.

Wakeman, D. R. [1], An application of topological dynamics to obtain a new invariance property for nonautonomous ordinary differential equations, J. Differential Equations 17 (1975), 259-295.

Webb, G. F. [1], Accretive operators and existence for nonlinear functional differential equations, J. Differential Equations 14 (1973), 57-69.

________ [2], Autonomous nonlinear functional differential equations and nonlinear semigroups, J. Math. Anal. Appl. 46 (1974), 1-12.

________ [3], An application of accretive operator theory to a nonlinear complex heat equation, Boll. U.M.I. (4) 11 (1975), 604-609.

________ [4], Asymptotic stability for abstract nonlinear functional differential equations, Proc. Amer. Math. Soc. 54 (1976), 225-230.

________ [5], Functional differential equations and nonlinear semigroups in L^p-spaces, J. Differential Equations 20 (1976), 71-89.

Whitney, H. [1], Regular families of curves II, Proc. Nat. Acad. Sci. U.S.A. 18 (1932), 340-342.

Winston, E.; Yorke, J. A. [1], Linear delay differential equations whose solutions become identically zero, Rev. Roumaine Math. Pures Appl. 14 (1969), 885-887.

Winston, E. [1], Uniqueness of the zero solution for delay differential equations with state dependence, J. Differential Equations 7 (1970), 395-405.

________ [2], Uniqueness of solutions of state dependent delay differential equations, J. Math. Anal. Appl. 47 (1974), 620-625.

Wonham, W. M. [1], Lyapunov criteria for weak stochastic stability, J. Differential Equations 2 (1966), 195-207.

Yen, C.-L. [1], On the rest points of a nonlinear nonexpansive semigroup, Pacific J. Math. 45 (1973), 699-706.

Yorke, J. A. [1], Invariance for ordinary differential equations, Math. Systems Theory 1 (1967), 353-372.

________ [2], A continuous differential equation in Hilbert space without existence, Funkcial. Ekvac. 13 (1970), 19-21.

________ [3], Selected topics in differential delay equations, "Japan-U.S. Seminar in Ordinary Differential and Functional Equations," pp. 16-28 in Lecture Notes in Mathematics 243, Springer-Verlag, New York, 1971.

Yoshizawa, T. [1], Asymptotic behavior of solutions of a system of differential equations, Contributions to Differential Equations, Vol. 1, Wiley, New York, 1963, 371-387.

________ [2], Asymptotically almost periodic solutions of an almost periodic system, Funkcial. Ekvac. 12 (1969), 23-40.

________ [3], "Stability Theory by Lyapunov's Second Method," The Mathematical Society of Japan, Tokyo, 1966.

________ [4], Stability and existence of a periodic solution, J. Differential Equations, 4 (1968), 121-129.

Yosida, K. [1], "Functional Analysis," 5th ed., Springer-Verlag, Berlin, 1978.

Zaidman, S. [1], Some asymptotic theorems for abstract differential equations, Proc. Amer. Math. Soc. 25 (1970), 521-525.

Zakai, M. [1], A Lyapunov criterion for the existence of stationary probability distributions for systems perturbed by noise, SIAM J. Control 7 (1969), 390-397.

Zakai, M.; Snyders, J. [1], Stationary probability measures for linear differential equations driven by white noise, J. Differential Equations 8 (1970), 27-33.

INDEX OF TERMS

Absolutely continuous function, 434

Accretive operator, 213
  examples, 213, 216, 232, 252, 293

Affine mapping, 255

Alaoglu theorem, 432

Almost periodic function, 438

Almost periodic motion, 111, 398

Ascoli theorem, 429

Asymptotic hull, 401

Asymptotically autonomous function, 184

Asymptotically periodic function, 184

Asymptotically stable set, 59, 169
  examples, 84, 87, 125, 128

Asymptotically stable solution
  of retarded functional differential equation, 311
  of an ordinary differential equation, 128, 175
  examples, 128, 172, 180, 186, 270, 312, 314, 319, 351

Atomic at 0, 338

Atomic reactor dynamics, 285, 317, 319

Attracted, 44, 56

Attracting solution, 175

Attraction, region of, 56, 168

Attractor, 58, 168

Autonomous ordinary differential equation, 130, 137, 184, 209, 244

Banach space, 429

Bidual space, 431

Bilaterally Lyapunov stable motion, 104

Bochner, 433

Bounded linear operator, 430

Brouwer fixed point theorem, 437

Brownian motion, 380, 445

Carathéodory conditions, 142, 143, 144

Cauchy net, 428
  problem, 244, 245, 258
  Schwarz inequality, 432

Chaos, 80, 88

Chapman-Kolmogorov equations, 372

Closed convex hull, 436

Closed operator, 437

Cluster point, 426

Compact motion, 37
  operator, 437
  positive orbit, 21
  process, 404
  set, 426, 428
  solution, 157
  support, 436

Compactly contained, 157

Compactness assumption, 63

Compatible, 194

Complete uniform space, 428

Conditional expectation, 442
Continuity property, 2
Continuous set valued map, 63
Contraction (Contracting)
  operator, 213, 437
  mapping principle, 437
Convergent net, 425
Converges in
  in mean square, 443
  in p^{th} mean, 443
  in probability (in pr.), 443
  strongly, 431
  weak*, 431
  weakly, 431
  with probability one (w.p.1), 443
Critical motion, 10
  orbit, 22
  point, 10
Diminishing, 184, 203
Diode, 353
Dirac measure, 377
Directed set, 425
Discrete dynamical system, 24
Discrete semidynamical system, 24
Discrete semidynamical system, examples, 48, 80
Distal, 132, 133
Distribution, 441
Domain, 429
Dual map, 225
Dual semigroup, 373
Dual space, 431
Dynamical system, 3
Dynkin's formula, 385
Elastic materials with memory, 314
Equation
  autonomous ordinary differential, 137, 184, 209, 244
[Equation]
  evolution, 209, 244, 245, 394
  functional difference, 339
  functional differential, 283
  kinetic reactor, 319
  Liénard, 142, 180
  limiting, 156, 173, 174, 181, 199, 416
  neutral functional differential, 285, 351
  nonautonomous ordinary differential, 137, 411
  renewal, 197
  retarded functional differential, 287, 326, 330
  stochastic differential, 380
  Volterra integral, 192
Equi-almost periodic set, 114, 398
Equicontinuity condition, 374
Equicontinuous, 133, 429
Equilibrium point, 10
Eventually, 425
Eventually stable set, 69
Eventually weakly stable set, 73
Evolution equation, 209, 244, 245, 394
Expectation, 441
Extension of a semidynamical system, 5
Extension of a solution, 8
Feller property, 373
Finite delay, 286
Flip-flop circuit, 285, 351
Flow, 3
Fourier series, 438
Frequently, 425

Functional difference equation, 339
Functional differential equation, 283
Fundamental period, 11
Generalized domain, 229
Generalized invariance principle, 44, 163
Generator
  infinitesimal, 219
  weak infinitesimal, 385
Global attractor, 169
  semidynamical system, 26
  solution, 126, 139, 190
  uniform attractor, 169
Globally asymptotically stable, 311
  Lipschitz, 437
  uniformly asymptotically stable, 169
Gronwall inequality, 438, 439
Group property, 3
Hahn-Banach theorem, 430
Harmonic solution, 126
Hereditary dependence, 283
Hilbert space, 432
History space, 289, 314
Homogeneous Markov process, 444
Hull, 401
Hull, asymptotic, 401
Infinite delay, 286
Infinitesimal generator, 219
Initial function, 285
Initial value problem, 139
  property, 2
Inner product space, 432
Instability of RFDE, 363, 364
Integrable function, 443
Integral equation, 192
Interaction of two species, 312
Invariance principle, 155, 165, 231, 310, 349
  for Markov processes, 384
Invariant measure, 378
  set, 36
Isomorphic semidynamical systems, 27
Issue, 10
Ito stochastic integral, 381, 446
Kamke, 191
Kinetic reactor equations, 319
Kōmura, 435
Krasnoselskii fixed point theorem, 437
ℓ-Lipschitz, 437
LSC, 62
Lagrange stable semidynamical system, 23, 54, 56
LaSalle invariance principle, 44, 78
Lebesgue dominated convergence, 434
Liénard equation, 142, 180
Limit set, 41
Limit set map, 62
Limiting equation, 156, 173, 174, 181
Linear functional, 430
  operator, 429
  viscoelasticity, 315
Lipschitz constant, 436
  operator, 436
Local semidynamical system, 26
Locally compact operator, 303
  compact space, 428
  integrable function, 434
  Lipschitz operator, 142, 436
Lower semicontinuous set valued map (LSC), 62

Lyapunov function, 77, 164
  examples, 173, 180, 226, 409
Lyapunov stable positive motion, 99
  semidynamical system, 100
  weak semidynamical system, 396
m-accretive, 216
Markov process, 369, 381
  semigroup, 372
  time, 385
  transition operator, 371
Markov-Kakutani fixed point theorem, 438
Martingale, 442
  convergence theorem, 387, 444
Materials with memory, 284, 314
Maximal interval of definition, 139
  solution, 8
Measurable function, 432
  space, 440
Measure,
  Dirac, 337
  support, of, 371
Merges, 19
Metric, 426
Metric space, 426
Minimal set, positively, 45
Minimality, 45
Motion
  almost periodic, 111, 398
  bilaterally Lyapunov stable, 104
  Brownian, 380, 445
  compact, 37
  critical, 10
  Lyapunov stable, 99
  orbitally asymptotically stable, 122
[Motion]
  periodic, 14
  positive, 7
  principal, 22
  recurrent, 105
  self-intersecting, 16
  uniformly recurrent, 109
NFDE, 337
Negative orbit, 32
  solution, 32
Net, 425
  Cauchy, 428
  convergent, 425
Neutral functional differential equation (NFDE), 288, 337
  examples, 285, 351
Nonautonomous Cauchy problem, 245
  evolutionary equation, 394
  ordinary differential equation, 137, 411
  retarded functional differential equation, 326, 330
Noncontinuable solution, 139
Nonlinear distributed network, 351
  renewal equation, 197
  semigroup, 209
Nonuniqueness, 189
  examples, 15, 304
Norm, 430
Nuclear reactor kinetics, 317
Null solution, 175
Operator, 429
  accretive, 213
  bounded, 430
  closed, 436
  contraction (contracting), 213, 437
  linear, 429
  Lipschitz, 436

[Operator]
  locally Lipschitz, 436
  m-accretive, 216
  quasi-contracting, 213
  weakly compact, 436
Orbit, 22
  compact positive orbit, 23, 37
  critical, 22
  periodic, 22
  positive, 20
  principal, 22
  self-intersecting, 22
Orbital attractor, 121
  asymptotic stability, 134
Orbitally attracted, 121
  asymptotically stable motion, 122
Ordinary differential equation,
  autonomous, 130, 137, 184, 209, 244
  nonautonomous, 137, 411
Orthogonal, 432
Period, 11
Period, fundamental, 11
Periodic motion, 14
  orbit, 22
  point, 11
  solution of an ordinary differential equation, 128
  solution of an RFDE, 331, 335, 364
Perturbation, 183, 187, 330
Phase map, 3
  space, 3
Population growth, 80, 283, 284, 312
Positive definite, 79
  hull, 37, 401
  limit set, 39, 157, 407
[Positive]
  limit set, examples, 88, 160, 181, 254, 259, 270, 392, 421
  motion, 7, 407
  orbit, 20, 407
  prolongation, 52
  prolongational limit set, 93
  trajectory, 156
Positively invariant set, 36, 163, 238
  minimal set, 45
Precompact space, 428
  positive orbit, examples, 128, 158, 232, 237, 240, 242, 265, 305, 325, 360
Preditor-prey system, 312
Principal negative orbit, 32
  orbit, 22, 41
  solution, 9
Probability measure, 440
  space, 440
Process, 400
Process, compact, 404
  stopped, 386
  uniform, 410
  uniformly compact, 405
Prolongation, positive, 52
Purely periodic point, 11
Quasi-contracting map, 213
Quasi-invariant set, 159, 329
RFDE, 287
Random variable, 369, 440
Range, 429
Realization, 444
Recur, 105
Recur uniformly, 109
Recurrence, 134, 187
Recurrent motion, 105

Recurrent motion, examples of, 106, 116, 189
Recurrent solution, 188
Reflexive space, 431
Region of attraction, 56, 168
  orbital attraction, 121
  uniform attraction, 169, 182
  weak attraction, 56
Regular conditional distribution, 442
Relatively dense set, 112
Renewal equation, 197
Reparametrization, 28
Resolvent, 216
Rest point, 10
Restriction of a semidynamical system, 4
Retarded functional differential equation (RFDE), 287
σ-algebra, 440
s-translate, 140
Sample path, 444
Schauder fixed point theorem, 437
Self-intersecting motion, 16
Self-intersecting orbit, 22
Semidynamical system, 2
Semidynamical system, Lagrange stable, 23, 54, 56
  Lyapunov stable, 100
  stable, 54
  examples, 5, 14, 48, 61, 69, 81, 106, 127, 150, 151, 215, 292, 339, 373, 403
Semiflow, 3
Semigroup, 209
Semigroup (strongly continuous) of operators, 212
  contraction, 213
  dual, 373
  linear, 220
  quasi-contraction, 213
Semigroup property, 2
Semi-scalar product, 225
Sequential closure, 395
Set
  asymptotically stable, 59, 169
  compact, 426, 428
  directed, 425
  equi-almost periodic, 114, 398
  eventually stable, 69
  eventually weakly stable, 73
  invariant, 36
  minimal, 46
  positive limit, 39, 157, 407
  positively invariant, 36
  positively minimal, 45
  prolongational limit, 93
  quasi-invariant, 159, 329
  relatively dense, 112
  stable, 52, 167, 175
  strongly invariant, 396
  weakly invariant, 37, 38, 306, 349
Skew product, 141
Skew product semidynamical system, 140, 151, 205
Sobolev embedding theorem, 435
Sobolev space, 435
Solution of
  a neutral functional differential equation, 337
  a retarded functional differential equation, 287
  a semidynamical system, 7
  a stochastic differential equation, 381
  an equation of evolution, 244, 245, 249
  an ordinary differential equation, 125, 138

Solution
  asymptotically stable, 128, 175, 311
  attracting, 175
  compact, 157
  compactly contained, 157
  extension of, 8
  noncontinuable, 139
  null, 175
  periodic, 128, 331, 335, 364
  principal, 9
  recurrent, 188
  stable, 175, 311
  uniformly asymptotically stable, 175
  uniformly attracting, 175
  uniformly recurrent, 188
  uniformly stable, 128, 175
  weak, 249
Space
  Banach, 429
  Hilbert, 432
  metric, 426
  of initial functions, 286, 288, 289
  of probability measures, 370
  uniform, 427
Stable difference operator, 342, 365
  semidynamical system, 54
  set, 52, 167, 175
  solution, 175, 311
Stability
  asymptotic, 59, 169
  eventual, 69
  Lyapunov, 100
  orbital asymptotic, 122
  uniform asymptotic, 169, 175
  weak eventual, 73
[Stability]
  examples, 125, 128, 172, 181, 265, 270, 312, 314, 319, 351, 388, 412
Start point, 9
State space, 442
Step function, 433
Stochastic continuity on compacta, 373
  differential equation, 380
  dynamical system, 369
  integral, 446
  process, 443
Stopped process, 386
Strictly convex, 255
Strong convergence, 431
  solution, 221, 245
  topology, 431
Strongly continuous semigroup of operators, 212
  invariant set, 396
Subbase for a uniformity, 427
Subnet, 425
Supermartingale, 444
Supp, 267, 271
Support, compact, 436
Support of a function, 267
  of a measure, 371
t-periodic, 14
t-transition, 3
t-translate, 286, 327
Topology
  strong, 431
  uniform, 427
  weak, 431
  weak*, 431
Tunnel diode, 352
USC, 62
Uniform attractor, 168
  process, 410

[Uniform]
  space, 427
  topology, 427
Uniformity, 427
Uniformly asymptotically stable, 169, 175
  examples, 173, 181, 186
Uniformly attracting, 175
  compact process, 405
  eventually stable, 69
  Lyapunov stable motion, 99
  recurrent motion, 109, 203
  recurrent solution, 188
  stable solution, 128, 167, 175
  tight, 371
Upper semicontinuous, 62
Variation norm, 379
Visco-elasticity, 284, 314
Volterra, 312
Volterra integral equation, 192
Weak attractor, 58
  convergence, 431
  dynamical system, 395
  infinitesimal generator, 385
  semidynamical system, 395
  solution, 249
  topology, 431
Weak* topology, 431
Weakly attracted to, 56
  compact operator, 436
  compact set, 431
  invariant set, 37, 38, 306, 349
  sequentially compact set, 431
White noise, 380, 381

INDEX OF SYMBOLS

$A^+(x)$, 56

$A_W^+(x)$, 56

$A_s(x)$, 94

A_T, 219

A_λ, 223

A^W, 385

$\mathscr{A}^+(x)$, 121

$AC(J;X)$, 216

$B_\varepsilon(x)$, 62

$\mathscr{B}^d$, 370

$B_b(\Omega; \mathbb{R}^d)$, 370

$C(\Omega; \mathbb{R}^d)$, 195

$C_b(\Omega; \mathbb{R}^d)$, 288

$C^k(\Omega; \mathbb{R})$, 160

co S, 257, 436

$\overline{\text{co}}$ S, 255, 436

γ_x, 2

$\gamma(x)$, 22

$\gamma^+(x)$, 20

$\Gamma_f^+(x_0)$, 156

D, 337

$D^+(x)$, 52

$\mathscr{D}(\phi)$, 7

$\hat{\mathscr{D}}(A)$, 229

Δ, 210

∇, 242

δ_x, 377

$\frac{\partial}{\partial n}$, 242

E, 382

E_x, 382

$E\{\xi(t)|\xi(0) = x\}$, 382

$\mathscr{F}$, 141

$\mathscr{F}_1$, $\mathscr{F}_2$, 142

$\mathscr{F}_1'$, $\mathscr{F}_2'$, 144

$(\mathscr{F}, \pi^*)$, 150

$(\mathscr{F} \times W, \pi)$, 151

$\mathscr{G}_p$, 194

H^+, 63

$H^+(x)$, 37

$\mathscr{H}[u]$, 401

$H^1(\Omega; \mathbb{R})$, 234, 242

$H_0^1(\Omega; \mathbb{R})$, 234

$H^2(\Omega; \mathbb{R})$, 235

i, 14

$J^+(x)$, 93

J_λ, 211

$\mathscr{K}_X$, 62

$\mathscr{K}_p$, 195

L^+, 63

$L^+(x)$, 39

$\mathscr{L}[u]$, 401
$L^1(J; \mathbb{R})$, 143
$L^2(J; \mathbb{R})$, 233
$L^p(J; \mathbb{R}^d)$, 195
$L^p_{loc}(J; \mathbb{R}^d)$, 201, 245
$L^p(J;X)$, 251, 413
$\mathscr{M}$, 370
$(\mathscr{M},U)$, 373
P_x, 382
$P\{\xi \in B | \Sigma_s\}$, 381
$P\{\xi(t) \in B | \xi(0) = x\}$, 381
$\mathscr{P}(t,x,B)$, 372
π, 2
π^t, 3
π_x, 7
π^*, 150
Q_ω, 213
$\mathscr{R}(\phi)$, 10
$\mathscr{R}^+(x)$, 105
$\rho(g,h)$, 144
S^1, 21
supp, 267, 371
$\overline{\text{supp}}$, 371
$\mathscr{S}(g)$, 191
$\mathscr{S}_t(g)$, 191
$T_\tau f$, 192
$T(t)$, 212
$T_\lambda(t)$, 223
$V'(x)$, 78
$V'(x,t)$, 164
$W^{1,p}(J;X)$, 251
$W^{1,p}_{loc}(J;X)$, 251
$\phi(x_0,t_0,t)$, 126
$\phi(f,x_0;t)$, 140
$\phi(f,g,k;t)$, 193
$\Omega^+_f(x_0)$, 157
$\Omega^+_u(\tau,x)$, 407
$o(\lambda)$, 247
$\|u\|_1$, 216
$\|u\|_2$, 236
$\|u\|_\infty$, 218
$|Ax|$, 229
$|\mu|(\mathbb{R}^d)$, 379
(x,x^*), 225
$\langle x,y\rangle$, 234
$\langle x,y\rangle_s$, 225
$\langle u,v\rangle_2$, 236
$\langle u,v\rangle_{H^1}$, 242
xt, 4
WT, 4
$t \wedge s$, 386
f_t, 140
x_t, 286
$(\mathscr{L},T)$, 232
X^*, 225
(X,π), 2
(X,d), 99
$(Y,\mathscr{U})$, 395
$(Y,\mathscr{U},\pi)$, 395

Applied Mathematical Sciences

1. John: **Partial Differential Equations,** 3rd ed.
2. Sirovich: **Techniques of Asymptotic Analysis.**
3. Hale: **Theory of Functional Differential Equations,** 2nd ed. (cloth)
4. Percus: **Combinatorial Methods.**
5. von Mises/Friedrichs: **Fluid Dynamics.**
6. Freiberger/Grenander: **A Course in Computational Probability and Statistics,** Rev. Ptg.
7. Pipkin: **Lectures on Viscoelasticity Theory.**
8. Giacaglia: **Perturbation Methods in Non-Linear Systems.**
9. Friedrichs: **Spectral Theory of Operators in Hilbert Space.**
10. Stroud: **Numerical Quadrature and Solutions of Ordinary Differential Equations.**
11. Wolovich: **Linear Multivariable Systems.**
12. Berkovitz: **Optimal Control Theory.**
13. Bluman/Cole: **Similarity Methods for Differential Equations.**
14. Yoshizawa: **Stability Theory and the Existence of Periodic Solutions and Almost Periodic Solutions.**
15. Braun: **Differential Equations and Their Applications,** 2nd ed. (cloth)
16. Lefschetz: **Applications of Algebraic Topology.**
17. Collatz/Wetterling: **Optimization Problems.**
18. Grenander: **Pattern Synthesis. Lectures in Pattern Theory,** vol. I.
19. Marsden/McCracken: **The Hopf Bifurcation and Its Applications.**
20. Driver: **Ordinary and Delay Differential Equations.**
21. Courant/Friedrichs: **Supersonic Flow and Shock Waves.** (cloth)
22. Rouche/Habets/Laloy: **Stability Theory by Liapunov's Direct Method.**
23. Lamperti: **Stochastic Processes. A Survey of the Mathematical Theory.**
24. Grenander: **Pattern Analysis. Lectures in Pattern Theory,** vol. II.
25. Davies: **Integral Transforms and Their Applications.**
26. Kushner/Clark: **Stochastic Approximation Methods for Constrained and Unconstrained Systems.**
27. de Boor: **A Practical Guide to Splines.**
28. Keilson: **Markov Chain Models—Rarity and Exponentiality.**
29. de Veubeke: **A Course in Elasticity.**
30. Sniatycki: **Geometric Quantization and Quantum Mechanics.**
31. Reid: **Sturmian Theory for Ordinary Differential Equations.**
32. Meis/Marcowitz: **Numerical Solution of Partial Differential Equations.**
33. Grenander: **Lectures in Pattern Theory,** vol. III.
34. Cole/Kevorkian: **Pertubation Methods in Applied Mathematics.** (cloth)
35. Carr: **Applications of Centre Manifold Theory.**
36. Bengtsson/Ghil/Källén: **Dynamic Meteorology**
37. Saperstone: **Semidynamical Systems in Infinite Dimensional Spaces**